# FROZEN TO LIFE

# FROZEN TO LIFE

## A PERSONAL MORTALITY EXPERIMENT

∞

D.J. MACLENNAN

www.djmaclennan.com

Published in Scotland by Anatta Books 2015
Email order@anattabooks.com for ordering information

10 9 8 7 6 5 4 3 2 1

British Library Cataloguing in Publication Data. A catalogue record for this book is available from the British Library.

ISBN 978-0-9933344-0-5

*For Inklings and Shatterlings*
*everywhere (and nowhere)*

# CONTENTS

# PREFACE

∞

This book escaped from my chest like a small alien and started to eat me. They say everybody has one book in them; they don't usually say that some of those books force exit. While writing it, a rash I had had on my back for years disappeared, only to emerge on my sternum – words said on the page, or from the mouth, sometimes write echoes in the body. Perhaps words *stifled*, even for the best of reasons, always do.

My good friend Smith, from whom I took authorial advice before starting, said that it was important to 'find your voice' when writing a book. He thought that the voice that came across in my blogs was coherent and would work for me. I have tried to follow his advice but have found that, compared to the voice in my short blog-posts, my book-writing voice sounds, by turns, more depressive, crazed, philosophical, and absurdist. I think it still has a certain coherence, and I hope you will hear it as one multi-timbral voice and not as several dissonant ones.

But, as you will see, there are good reasons why a person might find that he or she does not always seem to ring true with him, or her, self. There are times when the book examines themes in keeping with Smith's playfully-put charge that I am 'made-up'. I'm not yet sure to what extent he realised when he wrote it just how true that is – not just of me but of him and of everybody else. If it is true that I am more 'made-up' than most, then that is something I now not only embrace but also take as a compliment.

When opening the file that holds the words constituting this book, I sometimes got the Linux error message 'Invalid Argument'. While somewhat alarming at first, I soon discovered the cause: The OpenOffice text file for this book was stored on my network drive and not on the Ubuntu netbook upon which I am now typing. The time lag to call the file from the network was causing a 'pre-emptive' error. The error message became something of a regular reminder to me that I needed to keep my arguments clear and to back them up, as far as possible, with evidence. Opinions will vary about whether I heeded the message or simply compounded it with error messages of my own.

Throughout this book, you will see various references to the work of Derek Parfit, author of the books *Reasons and Persons* and *On What Matters*. His work – and the work of others such as Daniel Dennett, Douglas Hofstadter, Steve Grand, and Erich Harth – has been invaluable in helping me find ways to frame the sometimes-complex arguments that I have *had* to make in order to put across my case. I had not yet read *Reasons and Persons* at the time I made my decision, but had I done so by that point, it would have helped greatly in my attempts to rationalise my feelings about that decision.

I have many persons to thank for their help with this book, and various reasons to thank them. A few spring immediately to mind: Epsilon, for the useful spur of a particular drunken argument with me during which he both criticised the peer-review process and asked where I could *possibly* find scientific evidence to back up 'what you are doing'; Steve Grand, for providing fascinating responses to my rambling emails, despite the fact that he was extremely busy building androids and *Grandroids*, and trying to fix a hacked server; Liza Cleland, who spent many hours of a bleak Skye winter and spring copy editing and proofreading my work, performing the editorial equivalent of rinsing out crunchy sand and teasing out the byssal threads from the fibrillating mussel of my original text (partly by pressing me to think again about *how* we know what we think we know); and my wife, for all her love and encouragement, but also for her blanket failure to read any of the 'popular science' books I gave her. I hope that this book will begin to answer Epsilon's question, and provide my wife with a science-related book that she will actually read.

My other siblings have also been an inspiration. I can't imagine *not* being part of this sometimes motley but always fascinating, and curiously overlapping, group of selves. 'Middleishness' within this large family has, I think, helped to kindle my struggle for definition (of which this book now seems a logical product).

If I have inadvertently used the ideas of others without accreditation, I apologise for my omission(s) and ask that – if you still have a voice – you get in touch with me to discuss it. Face-to-face is good: While social media is wonderful, I find it hard to listen, develop arguments, and make human attachments without that special feedback loop that only kicks in with physical presence. I'll pour you a malt whisky and we'll talk it through.

# INTRODUCTION – INTO THE WHITE

∞

The cold sears my skin and *I* recede. An empty expanse of nothingness. A blank screen. Unforgiving but welcome. That is the point, after all: to stop the noise for just a while; to be aware of little or nothing but the burning cold and seaweed-ozone tang. I push out a little and test my depth. My toes just touch the stony bottom, and a brief panic thumps my heaving chest. Any minor swell or rogue wave will take me to drowning depth; at least that's how my body wants to see it. You are under threat, self. You are cold, self. Unprotected, self. I *will* learn this.

It hurts every time. The brain imbibes the bitter sensation, and the pain lessens as I gradually acclimatise. But it will always be beyond uncomfortable.

Sea swimming is a kind of addiction. It seems familiar on a level I cannot verbalise. Our ancestors must have spent so many of their waking hours wet and chilled to the bone. They walked for miles in drenching downpours; they bathed in icy streams; they forded great rivers, the stones pressing into their bare soles. Cold water was familiar, useful; dark, and deadly.

I got into danger when I began again to swim in the bay in front of the house. I set out from the ruined jetty, the midway rock just a few strokes distant. But the sudden awareness of my tiring arms and the depth of the water induced panic, and I called out to my wife on the shore before switching suddenly to survival mode and striking out for the rock with renewed vigour. I continue to swim, but I fear the sea. I am a fool, *risking* my life while only just beginning to discover what it is. Stupid. Cold. Ape.

The phone rings – my father asking if I will go with him to the funeral of a cousin who has died of a brain tumour. I decline. I did not know the man. He died aged fifty-two. I admit to my father that I am a coward when it comes to funerals. He bemoans the passing of an age when all the relatives would turn out to mark the death of even the most distant of kin: 'People would notice if you weren't there.' In the morning, he will cross the drear water to the ancestral island. There they will attend church for the verbal assault of the minister and the plaintive 'precenting' wail of the Gaelic psalms. Then they'll put my cousin in the ground – in the 'Elephants' Graveyard'. I'll never lie there.

This book is, at least in part, about death. But not about that kind of death. I chose something else. I have chosen to be cryonically preserved after my death – to be 'vitrified', head only, at liquid-nitrogen temperature, in a compartment of a huge stainless-steel container called a *dewar*. People sometimes ask me about this. Generally, they ask in a clipped, detached way that indicates to me that they want me to keep the explanation short, painless, clean, and death-free. Journalists demand more: They wish to root out the 'gory' details of cost, motivational triggers, and decapitation. Writers pen their cloying curlicues of love and loss with little or no regard for scientific accuracy.

It's not surprising that people don't want to simply *talk* about death, at least not outwith the bounds of accepted convention. But I find the lack of willingness to discuss, face-to-face, the worst thing that is ever going to happen to you, both illogical and depressing.

## THE DEAD PRINCESS – A MORTALITY TALE

The suitor looks down at the broken princess, her long golden hair bloodied from the impact of the fall. The suitor is from out-of-town and has never before seen this sort of thing. 'She won't speak to me,' he tells the court physician.

'She is dead, sire,' he intones.

'I'm sorry, I don't know what that means,' says the suitor. 'Can't you fix her?'

'Even the most powerful magic could not bring the princess back from the dead,' says the physician, stroking his goatee. 'Have you never seen death before?'

'I suppose I have had a sheltered upbringing,' admits the suitor, 'they never mentioned *broken* princesses. Can I still marry her?'

The physician is horrified. 'But she is *dead*, sire. We must put her in the ground. There will be a state funeral of appropriate grandeur for a princess so dear to our hearts.'

'Put her in the ground?' splutters the suitor, at least equally horrified, 'Down in the wet soil, where the worms will eat away her precious body and erase any semblance of the beauty and wit she once was. Can't I keep her? Perhaps *our* physicians have a casket where they could keep the princess without degradation, so I could at least look upon her and hope for a day when she may be fixed and re-animated.'

'Ugh, that's gross,' says the physician.

I find your death horrifying. Not just that you *will* die but what they will do with your precious configuration of atoms afterwards. They will burn you up, as if in dreadful mockery of the fires of the stars that created your constituent elements. Or they will put you in a box, down in the ground-rock and decayed vegetable-matter that swathes parts of this beautiful planet. They will allow you to dissipate. This is what happens. These are the conventions. This is respect.

But what else could they have done? They could have paused you. They could have waited.

Everything is made of atoms. Yes, there are smaller elementary particles (such as photons, electrons, quarks, and neutrinos) and there are certainly smaller units of measure, all the way down to the enigmatic *Planck length*, only properly describable in the language of mathematics, at around $1.6 \times 10^{-35}$ metres.[1] It is also true to say that, at root, everything is *information*. But understanding the arguments I will make in this book only requires the concept of atoms and their role as universal 'building blocks'. This will tell you much of what you need to know of the majestic, self-regarding structure that you call 'I'.

Atoms are good at waiting. They do not readily decompose.[2] We ingest them in various forms, as sustenance, throughout our lives, and they, in turn, become sustenance for other plants and creatures when we die – if, that is, we choose burial. What a wondrous and beautiful cycle of life we are part of. Plant a tree over my decomposing body, so that I may nourish it and bring forth new life after my death.[3] This idea has a certain poetic appeal, but it simply ignores the issue of *actual* death, and instead wraps the process in a kind of pleasant, Pagan, cyclic mysticism.

In my view, it is grisly and primitive to think of the death and decomposition of conscious entities in terms of simple recycling of raw materials. Perhaps you feel like raw material? I don't know. But you are so much more than that. You are a rare and special thing in this spartan universe: a complex, highly-evolved, self-aware being looking out at the Cosmos from your precious pale-blue marble of a planet and wondering how you came to be, what your future holds, and how long you will survive.

Is it selfish to wish for immortality while others are dropping like wilted petals all around? Some of the people I know wish for it; time is (with a few strange caveats) infinite, so why not have such a grand aspiration? But it's a concept I cannot (yet) internalise. Is that a failure of imagination? Will time tell? I am a practical person making provision for my own continued existence – the kind of provision that nobody else will make for me, not because they don't care but because they don't care as much as I do. It is unlikely that anybody cares as much about your impending death as you do. They will go on with memories of you encoded in the structures of their neuronal connections. You will live on, for a time, in their minds. Yet you will not live.

I claim that I am a practical and rational person, but we will need to see if the rest of this book bears that out. In order to get to the nub of my reasoning I must talk about some of the science involved – both current and speculative. I will need to discuss the thorny subject of 'the self'. I will look to the past, the future, 'the moment'. I must also go inward and look at my own thoughts, motivations, and mental state, weaving in *my* story (and the stories we tell ourselves are important). I will talk death – about what it is but mostly about what it is not. This will be disturbing for me and, perhaps, for you. I am not a philosopher or a psychologist, and I have, at least in the past, had limited time for the sometimes tortured (and often frustratingly *impractical*) arguments of some of them. Nor, despite the fact that I try to draw upon scientific evidence and to present my arguments in a considered way, am I scientist. I like to think that I make my decisions based upon realistic analysis of the available evidence, but I also know, from experience, how easy it is to mislead oneself and end up in blind alleys of pseudo-science.

I am an ordinary person; my journey is not unique, although the final destination is, I admit, somewhat unusual. You will find, in this book, much that is familiar: family, love, pain, joy, fear, and change. But you may find other strands less familiar, even alien. If you have an understanding of, and respect for, the importance of science, you will find common ground with my sense of elation – what physicist Richard Feynman called 'the pleasure of finding the thing out' – at discovering this sparkling realm.[4] If you have religious or supernatural beliefs then I think it would be fair to say that your voyage of self-discovery has not yet begun. You have much to look forward to, and to overcome, should you choose to open your eyes. And if, upon opening them, you find that they hurt, try to remember that it is not because the glare is bad for them; it's just that perhaps – to para-phrase the character Morpheus in *The Matrix*[5] – you have never really used them before.

# 1  THE BLOODY ARROW OF TIME

∞

> There's night and day, brother, both sweet things; sun, moon, and stars,
> brother, all sweet things; there's likewise a wind on the heath. Life is very
> sweet, brother; who would wish to die?
>
> —GEORGE BORROW, Lavengro

Time is a dimension, of sorts. Wrapped up with space, it forms a kind of
flexible surface called *spacetime* – a surface (sometimes called a *fabric*) that
can, under certain circumstances, become curved and warped. This is a
strange and difficult notion. The following thought experiment, based
upon Edwin A. Abbott's satirical novella *Flatland: A Romance of Many
Dimensions*, may help to illustrate the conceptual difficulty with dimensions.[1] Imagine you have lived your whole life only ever experiencing two
dimensions – length and width, within which all travel involves only
degrees of back or forward and left or right. Everything in your world –
other inhabitants, animals, trees, houses, vehicles – appears as lines. By
sight, you can tell these things apart only by subtle differences in their edge
shadings. This is your totality. One day, an enigmatic stranger leads you to
a new place that feels very different – it has a new property of 'up-ness' and
'down-ness', or as we call it, *height*. I would imagine that this would be a
disorientating and frightening, but also perhaps liberating, experience. I do
not mean to suggest, by this example, that you had not noticed time was
*there*; only – given that from our 'vantage point' we cannot fully conceive of

its dimensional nature – that you might have thought it was something else.

But what *is* time? Why do we perceive it as we do? Does it only go in one direction, rather than behaving in a poetic, cyclical manner? Grasping the concept of time requires that you think about decay, and recognise that everything you see around you is – universally – getting worse. We can, at least, take some solace from the fact that it is getting worse in an extremely interesting way.[2] We have a tendency to think decay happens *because* of the passage of time, but it is more appropriate to think of decay *as* the passage of time. How could you know that time was passing if nothing changed? 'Heaven is', in the facetious words of the Talking Heads song, 'a place where nothing ever happens'.[3] Our universe could not be more different to this; neither it nor we would exist if nothing ever happened. Unrelenting change is, and can only be, *all* that ever happens.

New stars are born in interstellar 'nursery' clouds called nebulae. Some of these clouds form by gradual accumulation of gas and dust; others form more dramatically, as happens when certain stars (the more massive ones) explode in supernovas, having first used up their supply of nuclear fuel and collapsed in on themselves. The force of gravity – really the warping of spacetime by massive objects[4] – can cause matter within a nebula to collapse inward and coalesce to form new stars whose nuclear fires are ignited under the immense pressures present. The gravitational field of a new star may trap other nearby swirling matter, which then clumps together to form gaseous or rocky planets. Is this decay? Yes, but decay of a specific kind. It makes mathematical sense to state that, before it exploded, the original star was in a less complex state than that of the resulting solar system. The decay of the original star has produced greater complexity in the universe. And so, despite the fact that the new solar system may appear superficially 'perfect', the *entropy* of the universe has increased.

Entropy is a concept found in the second law of thermodynamics. The German physicist Rudolf Clausius (1822–1888) first introduced it, in the 1850s. Drawing on the earlier work of Sadi Carnot (1796–1832), who had hypothesised that a *heat engine* could convert thermal energy into mechanical work, Clausius reasoned that a measure of the *loss* of energy from such as system was required.[5]

But what does a measure related to heat have to do with decay and the passage of time? What we perceive as heat is, we now know, the energetic (and somewhat ordered) jiggling of atoms. Left alone, such jiggling of atoms will, over time, become *less* energetic, never more: A piece of hot metal, for example, will radiate away its heat energy to the cooler air around it. The *thermal* energy of the hot object has not 'gone away', but it *has* taken on a more dispersed form in the now more-random jiggling of the atoms of the surrounding air. Although such a definition is disputed, you can see entropy as a measure of the *disorder* within a system. 'Quantum mechanic' Seth Lloyd prefers to call it 'invisible information'.[6] The second law states that the entropy of an isolated system always increases or remains constant. We could decrease the entropy of our 'system' containing the cooling metal, but only at the expense of its isolation – by dumping some of the 'randomness' outside of the now-warm room, by means of an air-conditioner, for example. This, however, would then serve to increase the entropy of a bigger system – the environment. The greater the entropy within a system the harder it becomes to put it back together exactly as it was before.

The concept of entropy can be confusing when we try to apply it to things that don't seem hot, or even warm. The *molecular kinetic* energy I mentioned above is, however, only one of the possible types. *Chemical* energy is a form of *potential* energy, stored in the bonds between atoms. Such bonds gradually change and break down (solids degrade over time by rusting, rotting, and so on), again changing/dispersing the previously-embodied energy of the solid and, again, increasing entropy.

We can also increase entropy by *applying* energy to break down a previously-isolated system. Think of an old chair (try to think of it as a chair-*system*) on a bonfire: It starts out resolutely solid and chair-shaped, but as the fire 'consumes' it, gradually becomes ash – some still smouldering on the ground and some carried away on the November wind. The chemical energy from the wood of the chair has changed and dispersed, increasing its entropy to a point where it can no longer be put back together. It's not like the time when one of its legs broke and you repaired it (although that also increased its entropy). This time the repair has become too complex for anyone to carry out. It could never again, in any case, be put back together in *exactly* the same way, because some tiny part of its mass is no longer available, having been removed from this system, now as more-disordered energy (heat, light, sound), in accordance with Einstein's famous $E=mc^2$.

There is a *direction* to the 'life' of the chair. It came from the heart of a tree; was 'released' and turned into a thing of utility, craftsmanship, and polished elegance; degraded over time through wear, tear, and breakage; then, ultimately, burned away to wind-blown and rain-soaked ash. The chair's 'life story' does not happen the other way around. This may seem an obvious and simple-minded thing to state, but it is, in fact, a profound truth. As long as (and only as long as) change keeps happening, it is possible to posit 'beforeness' and 'afterness' – past and future. Entropy gives a direction or 'arrow' to the unfolding of things. *We* interpret this direction as time.

So, we have a particular order in which things happen to us: we get born, we grow old, we die. Ho hum. The physics of this universe dictate that this is the order in which things happen, and nothing will change the fundamental order upon which such events rely. There are, however, special situations allowed by the physics of our universe that, while leaving the order unchanged, create some interesting and unexpected outcomes. Sitting in your chair (of polished elegance or otherwise) right now, you are travelling through time as fast as it is possible to travel through time (at a rate of one second per second[7]), though without something to measure this 'speed through time' *relative* to, it has little meaning. Travelling close to the maximum possible speed through space – usually referred to as *the speed of light* (the $c$ in $E=mc^2$) – has the effect of 'using up' some of your 'speed through time'. It might help to think of your 'route through time' as a straight vertical line; travelling through space relative to this line will pull you, to some extent, 'off course' towards the horizontal axis, thus making your route through time less straight, and therefore, longer. As a result, you travel more slowly through time from the point of view of a person who is not travelling through space at close to the speed of light.[8]

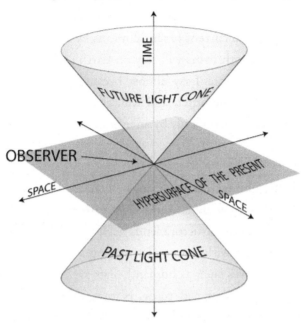

**Figure 1:** The 'future light cone' is the part accessible to us, and is bounded at the edge by the speed of light. From the point of view of an observer, a traveller accelerating out towards the edge of the future light cone travels more slowly through time. (Image: Aainsqatsi, *World_line.svg*.)

The 'twin paradox' is a useful illustration of this effect. One identical twin travels off across the Milky Way at close to the speed of light, while the other stays at home polishing the chair. When the travelling twin returns, she finds the stay-at-home twin much older than herself. Though they have both been travelling through *spacetime*, the traveller has moved more slowly through time as a result of her high-velocity journey through *space*, and vice versa, the non-traveller has travelled more quickly (relative to her twin) through *time* as a result of her *relative* close-to-zero-speed journey through space.[9] (The references to 'relative' here have nothing to do with the fact that they are twins, but relate [*sic*] instead to Einstein's special theory of relativity.) This effect is known as *time dilation*. The special and general theories of relativity also explain how length contracts and mass changes as a result of moving through space. All these effects become more pronounced the nearer to light speed one travels.

But interesting theoretical sleight-of-hand does not help with our decay problem, as we have no current practical means by which to realise such a situation. In reality, both twins will stay on Earth (or nearby) decaying at *almost* exactly the same rate. I say 'almost' because one twin may still have more wanderlust than the other, and will, as a result of her travels, age more slowly than the other, if only by an infinitesimally small amount.

## OUTRAGEOUS FORTUNE

All things decay, although some things, such as dense structures like diamond, seem to get off much more lightly than we do. A diamond's simple, lattice-like atomic structure lends itself to stability and endurance. Despite – in common with diamond – having a carbon basis, complex, wet, *biological* structures such as human beings are not particularly resilient. And they are not 'designed' to be. Natural selection has provided a method by which we can perpetuate our genes, and in some sense *survive*, without the need to actually live on indefinitely as unique individuals. This particular method of species-preservation works extremely well. We can argue about the finer details, and ascribe selfishness to one specific part of the system or another, but our existence demonstrates the effectiveness of the method. In *The Selfish Gene*, Richard Dawkins refers to us, in this context, as 'survival machines',[10] and it seems reasonable to see human beings (and all other plants and animals) – in view of our extremely limited lifespans – as vessels, or vehicles, for our 'immortal' genetic payloads.

The arrow of time will slay us all in short order. Not because of any malign purpose, but because that's the way we are built – weak, expendable, temporary. It will grind all to dust and keep on grinding; there is always enough time, just not enough for us. The cycle of life is both beautiful and terrifying, inspiring in me the kind of awe that others purport to feel about old Jewish gods. I recently saw one of those 'internet meme' images, in that style cribbed from work 'motivational' posters, emblazoned with the heading, 'Playing God'; and beneath the truly inspirational images of good people doing globally-beneficial hard science: 'Why not? The position was vacant.' Some of these 'God-players' will come up with solutions that may slow the ageing process or bring renewed health and vitality to the twilight years of senior-citizenship. In doing so, they will be 'defying' millions of years of genetic convention. They will not be slowing the arrow of time, but they *will* be bringing about a change in our relationship with it, for they may make it possible – for the first time – to extend our lives well beyond our genetically-allotted span.

Of course, countless human beings throughout history have got nowhere near living out their *genetically*-allotted span. They've been struck down by 'guns, germs and steel',[11] as well as a dazzling variety of other causes including accident and natural disaster. Death summarily deleted their individual richness – and if they did not reproduce, also their genetic richness – from its place in the pool, and life (in the wider sense) moved on. In the face of this profligate annihilation it is, surely, perverse to think of pursuing *artificial* survival for the privileged few. Shouldn't we accept death as an evolutionary price of adaptivity, facing up to both the flaws and the advantages of our sexual-reproduction-regulated existence? But just what is it that is artificial here? Why should we assume that this particular means of survival stumbled upon by natural selection is a satisfactory one, or even that it grants us a type and duration of existence worth having? It is currently 'in our nature' to die, but in the continued absence of a staying hand from God, it also seems to be in our nature to find innovative new ways to endure.

Moral relativism (or should that be moral relativity?) abounds in this territory. Person *A* should not survive, because billions of persons *B* have already died. Person *C* is dying of terminal cancer, and yet person *A* wants to have more than her normal lifespan. Person *D* died being born and never had a chance to experience life, yet person *A*, who has lived a rich and full life, thinks that she has the right to more of it. I will return to this subject later.

In the meantime, we can ponder what 'normal' or 'natural' lifespan actually means. Perhaps we call it 'natural' only because we feel we can do little or nothing about it. There is, however, another way of looking at this: We call it 'natural' simply because it *happens*. We impose *meaning* on this accidental circumstance through use of this word.

## RAILING AGAINST CONCEPTS

It will do you about as much good to rail against time *per se* as it will to pray for gods to intervene. The passage of time and the action of entropy bring about ever-greater complexity – a branching, blossoming tree of possibilities. Blossoming disorder (things getting worse), now unfolding within the constraints of the physics of our universe, creates novel opportunities for spontaneous ordered complexity to arise: Stars form from swirling debris; blind evolution is fuelled by energy scavenged from sunlight, plants, and animals; chaotic molecular processes in thinking brains become ideas that become created things. In the words of Professor Peter Atkins, 'Locally, there may be abatements of chaos. Collapse into chaos can be fantastically constructive.'[12] This is something to acknowledge and understand, if not to actively celebrate. Religions, particularly monotheistic ones, don't allow room for celebration (or understanding) of this kind of probabilistic exuberance. Obsessed with what happens to exist *now*, they assume an initial state of complexity in the universe: God creates everything from nothing, therefore he must be a mind-bogglingly advanced (and thus *complex*) entity. If we assume, as we should, an initial state of *simplicity*, then there is far less cognitive dissonance in the notion that the universe – and everything in it – came from a special kind of nothing, guided by no hand.

So, no railing means acceptance, right?

I don't do acceptance very well, particularly when it comes to the big issues. People like me are not protesting the implacable passage of time or the immutable laws of the universe, but we *are* striving for greater freedom from the bonds of our *accidental* nature. Human beings just turned out this way through the blind action of time and natural selection; evolution is a *drunkard's walk*[13] of randomness. There is no particular reason why we should live for only seventy or eighty years of human-measured time, with usually at least a handful of those years spent in decrepitude, senility, or physical pain. A red sea urchin at least gets two hundred years; baobab trees may live for several thousand; *Turritopsis nutricula* (a *hydrozoan*) can cycle indefinitely from immature *polyp* to mature *medusa* (jellyfish) and back again. Is it unacceptable for *Homo sapiens* to strive for a modest extension to that brief instant in the Stelliferous Era of this universe during which we get to make *our* benighted stand?

## THE CONSCIOUS STONE

A conscious stone was talking to his companion, the boulder, about human beings. 'I once had a dream that I was a human being,' he told the boulder.

'What did it feel like?' said the boulder, curiously.

'Well, there was a brief flash of light and a kind of strong thumping sensation, accompanied by a fleeting but dreadful sense of grief and loss,' said the stone.

'Doesn't sound very nice,' said the boulder, weathering somewhat. 'How long did it last?'

'About eighty years,' said the stone.

## PERCEPTIONS OF TIME

Our relationship with time is our perception of time. Because most of us are not physicists, and have not looked in detail at what it is and how it came to be, time really only exists for us as a perception. Even physicists in this highly specialised field struggle to understand what time *really* is and what (if anything) there was before there was time. The very concept of 'before' makes no sense without time.

Our everyday perceptions of time hinge on concepts like *past, present,* and *future*: I *went* to the dentist last week; I am at the dentist *now*; I *will be* at the dentist next week. That's perhaps too many trips to the dentist in quick succession, but it's a useful illustration because it infers the inclusion of a strong motivator that has an impact on our temporal perception of our lives – pain. We would generally accept that pain is a bad thing that we would wish, as far as possible, to avoid. If I were (like my wife) the sort of person who worried about going to the dentist, I might worry that my trip to the dentist next week would involve some degree of pain. I would be highly unlikely to worry that my trip to the dentist *last* week involved pain, because that trip is in the past. Perhaps the strongest reaction, of all possible temporal cases, would be to the one where the pain of the dentist's drill is hurting me *now*.

As obvious as this may all seem, it masks a philosophical quagmire about *degree* and *preference* in our attitudes towards *when* in the chronology of our lives things happen. For example, we might prefer to be in pain in one week's time in order to avoid pain now. But *how much* pain are we prepared to accept next week in order to avoid being in pain now? Would it definitely be the *same* amount or would we be prepared, for example, to take twice as much pain next week in order to avoid having half that amount of pain now? Or maybe it would be more – three times as much, five times, *ten* times. Could there ever be a correct response to such a question? What degree and preference would be *rational* here?

We are now in the realms of philosophy. But motivations (reasons) and chronology are important to the themes of this book, so I'll continue. It is fair to argue, as some philosophers do, that we have a *bias towards the future*[14] in our attitudes towards the chronology of events in our lives: We can prefer a particular good thing, such as a tasty meal, to be in our future than in our past (although the memory of the meal is good, it might not be *as* good as the anticipation of that meal in the future). We may prefer a particular bad thing, such as a trip to the dentist, *not* to be still in our future but to already have happened.

We also have a *bias towards the near*[15]: We prefer that future good things will happen as 'close' to the present as possible and that future bad things will happen as 'far away' from the present as possible. There are situations where we accept that we must experience one of the bad things sooner rather than later, to 'get it over with', but this again demonstrates a bias towards the near because we consider that 'getting it over with' *sooner* will be somehow *less bad* than waiting until *later*. This seems entirely subjective: Surely the same bad thing will contain the same degree of 'badness' whether we do it tomorrow or in a month's time.

The form and extent of our future/near bias varies (somewhat) from person to person. When we say that some people 'live in the past' we do not necessarily mean that they think that *all* the good things, in the whole span of their lives, were in their past. We just mean that they talk as if *more* good things happened to them in their past than are likely to happen to them in their future. We are quite used to expecting that 'old' people will have this perception, because (we rationalise) they have more life behind them than ahead of them, and the life behind them was of better 'quality'. But old people are just as likely to have a future-bias as anyone else: They too would probably look forward to that tasty meal later today more than they are enjoying the *memory* of the tasty meal they had last week. The future/near biases of people in constant pain, as opposed to those just remembering or anticipating pain, may be quite different. Perhaps their pleasures, blunted in the present by their ongoing experience of pain, are more fully experienced in memory. And, though it may be hard for us to imagine such people, what about the temporal biases of those experiencing constant pleasure?

It's easy to feel that we have an intuitive understanding of the 'normal' *desires* and *reasons* of the living because we are, ourselves, alive. But what about the desires of the dead? The dead cannot express a preference for any good things (or for the avoidance of any bad things). Yet we give weight to the desires of the dead by carrying out the wishes they expressed *before* they died, despite the fact that they will never know whether we – their survivors – carried out those wishes or not. It may seem perverse to question this tendency, but philosophers do question it. The 'common sense' answer is that we carry out the wishes of the dead out of respect for the wishes they expressed while they were still alive. It is entirely open to question whether respect is the main motivation or whether the true motive is to fulfil the requirements of the conventional expectations of other living people.

The wishes of the dead are relevant to my theme.

## I'M NEXTING

Our perception of time is, then, just that. Our brains perceive the passage of time in the way that evolution has dictated that they will. Specialised parts of our brains do an excellent job of 'making future', or 'nexting' as psychologist Daniel Gilbert has called it.[16] We can hypothesise (without getting too far into the architecture of brains at this point) that because the brains of earlier *hominins* had fewer neurons and were structured for different specialisations,[17] they were not as efficient at doing this as ours are. So perhaps time was different for them. And because of that differ-ence, maybe it was less painful. Maybe their limited ability to imagine the future saved them from the kind of anguish about what is yet to come that many (or most) modern humans experience. Perhaps our modern intuitive perception of time is actually quite bad for us.

A 'long view' of the future is certainly useful in various ways: It allows us to anticipate good things for a long time in advance; it allows us to make long-term plans for ourselves and our families; it even allows us to think about what might be right for future generations (although we seem to find it hard to place this very far up our list of priorities). But the long view also 'backfires' by giving us a strong awareness of our mortality and the 'mental space' to worry about what may or may not happen to ourselves, and to our loved ones, in the longer term. On balance, I consider my nexting capability more a blessing than a curse, but then again, I am not in a position to judge how it might feel not to have this capability. Some forms of meditation – for example, *mindfulness meditation* – engage with this concern by training participants to 'be in the moment'. Is this an attempt to switch off the *nexting*, or rather just a means to allow them to be less concerned about the future, while still being fully aware of its potential outcomes?

## PERSPECTIVES ON TIME

This kind of meditation is just one example of the ways in which we may try to shape our perception of our future (or past) to try to get our lives 'into perspective'. But who can say for sure what that means? The relevant time perspective we have changes from day to day and from situation to situation. It's fine to sense one's own startling insignificance in the vast sweep of time while standing under a breathtaking starry sky, but is it appropriate to feel insignificant within a sweep of a hundred years past to present, or a thousand years into the future? Should we be driven *insane* by our insignificance, as with Douglas Adams' 'Total Perspective Vortex'?[18] What is the temporal significance *threshold*? I don't know. I do know, however, that human beings tend to try all kinds of methods to avoid being insignificant. The Zen master might nod, knowingly, and say that the pupil needs to put this perception of time behind him. But who is the Zen master to say? He has made himself temporally significant by becoming a Zen master. If accepting insignificance were the goal then surely it would be better to do nothing, to say nothing, to care about nothing, to *be* nothing. Maybe that *is* the goal?

Regardless of what the Zen master, or anyone else, thinks, time's arrow will continue its flight. We could, with a nod to the Zen master, conclude that time is just an illusion. That may or may not be true, but it may be good for us to have that perspective. If we were like Derek Parfit's *Timeless* character, we could see all 'temporal states' as equal. We would be happier, because we could take just as much pleasure out of good things that happened to us in the past as good things that are happening in the present, or that are still to happen in the future.[19] That sounds like a very pleasant situation, but it starts to sound odd when it comes to the bad things. Can you imagine being just as distressed *now* about a painful visit to the dentist *last week* as a painful visit to the dentist that is happening *right now*? I do not mean to imply that just because we may not be able to imagine such a state of mind that it cannot exist. Most of us cannot even imagine constant pain or constant distress, despite that such states are hideously real for those experiencing them. States of constant pleasure might be harder to come by than states of constant pain, and 'achieving' them might require some radical cognitive re-engineering,[20] but such states are not impossible. Philosophising about different 'ground states' of being requires us to mentally deviate from our norms, to walk in some strange shoes.

Now walking in almost-inconceivably strange shoes, let us think about the attitude demanded of Timeless towards her own impending death. She is due to die tomorrow, but Timeless doesn't mind because she has an entire lifetime of memories of past good things to enjoy in the *just the same way* as if she was 'looking forward' to those good things in the future.

Parfit doesn't, of course, expect any *real* person to be able to actually hold this perspective. *Timeless* is merely a philosophical 'plot device' to illustrate a (clever) line of reasoning. Parfit concedes that, although the Timeless perspective might be good for us, evolution has not endowed us with brains that tend to think this way, and as a result, the fear of death is common in our species.

There are persons who strive to hold a state of mind something like the Timeless perspective. If we leave aside the mythological elements of their culture, we could say that the mindset demanded by what the indigenous peoples of Australia call 'the Dreaming' is one of timelessness: to live in the human world but always overlapping harmoniously on a continuum with the 'Dreamtime' of the ancestors and with the realm of what is yet to come.[21] And, as touched upon above, we know that the lifelong training of some types of Buddhists to 'be here now' demands a focus on the present so strong that temporal considerations melt away. Even religious traditions that do not demand a timeless perspective of their followers happily grant unfettered timelessness to their gods and goddesses.[22]

Am I afraid of death because I am unable to get my lifespan into the appropriate perspective? Do I have a distorted perception of what constitutes a reasonable lifetime? That is certainly a possibility. There are people who say that they have managed to get their lives 'in perspective', and perhaps I have something to learn from them. But only, I feel, if they aren't cheating; only if their reasonable defences have not been over-engineered to become elaborate walls of faith or other distraction, constructed chiefly as shields against genuine contemplation of the onrushing darkness.

Perhaps the most frustrating thing about time, for humans at least, is its damned *inscrutability*. We don't know what it is, but it affects everything we know and love, and we can't get back our pasts or undo the bad things. But there is absolutely no reason it should be 'scrutable' for us; the geometry of spacetime is – like the third dimension to a Flatlander – alien to us. The passage of time *is* the effect of entropy, and the effect of entropy *is* the passage of time – *ergo*, wonky philosophical Zen paradox. Or, to put it another way (in the words of Elbert Hubbard), 'Life is just one damn thing after another.'[23]

## THE ANTI-ENTROPIC DEVICE

A dictator ordered his best scientists to build him an anti-entropic device. When the device was ready, the dictator went to the lab for a demonstration. Before him stood a large wooden box, out of which an old man stepped.

'What's this?' shouted the dictator, greatly angered.

'I am an old man,' said the old man, 'I have led a good life. I have maintained my personal relationships and brightened the lives of those around me. I have contributed to environmental causes. I have cleared landmines. I have helped to reduce poverty, disease, and starvation. I have saved whales. I have always tidied up after myself. Compared to you, "General",' he said, 'I am an anti-entropic device.'

'You are a fool,' said the dictator. 'You are in no way *anti*-entropic. You cannot *unhappen* things. You cannot *undestroy* the things that have been destroyed. The very *best* that you could claim is that you are an entropy-*limiting* or *-slowing* device.'

'That's a fair point,' admitted the old man. 'It is true that I cannot, for example, *unhappen* the damage *you* have done. I *am* in fact more of an entropy-*limiting* device. And, as finely-tuned an instrument as I am, I still have unresolved questions in my mind.'

'Such as?' said the dictator.

'Will doing what I am about to do make me a better one?' said the old man.

The dictator laughed and shook his head, turning to his chief scientist.

'Now I see what this is. You are playing a joke on me. The device works as it should, as a *time machine*, and this old man is actually myself come back from some alternate future, as a demonstration of its effectiveness. I thought he looked familiar.'

'No,' said the old man, revealing a gun, 'That would be impossible. I am just an old man who has been sent to kill you.'

## 2 INKLINGS OF THE DEAD ZONE

∞

It was a difficult pregnancy: more than usually nauseating, and punctuated by bad dreams and forebodings that continued for some time after my birth. Hearing the screams from a fatal car accident on the road in front of our house unnerved my mother further. I was her fourth child, and the novelty must have worn thin. She had expected / hoped for another girl, but she got me. My parents named me after a pair of uncles on my father's side: a never-was-my uncle who died in infancy, and a living alcoholic-to-be one. If I were of a superstitious inclination, I might now imagine the 'birthing bed' in the old schoolhouse in Balavanich groaning under the weight of ill portent. Each of my four other brothers was born in hospital, but my sister and I can lay claim to having been born at home in the big birther – she in Glasgow, I in windswept Benbecula.

I was a difficult baby, and different-looking from the others. I spent a lot of my time 'girning' (crying) and projectile vomiting. My mother recounts a tale of taking me to a circus, where I cried throughout the show until, just before the end, my father took off his new spectacles and I, perplexingly, stopped crying. This may seem an unremarkable story, but it sticks with me because I can imagine that, even as a small child, my brain would find something oddly and uncomfortably out-of-whack in dad's new glasses. Genetic inheritance, *epigenetic* changes, early *cortisol* diffusion, and myriad other factors (including random quantum fluctuations[1]) had caused my brain to form that way.

At age two, my brother gave me paraquat pellets.[2] He, aged four years, ingested a small amount of the poison by licking the lid of the tin. I ate some unknown, and unknowable, quantity. My mother discovered us, and rushed us to the local army-base hospital to have our stomachs pumped. The scheduled flight turned back to collect us, and we were flown to Glasgow. At Yorkhill hospital, doctors pumped our stomachs again – tubing us with fuller's earth, and with syrup of ipecac to induce further vomiting.[3] They released us, around Christmas-time, after two weeks recuperation. The Scottish media lapped up the homegrown good-news story. A fifteen-year-old boy from one of the other islands who had accidentally drunk liquid paraquat from a Coke bottle a few weeks earlier was not so fortunate. He died of lung damage and other complications several months later.

We left Benbecula in 1975, when my father got the job of Director at the newly-established Gaelic-college in Skye. This looked like a great prospect for him and promised a new direction from the primary-school teaching posts he had held since leaving college. Skye was leafier and closer to civilisation, but still fairly 'wild west': Beat-up old cars without tax-discs plied the winding, narrow roads; opportunistic hippies piled in from England to stake their claims on the small, vulnerable crofting communities.

It was here, aged four years, that I first came to understand something of death. My mother was driving; kids bouncing, seatbelt-less, on the long leather couch of a back seat. She mentioned the death of some animal or other, and I piped up, 'Are you going to die one day, mum?'

'Of course,' she laughed. 'Everybody dies sometime, but it won't be for a long time yet.'

I cried inconsolably, the moisture from my tears causing the backs of my bare legs to squeak uncomfortably against the leatherette.

So far, so ridiculous – I don't remember my Santa Claus moment, but I remember my Grim Reaper one. What a miserable child. Yet I wasn't miserable. The old barn buildings that housed the 'college' were endlessly fascinating to explore. My mother worked tirelessly in the college kitchen preparing meals for beaded and bearded weirdos, while my brothers and I had plenty of time and opportunity to play on our own and, periodically, to steal food from the well-stocked catering larder. I remember discovering olives there – the ones stuffed with red pepper: 'What's this? Ugh, it tastes strange, and I still don't know what it is. Let's have another one.' Odd, bitter, an acquired taste, but moreish.

My favourite haunt was the dead wood on the hill behind the cottage where we lived. But I would never have considered going there alone. I was fearful, and sought the company of my older brother and, later, my younger. The three of us formed a tight caucus, demanding attention and often being disciplined as a direct result of my older brother's 'experimental' nature. Because I looked 'strange' and was less charismatic than the other two, I was frequently the subject of those experiments. I would only attain any feeling of boldness when spurred on by the others, and even then, I would take part in their dangerous escapades always with a burning knot of dread in my stomach. And the subject of burning knots reminds me of one of my brother's experiments. He once lit a piece of blue nylon rope and swung it around, laughing, as the fiery blobs zipped to the ground. A stray one hit my hand and stuck there like napalm, alight. I still have the scar. He was never a *bad* child, just wild and, it seemed to me, carefree.

His rabid adventurousness was not without consequence for him. One day, while playing in a tree by the cottage, he slipped and fell to the old concrete shed-base where I was standing. His head hit the edge of the base; he lay there, motionless at my feet, his pale face shrouded in the hood of his duffel coat. Blood oozed out onto the back seat of the car where he lay beside me as my distraught parents drove to the local hospital, ten miles distant. The doctor that stitched up his head said the hood of the duffel coat had probably saved his life, cushioning the force of the base-edge blow.

The brush with head-trauma didn't change him. His quest for cool and dominance only grew as he became more aware of his own infectious charisma. I loved to work together with him on small projects such as building toy boats and go-karts – watching his fevered concentration and commitment as he banged in the stubborn nails. But this joy was often short-lived. He often became frustrated with our burgeoning creations, and in fits of red rage, smashed them to smithereens. I couldn't comprehend these actions, and found myself wincing as if the hammer blows were falling upon me. Eventually, his behaviour broke some small component in me; I began to implore, to question, and to argue.

At least there was now some element of creation prior to his destructive acts. This was a kind of evolution from his younger days of smashing brand-new toy cars 'to get the driver out'.

My younger brother seemed carefree in a different way. Not for him the frenetic experimentation – he seemed to take life in his small stride, charming all who met him. Blame never stuck to him, and of the cause of one of his near-death experiences, he was, as usual, innocent. My mother kept goats, and my father had painted the enclosure fence with creosote.[4] My brother, aged only three or four, inhaled the fumes from (and probably ingested) the still-drying wood preservative, while visiting the goats. Later, back at the cottage, he took an unmoving epileptic seizure. His eyes drifted off to one side. Froth formed on his lips. My mother shook him, spoke to him, and stroked his face, but he simply stared, silent and unblinking. I cried into my mashed potato as we watched him in terrible fascination, awaiting the arrival of the doctor. Later, in hospital, he eventually regained full consciousness. He remembers nothing of the incident, bar the stench of creosote.

The goats were a source of trouble and amusement. We took a trip to the north end of Skye to collect the first one – my mother and our grandmother in the front of the Ford Cortina estate, my older and younger brothers along with me, as usual, on the back seat. The goat was deposited, loose, in the boot of the estate car. A sudden stop at a 'lollipop' road works signal caused the goat to fly, in Newtonian compliance, out of the boot and onto the back seat with us. The lollipop sagged to the ground as the road-worker doubled in uncontrollable laughter.

We were a ragtag band of hand-me-down children. This was the Seventies, and ill-matched scruffiness was almost *de rigueur*. The occasional new pair of wellies and gigantic, square, cotton 'sailing smock' were the only departures from the thrifty old-clothes culture that my mother espoused. In this context, I – a gap-toothed wurzel of a child – probably didn't look too out of place. Despite the similar attire, however, I couldn't help but feel that I stood out in marked contrast with the impish good looks of my peer brothers.

We spent many holidays on the tiny island of Raasay. My father had grown up there, and later met my mother there, while she was up on holiday from Devon with *her* father – himself a native Raasay man. The island stilled to a suffocating Presbyterian silence on Sundays, save for the unnerving sound of Gaelic psalm-singing emanating from the church, or the gravely muttered bible-readings from the old folk by the fire. I have warm, though indistinct, impressions and recollections of other times there: birds tweeting in resin-scented, waving pine trees; peat smoke and pipe smoke; calm waters and clinker-built boats. My father's parents, who lived there, are little more than vague sketches in my mind: a kind-faced Seanair[5] in cap and thick jumper; an old woman with a black shawl, seated in front of the whitewashed wall of a small stone cottage.

Despite what my father may have wished, religion did not become a major strand in the fabric of our home life. He tried, in a sometimes ill-tempered, sometimes gentle, mostly half-hearted way to make it a daily feature. My cohort brothers and I said a Gaelic prayer with him at bedtime, and I found this, back then, quite comforting if a little perturbing. At bedtimes when he was not available, I would pray a worry-mantra inside my little head, always afraid I would miss someone from my list of those to be kept safe by God, and by sin of omission, condemn them to harm. I also remember my father saying grace at the dinner table while we sat restless, chuckling heads only just bowed. As the prayer- and grace-saying faded out, mostly due to our growing embarrassment and lack of interest, his frustration about our absence of piety sometimes bubbled over, tumbling out as snarled condemnation. Years later, it became clear to me that he, at least in part, blamed my mother and her instinctive atheism for our failure to develop religious sensibilities. He didn't seem to consider the possibility that it might have been largely because we had developed minds of our own.

I later encountered a new tactic aimed at the religious indoctrination of children, when a small friend told me about Sunday school. There was no mention of God, just a collection of bright stickers showing a bearded man in flowing robes, and a big album to stick them in. I had seen this kind of sticker-collecting before: football players, sports cars, etc. I thought it looked like fun, and the boy assured me that I would get my *own* sticker album if I went along on Sunday. The reality of Sunday school was somewhat different, but the proselytising 'teacher' was cunning enough to keep me duped for a while. The charade ended when I repeated to my mother something – obviously something I found troubling – that the teacher had said. She, realising the tawdry way we had been drawn into an environment that no longer sat well with us, suggested that we stop going.

Primary school was enjoyable, despite the bun-haired religiousness of my first teacher. She was kind enough for the most part, and I keen to please. I rarely stepped out of line, but was always aware of a certain dark glee I would feel at the thought of breaking free of the bounds of the strictly-bordered boxes and shapes we were given to write and draw within. Generally, my desire to please and console overrode any notion of making waves. My earlier ambidextrousness faded and was soon forgotten; rapid pruning of redundant neural connections – use them or lose them. In that small battle of *sinister* and *dexter*, a right hand claimed the delimited colours.

My painful shyness continued to be a limiting factor, but my confidence grew as I began to realise I was good at some things: English, arithmetic, art, music. I loved to prove that I could spell long and complicated words, and understand their meanings. It didn't occur to me at the time that this was not, to my peers, a particularly endearing quality. I changed tack slightly and made more of an effort to make friends. I formed a strong bond with a solidly-built, cheerful boy, who radiated confident insouciance. Girls, though, were *fascinating*, and I took every opportunity to talk with them – to feel the warmth of their proximity and to listen to their calm, soft voices.

Although it was a very small rural primary school, I had other teachers. My strongest memories are of the headmaster, who taught me between the ages of nine and twelve years. He was a thin giant of a man — all wild curly hair and thick-rimmed square spectacles. He must have been quite young, as his tastes were very much of that era. An excellent and good-humoured teacher of 'the basics' (although we greatly feared his infrequent wrath), he was also keen to develop our artistic and literary sensibilities. He played us music from his large Jethro Tull collection; encouraged us to sing and to play instruments; worked with the visiting art teacher to realise grand, whole-class art projects; and read to us from books ranging from *The Lord of the Rings* to *The Ragged Trousered Philanthropists*.

He was also a fan of sci-fi literature, and I remember one particular story he read to us from an *Armada* compendium of science fiction tales. A man hit by a car looks certain to die of his injuries, but doctors give him the option of a new treatment — they can transplant his intact brain into a donor body. He accepts the treatment and begins his new life in the new body. However, he soon develops a relationship with his 'bereaved' wife. At first, she does not recognise 'him', but by the end of the story she has been able to tell, 'by his eyes', that he is her husband. I found this a bit of a weak ending. I was happy that 'he', after having been thought dead, had been reunited with the love of his life, but I couldn't see why, unless his wife knew of such pioneering brain-transplant treatments, she would ever jump to the conclusion that *this man* was her dead husband.

Maybe I have misremembered the story, but, if so, perhaps it is fitting that I have misremembered it in this particular way.

This inspirational teacher stayed on in Skye, for a time, after his wife died in a car accident. But it was clear that whatever dreams they had harboured and plans they had made for their life on the island had not survived her death. He left a few years later to teach at a school further north.

The sea played a large part in my childhood. It was always nearby: a meander through a field and scramble down a slope, from the cottage; then — a couple of years later when we moved a few miles along the road — a two-minute walk from the big house. We paddled, swam, and fished. Small 'cuddies'[6] could be had easily enough from the rocks in the bay, but we sought larger prey. In late summer, a hand-line terminated with feathered hooks and a chunky weight would usually catch you a bounty of mackerel, if you could get out in a boat to deeper water.

I remember watching the seabed as we struck out from the safety of the shore in my father's boat – the friendly pale sand giving way to deep turquoise then fathomless black. I enjoyed the camaraderie of those fishing trips; feeling the frenetic tugging on the line as we hit the mackerel shoal, and the excitement of hauling a silver-flashing line-full up from the depths and into the boat. We would usually gut them on board – slicing off the heads, then slitting the bellies to drag out the intestines, with rank little fingers.

But those trips were frightening too. Summer squalls often blew up, forcing us back towards shore, the underpowered Seagull engine over-revving as its propeller cleared the water on the crests of the largest waves. My other brothers were usually unperturbed, and mocked my strangled, panicked utterances. I tried to keep quiet and carry my fear inside. There were a couple of trips when, engine spluttered to a stop and father rattled, I saw anxiety on their faces too. I loved them more at those times. I understood.

My will to endure the terror of open water began to break. I feigned illness and stayed at home while my brothers chased the endorphin thrill of bouncing through the angry salt-spray in pursuit of darting shoals. I envied them and feared for them, but could not be with them.

It is difficult for me to get the flavour and texture of these memories across. The dark-weave stands out, but I am only emphasising those elements because they are in keeping with the themes of this book. They are, though, what I remember most vividly. And that being so, the emphasis is justified, because it says something about my personality. I, the 'case-study' of this book, need to reveal my personality in as honest a way as I can, so that you, the reader, can make judgements about my state of mind. 'Don't judge me,' is a staple TV-drama sound bite, but it's hollow. We all make judgements about one another. We also make judgements about ourselves and our *own* ability to make rational decisions. Am I rational? Has my fearful nature impaired my judgement? Am I being honest with you and with myself? Judge away.

I don't remember much of my relationship with my oldest brother or my sister from around this time. He is eight years older than I, and she six years older. With the age-gap emphasised, back then, by our youth and differing interests, there weren't many opportunities to make strong connections. We must have been, at times, something of a plague to them. Noisy, messy, and small, absorbing much of our parents' attention and breaking valued possessions, they must have wished for some peace to mature and to go about the important business of impressing their friends.

My sister's world was so different to mine that I didn't know what to make of it. Posters of horses, ballerinas, and David Cassidy adorned her bedroom walls. Her days seemed dominated by concerns about riding hats, saddle soap, and hairstyles. She often came across as haughty and aloof. But she was, I now realise, still smarting from having to leave behind her friends in Benbecula. She clearly found us infuriating at times, but she was not unkind. She kept her boundaries visible, to avoid infringements of the rules governing the tough – though princess-tinted – persona she was building.

My oldest brother was tall, rangy, and again, inscrutable to me. *His* bedroom was a shrine to rock music, specifically to the gods of the electric guitar. He looked an unlikely rocker, and it was clear that the music appealed to him on a level beyond guitar-hero strut and swagger. He learned to play the guitar, and as with other skills he tried his hand at mastering, became highly proficient, blaring his defiance against *something* through overdriven shoebox amplifier. But I think, now, that perhaps he was lonely. The move to Skye had affected his world too. He spent only a few months at the local primary school, where he had only his younger sister to relate to, and even that fragile attachment would very soon change when he was packed off to high school. We considered none of this at the time. We just looked up to him, both literally and metaphorically.

And I felt the music begin to seep into my consciousness: the sheet-metal vibrato of Ozzy Osbourne, the driving muffled-bass pound of Deep

Purple, among others. The sound and imagery of rock music seemed to stimulate my feeling of boundary-free dark glee. It held the promise of an adult world populated by powerful beings who had never restricted their colours to any set shape but let them flow, merge, and overflow. Here was freedom. Here was strength. Here was not me.

I shared the third upstairs bedroom with my two cohort brothers, and when we later moved from the cottage to the big house up the road, I again shared a room with them, though this time a vast high-ceilinged cathedral of a one. In later years, guitar-hero brother got the overhanging 'balcony' in the same space. We looked up to him more than ever.

The big house was astounding to us. Having lived on top of each other, jammed into the small cottage, for pretty well as long as I could remember, the change was dramatic. I think we wondered if it was a mistake. Was it temporary? How had our parents been able to afford such luxury? I half expected to come home from school to be told that we were moving back to the cottage. But it was real, and it was ours. And my parents began the painful, ongoing process of scraping together the relentless mortgage payments to keep it for us. They had overstretched themselves to buy the house, but my mother was strong, hard-working, and determined. She must have pledged to my father that, somehow, she would find the extra money required to keep it. My father resigned himself to the heavy burdens of his side of the bargain.

In our different ways, we thrived there. The location was perfect for new adventures up hill, over moor, and through river. And of course, there was the fractal wonder of a new shoreline to learn. My mother had the grand domestic canvas she had always wanted, and ample garden ground to grow food and to plant trees, shrubs, and flowers. My father had office space in the house and a double-door garage outside, to cram with car on one side and 'useful' pieces of timber and broken machinery on the other. From her base in the softly-appointed end-bedroom, my sister continued her drive towards independence and equine proficiency. My oldest brother blared away his weekends at home in frowning, world-weary concentration. High school, and the hostel where he had to stay during the week, was now his burdensome reality.

Doris Hurrell, my Devonian grandmother on my mother's side, was a woman of endless patience and unreasonable optimism. She enjoyed reading (and snoring beneath) novels of grandeur, mystery, and adventure; the first science fiction book I ever remember setting eyes upon was Frank Herbert's *Dune*, or maybe something by Ursula K. Le Guin, resting near the cigarettes and warfarin on her bedside cabinet. Gran was a radiant feature of our summer holidays: a source of unconditional support for my mother, a sympathetic ear for my oldest brother and sister, and a tireless provider of sunshine-trips and ice cream for my cohort brothers and me. My memories of her inextricably intertwine with feelings and impressions of all the best things about childhood. She was there the day we moved into the big house, standing on the doorstep, gathering us in with a beaming smile. I feel a keen stab of loss as I write this.

At school, I took up playing the chanter with a view to being able, some day, to play full Highland bagpipes. My piping teacher was a short, loud, scarlet-faced man, with hair to match. His body shape had obviously started out somewhat square-set, but middle age and beer intake had rounded him off in parts, so that he now looked like an over-inflated cube – an effect emphasised when he puffed up in the act of feeding air into his bagpipes. I had a sense that I was, somehow, ideologically ill-matched to this traditional musical instrument, but I persevered, bending my will to the task and to the demands of my relentless tutor. Later, when my younger brother took up the chanter, he quickly reached competency and then excelled. Not for him the silly worries about this not being the right instrument or not being 'any good' at it.

Daily life gradually became easier for me as I learned that primary school was actually a safe and pleasant place to spend time. I could push my dark forebodings to the back of my mind during the day. But they would, inevitably, emerge by night. Insomnia, punctuated by bad dreams, was usually my lot beyond the witching hour. I sought comfort from my brothers and developed methods for keeping them awake, to delay the onset of my restless vigil. Sometimes the night became a playground for all three of us: when we giggled together at the full moon's striping of our faces and bodies as it shone through the gaps in the Venetian blinds; when we ventured outside, in our pyjamas, to catch moths fluttering at the arched window. But my brothers were easy sleepers. They soon tired of my blatant tactics and of wakefulness, drifting away and leaving me at the mercy of the suspended-head lampshade and the wraiths in the toy box. Gamma's bed provided a measure of sanctuary on nights when dread trumped my shame and his chiding. Other nights, I got up and wandered the big house, glass of milk in hand – become a forlorn spectre myself, shying away from dark corners and glassy reflections.

My fears and worries were at least varied in nature: for the safety of my loved ones, nuclear holocaust, terminal illness, ghosts (a mortal terror of immortal revenants), werewolves, vampires, whatever scare-story had been on the TV news that evening. I can even remember feeling quite convinced that the Skylab space station was going to crush me to death upon its fiery re-entry into Earth's atmosphere. And it's not that I was a stupid boy; I understood that some of my fears were irrational and that I *should not have them*. But for some reason they lodged in my brain and tormented me – weakly in the daytime but strongly by night.

My fair-haired friend and his family were a source of warm-hearted distraction and refreshing pragmatism. I began spending whole weekends at their house. There were things to do there, and his father always seemed to have time available to take us out in a boat, for a brisk walk, or off to watch a rugby match. It was such a roller-coaster that I would forget my worries; the family were never gruff or pushy, so I felt that I could just blend into their harmonious and easygoing flow. I could forget myself. And, while my friend was confident, it was clear to me that he also had depth. It was a well-hidden quality, but it interested me and ensured that I never tired of his otherwise uniformly breezy personality.

It was he who first introduced me to computers. He had been tinkering with a fractious Sinclair ZX81, and I happily agreed to help him with typing into it some seemingly-endless string of code, via its awkward and unresponsive flat-plastic keyboard. It was a simple thing, but to me it was like modern magic. You typed in the code, ran the program, and it *did* something. I can't remember what that first bit of programming produced. It wouldn't surprise me if it had just been the word 'Hello' flashed up in glorious greyscale on the screen of the attached TV.

While I was distracting myself and looking for my place in the scheme of things, my mother busied herself with the task of working her fingers to the bone. She had a job as a cook, and in tandem with this, picked whelks (periwinkles) on the shore, when the tide was right. I have done this myself since, and I now know that it must have been a miserable way for her to make a living. The dealers sell the whelks to overseas buyers, making healthy profits on the deals. The pickers get a comparative pittance. We often went with her to the shore, dabbling along the tide-line while she bent her back to rooting the small brown sea-snails from their rock and seaweed habitat. I don't remember her complaining about the work, at least not in front of us.

My father's job at the college came to an acrimonious end, forcing him to look for work further afield. The only vacancy available that required his almost-unique skill-set was in Stornoway: a fifty-mile drive, three-hour ferry crossing, and then another forty-mile drive away. He faced the financial reality and took the work, and along with it, the difficult readjustment and the strain it put on the cohesiveness of the family.

Money became extremely tight as we awaited the first income from my father's new employment. We lived on homemade chips and tomato ketchup – no hardship for a child, but probably not very healthy. My mother switched off the expensive oil-fired heating, and we moved into the living room by night to benefit from the warmth of the open fire. She had an eclectic taste in television programmes, and so we (my two closest brothers and I) got our first taste of 'late' night TV. Carl Sagan's *Cosmos* was screening at that time. It had a deep impact on me: Sagan's soft and rational tones, ethereal music, weird sci-fi graphics; and of course, the mind-boggling (and -expanding) subject-matter, including black holes, evolution, time, and relativity.

The pay-cheque finally arrived, and our extended living-room sleepover ended. My parents somehow coped with the awkward working arrangement, for a while at least. There was even, at one point, talk of our moving out to Stornoway. They took us there 'on holiday' and insisted on showing us the school and some of the grim accommodation on offer. We weren't having any of it. We made it clear to them that we despised the town, the surrounding landscape, the whole island, and its entire people, and that we would never leave our school and friends to live 'out there'. They eventually conceded that it might not be best for us. They also accepted that my father would have to give up working there and come back home for good.

Over time, the tight gang consisting of me, my older brother, and my younger began to splinter. I still spent a lot of time with them, and we still shared a bedroom, but they were effortlessly finding plenty of new friends and interests of their own. My older brother continued to be drawn to exploits involving danger: riding motorbikes, trips out to deep water in tiny inflatable boats, smoking surreptitiously-acquired cigarettes. My younger spent a lot of his time, either on his own or with friends, by the water: on the shore exploring rock-pools, digging for lugworms, or worrying crabs; on the pier fishing, or hanging off the low steps to scan the seabed for lurking saithe.

I squabbled with him as he developed ever-more infuriating-brat-like qualities. He picked up new skills quickly and carried them with easy self-assurance. This alone would have been enough to affect my relationship with him – I found it hard to contain my envy – but it was also, I think, that I feared losing him to his shiny new interests and acquaintances.

I bottled up my rage against the injustice of these weakening bonds, and against the regular gut-wrenching pressure to take part in the mischievous escapades of my brothers and their peers. My desire to be like them and not to come across as 'weird' usually overrode my anxiety, but I sometimes gave myself away, unable to contain my fear and blurting it out in snivelling protestation. This did nothing to ease the true cause of my anger and frustration because, at that time, I was only tentatively in touch with my own emotions and sense of self. So I brooded, turning events over and over in my mind, and burning with a plaintive wish to find a certain deliverance – from feelings I could not fathom, and into the arms of something that I was entirely incapable of defining.

The arrival of a new addition, another boy, in 1981, brought fresh joy and consternation. My sister – having taking the pre-birth stance that it was 'disgusting' for mum to be having another baby 'at her age' – mellowed, and accepted the child once he became fact. I was delighted. For me, his tiny, happy, burbling innocence brightened the big house and made it a complete home. I gladly took opportunities to play with him and, when heavy-lidded weariness got the better of him despite his best efforts to thwart it, to put him to bed. I struggled with the prospect of leaving him when, two years later, high school and the hostel came to swallow *my* weekly life.

# 3  DEAD ROOTS

∞

Custom, then, is the great guide of human life. It is that principle alone which renders our experience useful to us, and makes us expect, for the future, a similar train of events with those which have appeared in the past.

—David Hume, An Enquiry Concerning Human Understanding

'Are you going?'

'I don't know. I've had enough of these things. Having to listen to all that hellfire and damnation shite – it's all wrong. It's horrible.'

'Me too. I wasn't going to go, but Dad's worried there won't be enough family there, you know, to carry the coffin.'

'... OK. I'll go. We can catch the first ferry back, so we don't have to hang around afterwards.'

I am a futurehead, and the past is not my domain. Context is important, though, so I feel it necessary to glance back over the history of death and establish some sense of my place within the chronology of ideas about how death and 'the dead' should be managed. Again, these have just been *conventions*, but conventions so entrenched that they have been defended down through the ages with the kind of zealous mystico-religious fervour that has often made it seem that the business of death/dying was/is far more important than the business of life/living.

As I have decided to break with the death conventions of my ancestors, it would also be worth having a look at what those conventions were. The Celts were, as far as we know, a proud, skilled, and culturally-vivid people, so looking at their customs should be interesting. I will also touch upon the customs of the culture that emerged here in the Highlands and Islands of Scotland after the introduction of Christianity. While I do not identify with this culture, it makes a certain kind of sense to search 'close to home' for clues to the emergence of beliefs about death, and to the practical consequences of those beliefs. Here will we find the conventions that my decision, in effect, defies.

But first, we need to look further back in time. A treatise on the death-rites of our amphibian ancestors might make for rather a short chapter, so I will start instead with species with which we can more readily identify: early modern *Homo sapiens*, and *Homo neanderthalensis*.

## RUST

Death happens a lot. Some estimates suggest that as many as 100 billion people have died in the last 10,000 years (since around the time of the demise of the last of the Neanderthals) alone.[1] Mystical conventions surrounding death were established some time in anatomically-modern-human (AMH) history. While good evidence of 'religious behaviour' in this regard does not appear until the Upper Palaeolithic era (approximately 50,000 up to 15,000 years ago),[2] behaviour such as the burial of bodies − seen by some as hinting at ritual veneration of the dead − may have begun as early as 300,000 years ago,[3] prior to the emergence of *Homo sapiens*. The earliest 'ceremonial burials' of any kind took place around 100,000 to 120,000 years ago.[4] I'm not sure what type of anthropological scalpel is generally used to separate 'religious behaviour' from 'ceremonial burial' as evidence of the holding/non-holding of *religious beliefs* by these humans, but I'm not going to debate that point, because, in the context of this book, it will add little to our understanding of the wider subject at hand. It is enough to state that early modern humans had begun to establish ceremonial *conventions* for dealing with their dead as early as 120,000 years ago.

What happened before 300,000 years ago is open to question. Chimpanzees generally take some interest in the bodies of their dead, poking and prodding them for a while before leaving them alone. There is evidence that chimps sometimes feed on the bodies; there have also been documented cases of chimpanzee mothers being extremely reluctant to give up the bodies of their dead offspring.[5] It is reasonable to assume that early *hominins* behaved in similar ways – taking some interest in the bodies of their dead, and perhaps pining over them, prior to leaving them to scavengers; or, sometimes, performing the scavenging role themselves, in eating the bodies. The surviving archaeological record of our genus provides evidence of an initially-gradual change in the manner in which corpses were disposed of – evidence of burial, though perhaps only to discourage scavengers, followed by a more rapid change after the arrival of species *Homo sapiens*.

Neanderthals too may have buried their dead, but this is still hotly debated.[6] It seems to me that some involved in this debate attempt to impose our values on Neanderthals. Though evidence is mounting to support a conclusion that they were not the kind of primitive near-apes they were once assumed to have been, clear-cut proof of *human*-style burial practices among them would not, to me, count as a key supporting factor.

Some archaeologists see the use of red ochre in human burial sites from at least 100,000 years ago as evidence of ritualistic behaviour connected with the disposal of dead bodies. There are varying hypotheses about the 'symbolic intent' behind this. It may have represented blood (and its connotations of menstruation, life, and death) and so may signify belief in a birth/rebirth cycle.[7] Pigments of red and other reddish/yellowish colours may have represented the sun, another important symbol of the cycle of life and the seasons. However, my preferred explanation is the practical – though still symbolic – one put forward by archaeologist Timothy Taylor and others: that the chemicals in ochre helped to stabilise and preserve the bodies.[8]

As language evolved (from origins as early as 200,000 years ago[9]) so other kinds of representation flourished. Rock carvings and cave art provide many of us with a visceral sense of connection to those early peoples, and they are often cited as evidence of the emergence of modern human aesthetic and ritualistic traits. Carved stone 'Venus figurines' are suggestive of reverential awe of life-creating female fertility. Red ochre and carbon black hand-prints from some 40,000 years ago evoke sacred moments frozen since the dawn of humanity. And although the subject matter of cave art often seems related to everyday concerns of life and survival – beasts of prey, fire, water, and so on – some anthropologists think that shamans in ritualistic trances undertook most cave paintings as a form of 'hunting magic'. Representations of the human form are less common in cave art, and themes of childbirth and burial are quite rare. Some *shaman* theorists take this as evidence that depictions of humans – particularly those at the beginning or end of life – may have been taboo in Palaeolithic belief systems. In the scramble for deep anthropological signi-ficance, simpler 'art for art's sake' explanations tend to get ignored.[10]

I suspect that, for many of us, our thoughts turn to ancient Egypt if we try to bring to mind the ways in which earlier civilisations dealt with their dead. The subject is interesting and evocative, set against a background of powerful pharaohs, beautiful queens, animal-headed gods and goddesses, and gigantic stone pyramids. I think, though, that there is also another reason for this 'comfortable' relationship with the death customs of this era: Those customs appear somehow 'clean' and 'respectful'. Or perhaps it is simply down to familiarity: We hear about these subjects as early as primary school, where we learn that, while different from us, these peoples were 'civilised' and handled their dead with due care and attention. (Generally, ritual sacrifice elements of those customs are swept to the corner of the discourse, under the reed matting.)

Given that the series of dynasties we refer to as 'ancient Egypt' spanned some 3000 years, we should not be surprised that the death traditions of those societies 'evolved' over time. We know, from what was recorded in sources such as *The Book of the Dead*, that Egyptians of this period believed in an afterlife, which they called the *Duat*. The soul that developed in the imaginations of these peoples became a tripartite one, consisting of a 'life force' (the *ka*), a unique 'personal spirit' associated with the 'strength of character' of the person in life (the *ba*), and – for the 'worthy' – an after-death immortal union of these two (the *akh*).[11] Along with belief in these various facets came the baggage: elaborate ritual aimed at cementing the *ka*/*ba* union, and religious elite placed to oversee and enforce the various proceedings.

As Ancient Egyptians found, imagining an afterlife running in parallel with, and looking quite similar to, the living world seems to come quite naturally. Placing the dead within such a 'landscape' is more complicated; in this regard, the body presents (and represents) something of a problem.

For me, one of the important aspects of the significance of Ancient Egyptian death customs is the indication of a belief in a *physical* connection between the world of the living and the world of the dead. The reverence with which the corpses are handled, and the way that they are preserved, shows that these people believed that the physical body would still be, in some way, *required* in the afterlife. This is an interestingly paradoxical belief, and one that the 'tripartite soul' idea was used to explain. The physical body is embalmed, internal organs removed,[12] and sealed in a tomb. Every effort is made to keep it dry and safe from attack by vermin. Yet, at the same time, they believed that this person had an incorporeal 'spirit'; what possible use, then, could this *akh* have had for a dried-out husk of a physical body? The prevalent explanation has been that Ancient Egyptians thought that the spiritual elements of the deceased would still need a physical 'home'. I am not so sure.

I realise that the questioning of paradoxical religious beliefs is a subject that would (and does) take up a vast amount of print space. I have only highlighted this particular one to illustrate how belief in a physical connection between this life and 'the next' has changed.

However, part of the paradox we see in such behaviour comes from a modern misunderstanding of the purpose of mummification and entombment in ancient Egypt. The world of the living was replete with terrors, so was the land of the dead. Mighty individuals hosted mighty *bas*. The veneration of those rulers is apparent in the treatment of their remains, and it is understandable that we wish to see pyramids (especially the early, stepped ones) as their 'stairways to heaven'. But, in so doing, we may miss the more ominous significance of such extravagant entombment: that these people believed that powerful and potentially-dangerous souls needed to be *contained*, lest they escape and terrorise the living.[13]

## SOUL MINING

The devout peoples of today that purport to possess an immaterial soul might struggle to recognise the version believed in by earlier human civilisations. And most would be justly horrified by the prospect of ritual human sacrifice as a means of giving that soul the best possible start in its 'new life'. Certain Iron Age peoples, for example, apparently had no such qualms. Personal servants were ritually slaughtered that they might continue to serve their dead Scythian kings in the afterlife. Thracians slit the throats of the favourite wives of their dead warriors, so that those high-status males would each have a companion there.[14] Norse ritual practices involved the sacrifice of slave girls and boys initially led to believe that they would join their dead chieftains in *Valhalla*, only to find out at the last – from the cold lips of the 'Angel of Death' (*Malak al-Maut*) – that they would not be admitted there.[15]

It is all too easy for people of faith today to feel that *their* religions swept away such 'barbarous' traditions. The Old Testament story of Jephthah's vow to burn his daughter (in the Book of Judges) is less well known than the tale of the 'mercy' of God in deciding that Abraham did not, after all, have to kill his son. For Abraham, the instruction from God turned out only to be a 'test'. Jephthah's vow, on the other hand, can be read as having been carried out as promised.[16] Although the interpretation of the Jephthah story is contested, I see both examples as showing willingness, on the parts of these characters, to commit infanticide based only upon terrible whims of belief viewable today as signatures of a most psychopathic form of OCD. Never mind constant washing of the hands or checking of the cooker switch – this is full-scale, grandiose and lethal intrusive thought: God *told* me to do it.

Whatever our interpretation, it cannot be claimed that these stories are not set in times and societies familiar with extreme violence – including sexual violence – and ritual human-sacrifice.

In the New Testament, Jesus is judged and slain primarily by foreign aggressors (the Romans). Perhaps, by distancing them from the ritual brutality, this makes the story of his death more palatable to modern Christians. But the 'consumption' of Jesus by his followers, including present day ones, is, of course, ritualistic in the extreme. He is portrayed as a *scapegoat*,[17] washing away the sins of the masses in the very act of dying – bloodily, painfully, and ignominiously – on a wooden cross. The frankly disturbing rite of the Eucharist (known in these parts as Communion) alludes, on all but the most doctrinaire of interpretations, to older religious traditions that would have involved ritual cannibalism and human sacrifice.[18]

I could go on, but you get the grim picture. Ritual slaughter continues, though to a limited extent, to this day: in *muti* murders, in 'honour killings', in flying aeroplanes into tall buildings for 'martyr' reward in the afterlife,[19] and so on.

We can read into the examples given, and into many others, that 'the soul' was, and still is, of the utmost importance to some. The idea of a 'duality' – the physical body and the incorporeal spirit – has been developing since the evolution of the human capacity to think in abstract terms and to weave mystical concepts from threads of fevered, often drug-inspired, imaginings. But empirical observation no doubt also played a role. For example, sensations associated with what we now call a 'phantom limb' may, in earlier times, have been taken at face value as proof that cutting away part of the physical body would not necessarily sever its spiritual counterpart.

The notion of some form of physical connection to the afterlife is evident from Celtic archaeological remains such as those found at Vix in Burgundy, France, and at Hochdorf, Germany. High-status individuals were sometimes buried wearing gold or bronze adornments including armlets, belts, and torcs.[20] This imagined afterlife again seems a curiously materialistic place – one where precious metals are still valued and where the status of the wearer is preserved by continuing to sport them after death.

We should not assume, however, that the intention behind burying 'grave goods' was in all ancient societies primarily about furnishing dead individuals with afterlife chattels. Among those with more worldly views of death – such as the ancient Greeks – burial of an elaborately-painted *amphora* or other valuable object in a grave may have meant little more in symbolic terms than does the inclusion of an expensively-ornate coffin today.

While modern-day peoples have preserved the custom of trying to ensure that corpses are well-presented when they enter the grave, we have lost, at least in the West, the tradition of including grave goods. The modern-day soul may require some clothing but it does not, apparently, have any further use for its personal high-tech status objects: its tablet computer, its digital camera, and its smartphone. Apart from wedding rings, the inclusion of enduringly valuable objects – such as gold watches – is rare, and the deliberate inclusion of data/paperwork conferring great wealth – such as Swiss bank-account access numbers – virtually unheard of. And although some colourful Louisiana-style funerals feature non-traditional coffins, including some shaped like cars, we would be extremely surprised to hear of a body formally buried inside a real car. This contrasts sharply with certain historical burials, such as the one at Vix, where tribes often buried horses and/or chariots along with the wealthy and important dead.

In that view, then, the soul must have been decidedly person-shaped to be able to utilise grave goods, spend its wealth, travel in its chariot, and enjoy the ministrations of its servants and concubine(s). While apparently settled on the idea of the soul as a singular entity, it is questionable whether the modern view adheres fully to the tradition of the anthropomorphic soul. Christians imagine heaven as a clean and billowy place in the sky where a father-figure God tends to his 'flock' of human-shaped souls; Allah's presence in the after-death paradise seems to be less important to some Muslims than the promised clutch of pleasingly virgin-shaped virgins. However, the view widely held by those with 'spiritual', if not conventionally *religious*, beliefs seems to be of a soul more wispy, less weighty, and altogether less substantial than the traditional kind.

The perceived format of the soul is an important consideration when looking at the ways cultures of the past and present dealt/deal with the body after death. The means by which a soul is contained or released must be compatible with that format, lest the soul end up in some terrible limbo, or at very least, some slightly-embarrassing quandary. The volume of blood spilled over these questions of soul etiquette indicates the high importance, to believers, of getting it just right.

To other early civilizations, *efficiency* was the name of the ritual-death game. For example, the poor souls of the hundreds of thousands sacrificed at Tenochtitlán had not even limbo to look forward to; only quick 'processing', after the long wait on the blood-caked pyramidal-temple steps, prior to rapid consumption in the maw of Huitzilopochtli or one of the Aztecs' many other baleful gods.[21]

The night is chill and moonless. Candles burn to hold back the demons, as the mourners raise their 'caoine' (keening) over the gelid body laid out on the bier. The knife wound in the dead warrior's side has been cleaned – the druid ensuring that all was done in accordance with the teachings – and he has been dressed in fighting garb, adorned with his most-valued bronze possessions. The shroud will be cut open, so that his movement will be free and fluid as it was in life. He will face east in the grave-pit; should he rise, he will see the invaders coming and, this time, hack them down. His beloved dogs will go with him. They will stand by his side as they did so often at watch on the green knoll; they will tear the throats of his enemies, should they dare to return.

Before the arrival of Christianity, Celtic customs held that the dead would continue on *bodily*.[22] I am not adding 'in the afterworld', as evidence for their believing in such a 'place' is scant. Granted many believed in 'other-worlds' – ambiguous places beyond the edge of the map such as the Irish-Celtic *Tir nan Og* – only reachable by strange and often terrible journeys, but even these appear to have been conceived of as corporeal realms. To some extent, they believed that life after death was 'business as usual', with the deceased putting in notable (if infrequent) appearances in the societies of the living: voyaging from place to place; fighting; attending feasts; and perhaps, for the males, even impregnating a live female or two.

We can see how the 'new' *Christian* template for life after death overlaid onto the customs of early Christianised peoples: The soul ended up somehow 'in between' for some time after the death of the body, before (all being well) finding its place in a new metaphysical homeland of one extreme or another from which, under 'normal' circumstances, it would not return. Christianity, as always, did an efficient job of incorporating existing traditions into its belief-framework, and so we see that Celtic beliefs about life after death, for example, did not change abruptly but became gradually less materialist over time. The old Irish-Catholic laying-out tradition of opening a window to let the soul of the dead person exit would make little sense in a context of belief in a *material* afterlife.

In the Christian tradition, bodily resurrection is reserved almost exclusively for the Jesus character, with other souls demoted, at least temporarily, to a state of patient incorporeality. Only Judgement Day will reunite those other dead souls with their long-lost bodies, with God performing the nanoengineering marvel of patching them all up in readiness for the last hurrah. But even this slim hope of a bodily rising from the dead was vitally [sic] important to Christians, and the idea persisted virtually unmodified well into the nineteenth century, strongly colouring views about post-mortem scientific practices. As Mary Roach points out in her immensely-entertaining book *Stiff: The Curious Lives of Human Cadavers*:

> Back then no one donated his body to science. The churchgoing masses believed in a literal, corporal rising from the grave, and dissection was thought of as pretty much spoiling your chances of resurrection: Who's going to open the gates of heaven to some slob with his entrails all hanging out and dripping on the carpeting?[23]

The cleanliness and general appearance of 'the soul' is still important to believers today. Some readily criticise those who wish to hang on to life for as long as possible, perhaps seeing that desire to remain at very least alive, if not young and beautiful, as a grotesque *Dorian Gray* type vexation of the immortal soul. But the attitude of the *believer* demonstrates a kind of *reverse* Dorian Gray scenario that could be seen as equally grotesque: The *body* must putrefy in the ground so that the *soul* gets to emerge sprightly, beautiful, and eternal.

## DWELLING ON THRESHOLDS

The idea of the 'in between' I mentioned above has been contentious in Christian circles, and it seems that this faith and certain others have never managed to fully incorporate the earlier notion of *liminality*. Catholic 'purgatory' does a fair job of acting as a holding-place for souls in transition, but other Christian traditions say little or nothing about what happens between the death of the body and the moment of entrance of the soul into heaven (or hell), leaving the faithful to contrive various means of floating or tunnel-walking to reach their final destinations. This religious reticence about dealing with this *liminal* (from the Latin word for 'threshold') period of the soul may be to do with earlier associations between liminality and dangerousness or uncleanliness: The soul is, at this time, neither one thing nor the other, neither in one place nor the other; and so is, in some way, *volatile*. Upon encountering Christianity, the peoples of earlier civilisations might well have had incisive questions to ask about this blurred area: So when I die I go to heaven, but exactly *how* do I go to heaven; and how long does it take to get there?

We could reasonably argue that, for Jesus, the liminal period begins when he dies on the cross and ends when he rises from the dead in his tomb, and then, finding the stone rolled away by an angel, makes his exit from the world to return to heaven. We could equally view his liminality, in the sense of ignominy and social exclusion, as beginning even prior to his condemnation. Crucially, he is *accepted* back into heaven by his 'father' only after this time of terrible suffering, followed by a limbo-like waiting period after death.

In other religious traditions the liminal period lasted much longer – sometimes for several weeks or months – and for the famously well-preserved 'bog bodies' such as Tollund Man and Lindow Man, it never ended. There is strong evidence to suggest that their violent deaths, prior to burial in the supposed-liminal environment of sphagnum-moss bogs, was designed to confuse, confound, and generally *vex* their powerful souls to a point where they ended up forever trapped inside their perpetually-recognisable bodies.[24] The absence of industrial-scale peat removal on the islands of the Hebrides has meant that the leathery remains of vexed precursors have a chance to tan a while longer. To date, only at Cladh Hallan in South Uist has evidence suggestive of liminal-bog burial been uncovered here.[25]

This may be uncomfortable subject matter, but it is important to recognise the recurring theme, throughout the history of religion, of periods of transition or *rites of passage*[26] between life and death.

The perception of the existential state of the dead person still differs among religions, though perhaps not as dramatically as in the past. The belief in a *material* bodily existence after the cessation of life (such as that suggested by what we know of the religion of the Celts) has faded, though the reverence of dead bodies, in terms of attempts to make them presentable after death, continues to this day. The soul has become insubstantial and composed of something entirely and indefinably *other* or *extra*. Many (perhaps the majority of) religious people now believe that this soul exits the body immediately at the point of death and proceeds straight to the afterlife, with little or no interim 'quarantine' period.

Despite the growing acceptance of spiritual quick-exit, the idea of the soul having *substance* – even to the extent of having a measurable *weight* – hung around well into the twentieth century. Roach cites the experiments of a Dr Duncan Macdougall of Massachusetts, in which he tried to determine what precise change in body weight might occur at the moment of a person's death. Any such change, he reasoned, must be caused by the loss of the slight but calculable weight of the soul. Macdougall's conclusion of a soul-weight of 'three fourths of an ounce' was, upon its publication in a 1907 issue of *American Medicine*, disputed on various grounds, including one that the loss could be wholly explained by moisture evaporation.[27]

## SULPHUR

Rather than examining, in detail, the mysteries of soul-transition – a preoccupation so important to earlier religions (and later religio-scientific cranks) – *Protestantism* focussed on the mode of worship. In this belief system – within Scotland, under the dire spell of the *Calvinist* teachings of John Knox (*c.* 1514–1572)[28] – the church became a gathering place where the faithful would commune *directly* with God, instead of via elite priestly interlocutors. With no role left for the elite, their gaudy trappings of wealth and power could be swept away – or at least out of sight. This dramatic change may have felt (initially) *empowering* to the ordinary folk of the sixteenth century, suffering as they were under the coshes of violent dictators and untouchable, bureaucratic, and hierarchical churches.

Calvinism lends itself well to the modes of understanding of hard-bitten peoples living in harsh environments: the simple and unadorned places of worship; the lack of ostentation in praise; the manual-labour ethic; the resignation to punishment, suffering, and death. It is *dour* in every sense of the word. The focus, inside this stripped-down vehicle of faith, is firmly upon sin; more specifically, upon what these believers call 'original sin'. According to their reading of the bible, we are all born sinful because of the original sin of Adam and Eve in the Garden of Eden. This is an odd and unforgiving belief, but crucially, it is one that allows a certain kind of *license*. The license is firmly in the hands of the groups – very often groups of elderly men – in charge of applying the perceived laws of the church. Paradoxically, it allows them the flexibility to judge all situations through this same lens but to apply their laws stringently or leniently as they see fit. An infant that has died can be adjudged either low on sin while still sinful (because of its short time on Earth within which to transgress), or sin-laden because of the compounding of the sins of its parents *on top* of its original sin. A 'fornicator' can be reviled but partially forgiven (because he/she was born sinful but has made a poor effort at sticking to the righteous path), or excommunicated (because his/her entire sin quota has been exhausted).

Addictions, such as alcoholism, arise, almost naturally, as a self-medicative response to the strictures of such religions. The self-loathing inherent in addictive behaviour provides, in turn, a deep well of *aithreachas* (repentance) for the controllers of the church to draw upon. The cycle of transgression-repentance-punishment-forgiveness-transgression feeds the system by ensuring the regular return of its victims, who become examples of the results of straying from the path of 'righteousness', and indeed, of finding it again.

My grandfather on my mother's side and both my grandparents on my father's side were raised in *Presbyterian* religious traditions, in line with what I have outlined above. The ultimate penalty for those failing to live in accordance with the laws of this faith would be to burn in hell after death. This, for people who truly believe in it and for those who cannot shake the fear that it *might* be true, provides a powerful incentive to stick to the rules.

While the structure of strict Protestant-format religions was different to what had gone before, the means of ultimate control were (and still are) the same. The prospect of death, with all its hellish associations, loomed large in the minds of the faithful, and so their lives were controlled and coloured by their perception of their eventual deaths and 'afterlives', as communicated to them via the medium of their in-groups and church 'elders'. The prospect of punishment and humiliation in *life* would also have a strong bearing on social norms, but death was the great arbiter.

I have asked my father, who was also raised with a Presbyterian-style Christian religion, about the curious focus of the Protestant churches of the Highlands and Islands on the Old Testament and its brutal doctrines. I suggested to him that the central tenets of the Old and New Testaments are mutually incompatible. He simply told me that the Old Testament was known as 'the forerunner' to the New.

There is no way around the gaping tautologies with which this type (and all other types) of religion is littered. And this is by design, so that there is no way *in*: no way in for people like me who would only make mischief with the wide-open non sequiturs, but equally, no way in for people like my father who have been expected to *identify* with this set of beliefs. It is a religious culture about as inviting as the heavy leather-bound Gaelic bibles that stand guard – impenetrable and long since last read – upon his living-room shelves.

In common parlance we sometimes call such cultures 'God-fearing', but it may be more accurate to say that they are *afterlife*-fearing. These persons seem to have a strong desire to continue on after death but spend their lives worrying about how to ensure that their received afterlives are of the right kind. In other ways, however, consensual belief in a fiercely-circumscribed afterlife may act as a strangely-utilitarian forum for contemplation of the demoralising insecurities of life. The *judgement* aspect of this process is, in many ways, incidental: God 'works in mysterious ways', so his specific criteria for judging the goodness or badness of a particular life cannot be known. Because of this, the morality of an individual's life is always open to interpretation both by God and by the group. Fear of the eventual outcome is sustained, and structural control is maintained.

By the time of the introduction of Protestantism, the idea of the existence of an immortal soul reigned, at least in Western cultures, virtually undisputed. This soul, while no longer as material as in the sense of the earlier Celtic belief, can travel to a narrow range of metaphysical destinations and can certainly suffer pain. It must be aware that it is suffering pain (or feeling pleasure) and it must know the reasons *why* it is receiving those sensations. Therefore, it must be continuous with the original living person and must remember that it *was* that particular living person, otherwise the pleasure of God's eternal proximity would not serve as reward, nor the unending pain of hellfire as punishment.

The rampage across Europe of the Black Death had earlier created the wild-eyed context for the systematisation of this new imagery of doom: hell with its burning piles of diseased souls; the Grim Reaper cutting down the unrighteous, even in the prime of their lives; the Four Horsemen entering stage left, skulls agleam, for what must surely be the End of Days.[29] Old Norse and European mythologies blended with the darkest predictions of Revelation to create a toxic brew of fear and torment. Cue the opportunistic 'logic' of Puritanism: Unclean souls must burn like the disease-ridden bodies of the dead; the sin-infected living must be cleansed in fire at the stake.

The Plague devastated the old order, as people saw monks dying in numbers, and church-leaders fleeing in terror. Hell had come to Earth, smiting elite and peasant alike. This great levelling had cleared the way for the common man to negotiate directly with his maker.

In modern times, the notion of hell has gone out of fashion, at least in the West. The clear-cut balance of one afterlife zone for the good (heaven) and one for the evil (hell) has been lost. The faithful now expect their souls to be 'saved' if they do good, but make far less mention of what they expect to happen if they fail. While this fudging of the hell issue has made religion more palatable for some, it has done nothing to check the decline in church attendance. On the other hand, however, the Presbyterian-style sects that have stuck to the old interpretations have seen their attendance figures utterly decimated.

## SHADES

Religion in the West is *itself* in a liminal state. Non-belief is becoming more widespread, religious observance is slipping (in an TV interview filmed not long before his death, author Iain Banks referred to religion today as a 'minority sport'), and religious beliefs are becoming less literal. Despite this, the notion of the immaterial immortal soul seems quite entrenched. I recently watched the film *The Lovely Bones*.[30] The theme is a popular one, but one that I have always found unsettling. Balance is brought to the grisly murder of a young girl by the fact that her ghost lives on, allowing her to observe the fate of her killer. The story culminates in her gaining entry to the afterlife paradise, along with a merry band of other infanticide victims.

*Must* we keep doing this – mythologising death to the extent that even stories of child-murder can have happy endings?

The character of the girl in the film has emerged from the lineage of mythology of the 'unquiet dead'. She is a tormented soul, and cannot rest until her murder is avenged. She is dead, and she is unquiet, but she is not *un*dead. Traditionally, that fate has been reserved for those considered to have been so evil in life that they flatly refuse to die properly. The belief in vampires and lycanthropes has also gone out of fashion, but it was once an important outlet for poor, ignorant folk lacking an explanation for the failure of some dead bodies to decompose as expected. Pulling out the teeth, cutting out the hearts, or lopping off the heads of these night-stalking miscreants was expected to put an end to their death-defying effrontery.

Religions are only just beginning to respond to the perceived challenge to their beliefs posed by cryonics. We know, both from our own experience of religious views and from what I have outlined above, what form the objections of believers will take. Even though they may well not know the word, the general perception is that a cryonically-preserved person is liminal: Trapped in this 'unnatural' way, the soul cannot make exit to the afterlife. Some might take an extreme view – that those in cryonic suspension are *undead*. Such views, while utterly misguided, are certainly deep-rooted. Moreover, we should not underestimate the power of the notion of liminality to creep into the minds of even those who do not profess to hold religious views.

Later in this book, I will discuss further the reasons why people either *really* believe they have a soul or *sort of* think that they have *something like* a soul. While it is fair to take issue with those who think that such beliefs have been, somehow, *immanent* in human beings,[31] we cannot dismiss the idea that their most tentative roots stretch back towards the primordial origins of life. What remains of the hominin fossil and artefact record shows us that burial practices have 'evolved' in complexity over time, and we project onto this apparent development a philosophical significance that is recognisable to *us*. Such projection serves the needs of religious and non-religious alike: The notion of proto-humans having only 'proto-souls' might sit well with those people of faith who employ tortured means to try to incorporate evolution into their bolt-on belief structures; the idea of 'progress' away from 'primitivism' and towards 'enlightenment' fits the narrative of ever-increasing human rationality.

It may not be possible to draw useful conclusions about the emergence of ideas of 'immortality' or 'the soul' from the historical record, because our ideas are so coloured by modern, individualistic, materialistic interpretations of what 'survival' means. We have, perhaps, lost sight of what a death meant to ancient communities in terms of common loss of an active, integrated contributor who had been raised to that status at great cost to the group. In that context, transmission of the stories, deeds, and *role* of that lost contributor to other group members may have carried a significance that was 'alive' to them in ways that we simply cannot grasp.

So, notions of *personal* survival in some form of 'afterlife' may have been the exception rather than the rule. Elements of Abrahamic religions have been instrumental in collapsing down the panoply of human responses to loss and continuity of life towards a focus on the spiritual persistence of individual punishable or rewardable agents; billions of us have inherited fragments of this perspective. We must accept, then, that we run the risk of doing battle with a straw man with a straw soul when we argue against the spiritualistic interpretations of 'self' and 'soul' of 'the past'.

On the face of it, it would not be difficult for an archaeologist or anthropologist to formulate a way of placing my own views about death into a historical context. Given that my views appear new, and entail modern science and technology, some would place me right at the modern-day end of their timeline. Others would disagree. Timothy Taylor argues that

> there is a law of conservation of questions. 'How can I stop the soul of the deceased reanimating the body?' is now being replaced with 'How can I live so long that my life becomes indefinite?', a question previously only asked by the most arrogant pharaohs and emperors.[32]

As someone who studies the evidence of past human-behaviour, as laid down in the fossil and artefact record, it is normal for Taylor to seek clues to the reasons for today's human behaviour within the context of historical practices. But is that the correct approach to apply where the past is not the issue? Religious dogma has dominated death for so long that we have come to see death *as* custom. When it is not – when the new practices involved are in fact intended as an extension of *medicine* – some historians still insist on shoe-horning in as 'comparisons' what many modern life-extensionists see only as frustratingly anachronistic non sequiturs.

Nevertheless, it is not unreasonable to see cryonic preservation as an elite burial practice, in that it is currently only available to the few. If, like pharaohs, we are simply hogging resources that would be better distributed to the many merely for the sake of a grandiose *custom*, then we are indeed arrogant. If, however, we are onto something – a *medical* technique that could some day become cheap and commonplace, ultimately rendering permanent death *optional* for each of us – then our reasoning is rather more nuanced.

I accept my tenuous fragility but try, with the scarce and rudimentary tools at my disposal, to mitigate it. Is that arrogant of me?

He is gone from this world and they do not speak his name. We hold the grey tassels, while the older men take the real weight of the coffin with the heavy straps passed underneath it and threaded through the gold-coloured plastic handles. They lower him through the stark hole in the green-baise surround, and we drop the tassels down onto the varnished wood below, as the minister – like brooding, latter-day *Malak al-Maut* – looks on. Time now to speak of him and his vibrant life? No. Time instead for the grimmest of sales-pitches: We are sinners and our only hope, our only salvation, and only ever achieved in the next life, is through the 'straight gate' of Jesus Christ.

And it always sounded wrong to me, even when I was a very small child. It rang with suffering, with hopelessness, and with blank resignation. It rang, like a cracked but still-sonorous bell, with evil.

# 4 MONKEY IN THE MACHINERY

∞

The bus always smelled rank – diesel fumes and Monday-morning damp mixed with body-odour, stale cigarette-smoke, and flatus. The early rise followed by the bone-jarring forty-mile trip north to high school was, to my mind, a finely-honed instrument of demoralisation. Time to think; time to worry.

It must have been grim for my parents too – stumbling out of bed in the black winter mornings to prepare porridge, or fried-egg sandwiches, for their tired, embittered, and forlorn offspring. We didn't make it easy for them. Still groggily sleep-deprived, and racked by teenage forebodings, we usually sat in silence. The voice from the static-ridden radio seldom penetrated my bubble of anxiety, as I tried in vain not to *think* about what lay ahead.

I remember little specific about Gamma's, and later Epsilon's,[1] demeanour on those mornings. We had usually managed a few terse exchanges by the time the bus arrived, but often only because it was late, and one of us (usually me) would say something to break the tense monotony. Once on board the bus, we separated immediately into our factional seating-positions: Gamma behind me, progressing towards the 'senior' seats at the back; Epsilon in front, along with the rest of the diminutive, giggling first-years.

I still have bad dreams about losing my school timetable and ending up in the wrong class, in the wrong school, arriving on the wrong day. These types of dreams are common: Great changes are happening in the brain during those early high-school years, so anxious memories from those times can stick and become a subject of constant reinterpretation by the subconscious. But I *could* be quite disorganised. In the absence of a timetable or my peers, I sometimes panicked and could not remember where I was supposed to be. I often felt panic, but internalised much of it.

The hostel was profoundly hierarchical, and occasionally brutal. Later in my high-school life I read Golding's *Lord of the Flies*, and in my mind, applied the scenario to the hostel. It was not difficult to imagine how, in a hostel cut loose from societal and situational constraints, bloody havoc might ensue. It would have been easy for me to make a list of the resultant killers, appeasers, and victims.

As the seniors often reminded us, however, we had it easy compared to the days 'back in the old hostel'. Dark hints of a past of extreme bullying and full-blown violence permeated their sneering put-downs. And I had good reason to believe them. Alpha had done some time in the old hostel, and while he wasn't overt about it, it was clear that the experience had left its mental scars.

A combed-over ex-forensic-policeman and his drunken harridan of a wife ran our hostel. Despite the acoustic-dampening of the polystyrene ceiling-tiles, her grating Western-Isles-accented screeching *reverberated* along the corridors as she lurched, each evening, brown paper bag clutched to disproportionate bosom, towards her staff quarters. The male warden continued at the hostel after the marriage broke down and she left for good. The authorities assisted him in fudging the fact that they had dismissed her after sustained complaints by parents.

When I first arrived at high school, I was relatively tall for my age. This helped me to avoid bullying by kids in my age group. I experimented with my own aptitude for violence by picking a couple of fights. But I was a hopeless failure on this count, poor co-ordination and fear of pain letting me down badly. I suppose this was another spur to develop a new persona: A competently aggressive me would have had no need of such psychological armour.

This also held true for my efforts at sport. In hindsight, it was naïve of me to imagine that I might be good at shinty. My father waxed lyrical about the sport, so I thought he might be pleased if I learned to play. Shinty (*caman* in Gaelic) is a ridiculous game: Testosterone-fuelled Highland gorillas slash at a tiny leather ball with solid, hooked, triangular-section sticks. The game looks a little like hockey but appears to have fewer rules. One main difference is that, unlike in hockey, you must always *hit* the ball rather than scooping it. I always seemed to scoop rather than hit, for fear of smashing somebody in the face with my stick. The other players held no such reservations.

As well as an array of other delusions, I initially clung to the idea that my 'culture' was important to me, hence the shinty and my persistence (for a while) with the bagpipes. This fallacy also led me to choose Gaelic over French as my only language-subject. After my third year at high school, I was allowed to take German as an additional language, but I regret not having had the self-knowledge and strength of will to choose French in the first place. I had been too eager to please, and had just done what I thought was expected of me. There was no *me* in those early decisions, only slack-jawed acceptance.

It's not as if the Gaelic classes taught me anything worthwhile about my culture. And when the history of the Gaels did arise, the teachers put it across to us in such stultifying terms that it sounded as dry and dead as withered, wind-scattered leaves. The language seemed heavy with loss and arcane in structure. I could read and pronounce the words with little effort, but my brain seemed to reject the notion of attaching *meaning* to them. And, I suppose, my feelings of guilt did not help me to relax into the cadences of the language – guilt that I wasn't really trying, guilt that I felt no connection to it, guilt that I hadn't already learned it direct from my father.

But other classes began to give me opportunities to meet people the like of whom I had never come across before – people my age but with greater experience of life and with similar interests in music. I worried about coming across as a naïve backwater boy, so I didn't find it easy to make new attachments at first.

Friends from primary school days were with me in the hostel, but the suddenly-greater pool of like-minded peers quickly altered old allegiances. I seemed unable to seal new friendships and was losing interest in most of my old ones. Associating with siblings was considered taboo, as it smacked of weak, 'no mates' reliance. As a result, I felt torn and alienated, but still clung to a sense that it would all be OK, that someone would help me.

An experience at a fairground came as a bitter wake-up. By special dispensation, the junior hostel-residents were allowed to go down to 'the village' to visit the fair. It only came round every few years, so, the hostel warden assured us, this was a great and rare privilege, and one we should not take lightly. The fair was simultaneously exhilarating and frightening: lights, pounding music, loud and leathery-faced stallholders, 'waltzers' and other huge rides, candy floss, fruit machines, *Space Invaders*. Then there were the other kids: nervous-looking ones my age, brash and confident senior boys, beautiful and terrifying senior girls, tight groups of second- and third-years.

I leaned against a post to the side of the tent where the slot machines and video games beeped and rang their enticing melodies. A boy detached himself from a group of second-years and sauntered over. I recognised him from school, and he had acknowledged my presence before, so I had no reason to suspect malign intent. He fiddled with the video game beside me, then turned and offered me his can of Coke. I accepted and took a swig – before gagging as I realised that something was wrong. Laughter and jeering rang out from the knot of second-years as I coughed up the urine I had swallowed. 'It's pish,' they shouted, 'you just drank pish.'

Gamma helped me to root him out, and I was genuinely grateful. But Gamma had a 'cool' reputation to protect, so he would have had his own vested interests in helping me to find the poisoner. I worried that I had embarrassed Gamma by being such a gullible fool. At first, I felt stupid and ashamed; later, I felt only anger.

Days later, the school machine closed in on the offender, and he was punished, but not before I had landed a few frenzied punches on him during a poorly-staged fight. My aggression rang true for once, but he didn't really fight back. He had known this was coming, and accepted the blows; he had been a dupe, after all, goaded into the foul trick by his foul peers.

Bullying is routine in high schools. With the abundance of available victims, pathological cruelty always finds an outlet there. I cannot even begin to imagine what it must be like for kids nowadays. Word of my distasteful experience spreading, cancer-like, through social-media websites and instant messaging might have broken me. But, back then, the past was easier to delete, and the furore quickly died away, leaving little but the odd chuckle behind my back, or muttered jibe about 'feeling thirsty'.

I changed. It wasn't the work of a moment, but the exponential nature of the change seemed to go far beyond what we had been warned to expect from the emotional effects of puberty. I pursued interesting allegiances with new-found vigour, and found reward in the discovery that there were others possessed of a dark humour and fractious drive similar to my own. Nevertheless, musical taste was always the key gauge and attractor. We began to exchange vinyl albums and compilation tapes; we greedily gobbled up the latest music-industry gossip from *NME*; we bragged about our latest band T-shirt acquisitions.

The music of The Stranglers was an early passion. I was familiar with some of their albums from Alpha's collection, but finding that my new peers shared this taste was the spur I needed to begin exploring their music in more detail. Alpha's collection turned out to be a goldmine for impressing my new friends: The Police, Pink Floyd, Genesis, Ultravox, and many others. We didn't greatly discriminate between this muddle of genres back then; we just listened to what our ears liked. And I learned about raucous, exciting (and dangerous) bands of which I had little or no previous knowledge – The Clash, The Sex Pistols, New Model Army. But the band that struck the blackest, most-ringing chord with me was Killing Joke.

Knowing, as I do, a couple of excellent music journalists, I am trepidatious about *describing* the feelings generated in me, at that formative stage in my life, by Killing Joke's music. In that regard, my cynical-musician streak (and over-exposure to *NME*) used to attract me to the Frank

Zappa 'Writing about music is like dancing about architecture'[2] school of thought. However, having since experienced both soaringly-sincere music journalism and symphonically-beautiful architecture, I am now more inclined to defend such cross-disciplinary effusiveness.

I can honestly say, then, that it evoked in me feelings akin to the dark glee I had sometimes felt during my primary-school years. But there was now an added element – a jagged edge of cold-blue anger looking for an outlet. And this music was just such an outlet. It towered above me, then soared around me with a joy, a pain, and a rage that transformed it from a wall of bass-thud and devil-chord into something approaching a cross between battle cry, terrifying sermon, and plaintive *mea culpa*. I filled myself up with it, and chimed with its mighty resonance.

It wasn't long before my new friends and I were plotting for our own imagined musical futures. We talked instruments, invented band names and song titles, and decided on line-ups. I had always assumed that I would be the front-man. It seemed logical to me that my barely-contained bitterness and confusion would metamorphose into a lyrical splendour, which I would impart to adoring fans with a surging, mellifluous howl. After all, I was the only one of us that claimed any vocal ability. In truth, I had little evidence for such a claim, my voice so far having only been tested at Gaelic 'Mods' (traditional music competitions) and primary-school carol concerts. Determined that neither Gaelic Mod nor religious singing would feature in my high school life, I had abandoned all crooning opportunities. Nevertheless, despite never having committed lyrics to paper, I imagined myself a singer-songwriter, and by extension, the newly-charismatic lead vocalist of our putative band.

And, in time, the lyrics came – stilted, prosaic, mawkish ones, but lyrics nonetheless. *How* could it be, I wondered, that my 'special' thoughts and feelings, once transferred to paper, looked like *this*? I persevered. Over time there was some improvement, but they always fell far short of my expectations. The few times they *seemed* to ring true were when I wrote words for hypothetical slow piano 'ballads'. Even then, however, they often came across more suicidal than sweepingly sincere.

The friend who I had assumed would, one day, become the lead guitarist, turned out to be a fascinatingly (often infuriatingly) slippery customer. He had his own grand – and I thought at the time rather fixed – ideas about how 'the band' would look and sound. To my disappointment, he quickly slipped out of the running, forcing me to look elsewhere. The new contender was a sporty high-school heartthrob, but one with an interesting extra dimension of guitar-hero potential. The 'audition' went better than expected, his battle with wayward tuning more than compensated for by his overdrive pedal and confident swagger. He also knew a drummer – one who had recently scaled up from pipe-band-snare to full kit. As a serendipitous result, the line-up was almost complete.

I left Epsilon with little choice about his new role. Having purchased the bass guitar with the sole intention of hanging it from his slight shoulders, I would have struggled to take 'no' for an answer. He accepted with his customary shrug of blasé resignation, and as expected, began a rapid ascent to virtuosity.

Early practices were enthusiastically shambolic: Epsilon competently holding the rhythm while trying with dogged determination to impart some of it to the drummer. I seemed to spend much of my time attempting to help the lead guitarist to get in tune. He insisted on making liberal use of his 'whammy bar', which, more often than not, either detuned the instrument or snapped a string. My keyboard playing consisted mostly of simple chord-washes, which, with a following wind, would help to blend out the screaming dissonance of the electric guitar. My singing was strained and lacked power, but I put my all into it, buzzing with a new and confident self-assurance that this was what I had been waiting for.

Outside of the milieu of music, there was also a certain amount of strain and atonality. My early proficiency at maths had given way to puzzled discomfiture, as I sought (albeit half-heartedly) to internalise the formal symbolism of algebraic 'curly brackets' and Venn diagrams. My maths teacher – a poor, D.T.-racked sot – singularly failed to mention *any* relationship between this abstruse discipline and the computational magic I had experienced as a pre-teen. 'Science' classes also lacked inspirational content. When I look back, I really do wonder how it was possible for those teachers to render subjects such as physics so uninteresting. One sniff of the tantalising world of the quantum, for instance, might have had me hooked, even head-over-heels. But, in truth, I wasn't *looking* for anything else by that point. My course seemed set fair towards a land of musical self-realisation (with some fame for good measure), and there simply didn't seem to be space left in my brain for much other than this, and the growing demands of coursing testosterone.

There were exceptions. A classics-teacher with a taste for strong black coffee and dry put-downs made darting forays into my world. His attempt to show me something of his own world fizzled, due to my failure to appreciate his extensive Bob Dylan collection. He later taught me classical studies, when I took it as an add-on subject in my fifth year at high school. The class was enjoyable but, to me, lacking in relevance. His Latin class could have become a goldmine of relevance, had I imagined that I might, some day, wish to acquaint myself with the precision labelling of brain regions and of other scientific and philosophical concepts.

I made headway in English class. Seeing some connection between this subject and my desire to write songs, I worked hard to produce decent creative writing. The formalities of grammar were a problem for me, but every other element of the subject seemed obvious, even easy. I sneaked sci-fi into short stories and shoehorned plagiarised song-lyrics into my poetry. The diminutive tweed-lined teacher handed assignments back to me with a quizzical look and a softly spoken, 'Well done' or, sometimes, a quizzical 'Hmm...'

The books they gave us to study in English class seemed, for the most part, stuffy and boring. Some English teachers took it upon themselves to educate us about our culture, insisting that we read about the lost communities of the North of Scotland. Those books, though beautifully and poetically written, gave me no feeling of connection to the characters; I failed to engage with the rain-sodden hopelessness of their daily struggles for survival. This was not, *per se*, because these books dealt with Scottish Highland history but more to do with my lack of feeling about *any* kind of history; I don't remember anything at all that I learned in my early history classes.

Iain Banks saved me from my pit of literary despair. His modern High-land-gothic tale *The Wasp Factory* opened my eyes to the tantalising strangeness of life in remote parts. The twisted protagonist of the book – a weird misfit living on an isolated Scottish island – awaits the imminent return of his crazed brother, Eric. The book was bleak and frightening, but it was also ghoulishly cool. And it spoke to me as much about the playful Scottish anarchism of the author as it did about the creative insanity of the characters. Banks' second novel, *Walking on Glass*, has a Gormenghast-like quality, adopting an interestingly omnipotent viewpoint to watch several lives collide.[3] In *The Bridge*, the collision is a literal one: After a car-crash on the Forth Road Bridge, a man known as John Orr finds himself inhab-iting, along with countless others, a vast bridge – a structure within which he feels safe but from which he senses a growing compulsion to escape.[4] (The influence of Alasdair Gray's dark vision of the city of Unthank, in his brilliant *Lanark*, on the tone of *The Bridge* did not occur to me until recently.[5]) Both *Walking on Glass* and *The Bridge* have strong elements of sci-fi, and this led me neatly into his 'Iain M. Banks' 'hard' science fiction books.

Banks' 'Culture' novels are set in a universe where science and techno-logy have become so advanced as to appear (to borrow the words of Arthur C. Clarke) 'indistinguishable from magic'; they re-ignited my passion for sci-fi. His ideas, though often disturbing, had meat on them into which I could sink my teeth. One particular image – a freshly-severed head bouncing down a hill before drone recovery, within seconds, to a Culture mother-ship – sticks with me. The unfortunate (or fortunate, depending on how you look at it) victim spends a rather boring few months on board – a head on life-support awaiting a new body, which is being grown, in a vat, from his own cloned cells.

Back at the alt-rock face, I introduced edgy cover-versions into the band's growing repertoire. Gary Numan's early work was a good fit for our instrument line-up, our musical abilities, and my desire to achieve an air of detached freaky-cool. But my own songs, in all their ill-starred and jumbled perplexity, dominated our set-list. I had succeeded in finding some kind of writing style, but unfortunately, it had not resulted in songs that I *liked* very much. So I persisted in pushing those songs away, in the hope of developing a sound that was more in keeping with the style of the bands I actually listened to. We got a few gigs, and with great enthusiasm stumbled noisily through our ragbag of tunes, the upturned faces of admiring pretty girls periodically framed in the glow of the disco lights as they smilingly worshipped the guitarist.

It gradually became clear that the drummer and lead guitarist were getting sick of my increasingly demanding nature and of my refusal to introduce any of their songs or suggested cover-versions. They left the band, and I did not care. Epsilon and I continued to play together, though reduced to a state of lonely new-wave. A friend who had been on the periphery of my initial high-school tribe, as a result of absence through long illness, later entered the frame and became our new and close-fitting guitar ally. We dispensed with the idea of getting a new drummer (mostly because we couldn't find one), opting instead for the metronomic certainty of a drum machine named Freddy. We rejoiced in our new and genre-busting sound, and, under duress from Epsilon and the new guitarist (I will call him Guitar from now on), we developed a clattering sense of humour.

Word of a small recording-studio reached us. We went along to meet the man who owned it, to plead our case. He was an expressive-faced, lantern-jawed Glaswegian with a love of computers, guitars, and filthy jokes. At first, we were daunted by him – he had played in *real* bands, he was an electric-guitar virtuoso. But his gently-cajoling style and willingness to help us out quickly endeared him to us. The studio – 'the shack' as he called it – was a sound-insulated stone shed in his garden. There we recorded a cassette album, which we later touted, with great pride and little success, around our high school.

An opportunity to appear on Gaelic television brought much hilarity and even a little welcome controversy. The people behind the 'youth orientated' show in question were keen to give the impression that a vibrant Gaelic rock/pop music scene existed. We had never, as a band, written or played any form of Gaelic music, but there was no question of our resisting the lure of the publicity and BBC fees. We concocted a rambunctious 'Scottish-Cajun' style ditty, and after recording it in an Edinburgh studio (its original essence somewhat squeezed out by the programme's music producers), made a video for it at Dunvegan castle, in Skye. Guitar's notional version of the video would have been, if realised, vastly superior to the end result; he would have had us snarling our stuff amidst the wrecked cars and mouldering black bags of Portree rubbish-dump, rather than casting moody glances at the camera through the battlements of a local tourist-attraction.

The programme's producers took great exception to our later 'shock' revelation that we had 'only done it for the money', and that we were not really interested in Gaelic at all.

Guitar, Epsilon, and I spent a great deal of time together – initially at after-school practice sessions and then at our respective homes. This would often involve Guitar coming to stay at the big house for entire weekends. We laughed together, we brooded together, and later, learned to drink and smoke together. Whisky-laced black coffee served as our 'song fuel'. Epsilon popped in and out of this tight group, often like a fleet-footed dilettante. But we did not hold this against him, now realising that his talent and popularity would always keep him in demand in other quarters.

Guitar's family were, like Guitar himself, intellectual and artistic. I had expected to feel uncomfortable around them; as often proved the case, my fears were unfounded. They crackled with nervous energy (which was particularly remarkable in the case of Guitar's father, given his poor health) and supported our musical ambitions with boundless enthusiasm. The multitudinous shelves of their house sagged under the weight of classic novels, art and philosophy books; and the walls carried immense (and, to me, somewhat disturbing) paintings by Guitar's father, by his brother, and some by Guitar himself. Their creative talents were manifold and their humour quick, clever, and often deeply abstract. I loved to spend time with this Bohemian clan, knowing that they were good people and feeling that their influence on me was unreservedly beneficial.

Life with my own family had receded into the background. I saw my parents only at weekends, and spent little time with them apart from when my mother was dutifully ferrying us to some gig or practice venue. Seeing little Zeta was always refreshing, and would, for the few hours a week I was around to play with him, take me out of my bubble of troubled self-obsession. We saw Alpha from time to time. He had gone to University in Glasgow but returned to Skye after his studies there. He had his own band and, evidently, plenty of concerns of his own, now being under pressure to find regular sources of income. Beta married a man she had been going out with since the age of sixteen. They were married in Skye – Epsilon and I proudly playing the bagpipes at the wedding. I misliked her overbearing braggart of a husband. I think Alpha positively hated him. The marriage did not last.

I began to see less of Gamma at weekends. His love of cars and motorbikes had brought him into contact with a different crowd. He also had 'serious' girlfriends. I had assumed that he – being the handsome, charismatic centre of gravity – would not be subject to the kind of teenage-relationship angst from which I suffered. But I had underestimated his capacity for introspection, and there were times when he was plainly struggling with emotional pain.

Staying in the hostel brought opportunities (usually carefully monitored) to associate with girls. Stolen moments in girls-hostel outbuildings were the stuff of boys-dorm banter. Such 'relationships' were, of necessity and design, usually quick and shallow. I experimented a little, but found, after a fevered while, that I wasn't particularly suited to the 'playing the field' mentality. The idea of a *proper* relationship – one with someone I really cared about – was immensely appealing to me. Because of this, I sometimes over-committed and ended up feeling like an idiot. I wasn't an easy sell anyway – all weird angles, bad attitudes, and black-dyed hair.

Holiday work at a local hotel brought me sharply into the adult world. Sweltering shifts at the gigantic kitchen sink were often rounded off with several refreshing pints of lager-and-lime. The other members of staff were friendly but seemed in an inordinate hurry to make me grow up. I didn't object, and relished the chance to share my slurred and opaque 'wisdom' with people older than me.

One summer, at a beach party on the shore below the hotel, I met a girl who I came to care about in a way that I had not experienced before. She, like me, was on her school holidays. The fire from the party had long since died out, but we stayed, sitting side-by-side on the rocky shore, pouring out our young hearts as we watched the sun rise. We struck up a romance-by-mail, which, after a while, and to my stinging disappointment, faded into friendship-by-mail. I did see her again, but I made my usual stupid mistake of coming on too strong (at least in the emotional sense), as well as some other new mistakes. She was more of a realist than I was: we were just kids, we lived hundreds of miles apart, it could never work out.

My memory of that summer is awash with The Waterboys. That was the river. This is the sea.

My customary gloom, and my taste in music and literature, had inspired an interest in the 'supernatural'. I enjoyed being scared rigid by tales of Ouija boards (I had not yet heard of *ideomotor action*[6]) gone wrong, and of teenage psychokinesis and telepathy. I genuinely wanted to believe that it might be possible to read minds and to move objects around just by *thinking*. I now realise that I had an ulterior reason for swallowing this kind of nonsense: The gaping hole in my sense of self and in my understanding of the universe needing filling up, and the occult seemed obvious candidate material for the job. My confused philosophy at that time tended toward the notion that paranormal occurrences were regular but that they were all human-centred and would, one day, be proved to result from the untapped potential of the human brain. This felt like a good fit with my otherwise uniformly-atheist outlook.

Authority and I were never going to sit well together. The headmaster insisted on wearing one of those academic cloaks – a black one that billowed out behind him as he crossed the courtyard in self-important haste. How could my friends and I *not* consider this an open invitation for *Batman* gags? I despised the inequitable systems of high-school life, and made myself heard at every opportunity that arose to question and condemn. The 'gold'-braided 'prefects' were a favourite target for my bitter scorn. The headmaster was adept at handling 'traditional' rebellious behaviour: smoking, bad language, petty vandalism; but personal insults and rabble-rousing were in a new and intolerable league. He began to monitor us closely, and to turn the crushing screw of rule-bound dominance. His special relationship with the hostel warden meant that there was no let-up for me at the end of the school day.

In reading over some of what I have, so far, written in this chapter, it now seems amazing to me that I did not realise at the time that I was living the life of a walking, talking, moping, and wailing teenage cliché. All the elements are there, and we could concoct, if we wished, a simple elemental recipe for such a period of bewildered angst. The neurochemical foment of the teenage brain does not, however, allow the unhappy subject the benefit of an external perspective. Trapped in an emotionally-bruising state of rapid neurological change, the teenage sense of self can become stretched and warped in new, frightening – and sometimes life-threatening – ways. Few teenagers are ever given a proper explanation of what is happening to them; the standard 'hormones' interpretation just doesn't cut it.

This state of affairs had rendered me a kind of explosive introvert – naturally inclined to shy away but driven to displays of riotous self-importance and self-promotion. It's no wonder that I wanted to be a front-man. My friends and I subverted the school-uniform code by wearing our own smart but entirely monotone versions. I grew my dyed hair ever longer and hid behind a preposterous mat of fringe. A developing interest in the music of The Clash had me adopting a more 'punk' attitude, but I feared, in my bleaker moments, that I was fooling no one.

A stage name, to cement my new identity, seemed an obvious requirement. I cobbled one together from the first name of Killing Joke's lead singer, and the protagonist of a cult Sixties TV show. This rebirth invoked much ridicule, but, while there were many soft areas in my underlying personality, the stage name was part of my wall, and was, as a result, unassailable. And it really helped me. I felt *in control*; this was a special thing that I had chosen for myself. Plated with this clanking self-assurance, I pushed myself forward.

Art classes allowed me some little self-expression within school boundaries. Chalk pastels suited my blended-out style better than the stroke-by-stroke discipline of paint. And 'graphic design' was always a great excuse to mock up album covers. But Guitar and my inscrutable friend Smith (as I shall now call him) were on an entirely different level of artistic ability. Their sketching was quick and fluid, and their brush-strokes brightly confident. While I laboured over unimportant details, they fleshed out their life-drawings to reveal skin tone and bone structure. But, at least in part because I did not wish to be separated from them, I continued with my less-than-artful art.

As neither Guitar nor Smith stayed in the hostel, they must have found my endless complaints about it tedious. But they never showed it, and were always willing to sympathise, understanding, at least, that the draconian constraints of the place were wearing me down. Perhaps I had never really gotten over my initial homesickness. I did miss home, but also began to find that by Sunday nights I was more than ready to leave it again. The visits there were too brief to allow me time to settle.

By fourth year, we were allowed 'special' dorms in the hostel, which we each had only to share with one other resident. I enjoyed rooming with Thor, my fair-haired friend from primary school. He had a knack of lightening [sic] the atmosphere, although it was clear that he was now carrying some substantial burdens of his own. He had not found his true niche and, at times, was something of a loner himself. Nevertheless, his sporting ability, his powerful frame, and his patent good nature meant that he could usually find ways to connect into otherwise disparate groupings. We had come to share some musical tastes, and Big Audio Dynamite frequently blared from a ghetto-blaster on our dorm windowsill. Thor's self-reliance and physical prowess manifested in a love of climbing. He drove himself ever harder, mounting expeditions (usually solo) up the mountains of the Cuillin Range and further afield. I went with him a couple of times; I found the experience both exhilarating and terrifying, though not remotely in equal measure.

My grandmother, Doris, died. 'In her sleep,' they said. This had not been expected. My mother, and indeed the whole family, must have been distraught. But, as I did not travel to Plymouth to attend the funeral, there is a hole in my emotional memory of that time. To my way of dealing with grief, perhaps it's better that way – I remember only her life.

Pressure to decide on some academic path grew. Our 'careers advisor' was a myopic boor with no concept of what 'career in the music industry' might mean. He would lean back in his chair, exasperated, before returning the conversation to subject areas with which he was more comfortable: utilities-industry work, Gaelic, and straightforward manual labour. He bristled, also, at any mention of art-college. I think, to him, a desire for creative expression smacked of latent homosexuality, and that a 'creative streak' was something best purged with a proper dose of shinty, cold showers, and dirt-shovelling. I had hoped, honestly but naïvely, to find a career that would fulfil some element of my thirst for stability while still allowing me the creative freedom I craved. It was clear, however, that I had come to the wrong place for such flexible thinking. My sloppy efforts in maths and physics had hobbled me, and opportunities to learn more about computing at school had been almost non-existent, so there was no prospect of a career in science or technology.

On a school-arranged minibus tour around the art colleges of Scotland, Guitar, Smith, and I somehow managed to get permission to stay with Guitar's brother for a night. He, a grinning paint-spattered giant, had already studied at art college for a couple of years. He kindly admonished our drunken trashing of his tiny (and admittedly quite trashed before we arrived) residence. The trip was a learning experience of sorts, but I gleaned from it none of the artistic inspiration and aspiration that I saw on the faces of Guitar, and to a lesser degree, Smith.

But I now saw a clearer path ahead. I had heard of a 'rock school' based at Perth College. Could this be my salvation: learning the disciplines of music; writing with like-minded lyricists; blasting out edgy, alternative songs, in an environment of majestic fraternity? I didn't want for imagination.

Mooching after girls, playing music, and finding ways to get hold of alcohol – late-teenage years are so clear-cut, aren't they? Guitar and I looked older than our years, so we usually managed to get a few drinks, or a Red Stripe carry-out, at the local pub. Our front of confidence even extended to noisy jamming-sessions there. And, while most of our mooching was entirely unproductive, our unsubtle presence in the pub sometimes gave us a better-than-average chance of female company.

While in that bar – still under drinking age – with Beta one evening, her now-estranged husband loomed large. 'You think your little brother is going to protect you?' he shouted after us, as we left. *Protect her? From him, that overgrown boarding-school bully? And from what kinds of habitual words and deeds?* I dreaded to think, as Beta turned her back on him and we walked away.

Gifts I received for my eighteenth birthday included *The Flowers of Romance* by PiL, smoke bombs (one of which I ate at my party, thinking it was a sweet), and several intoxicating puffs of an expertly-rolled spliff. I sang with the band: I remember a particularly best-china-rattling cover of 'Brand New Cadillac'. We had packed my parents off to a hotel for the evening, so that we could hold the party in their house. Perhaps the presence, at the party, of Beta and her new man had reassured my mother that no major damage would be done. But Beta couldn't be everywhere at once.

Having intended to leave high school, a few months later, in a blaze of (self) righteous glory, I had to make do with lukewarm ignominy, when Guitar and I were caught, during lunch-hour, having an under-age pint with our meal in a Portree pub. I have never been entirely clear on this, but I think I may have been expelled a few days before I was due to leave anyway. It was hardly a 'rock 'n' roll' end to my fractured schooling.

And how did I feel, awake in bed in the early hours of the morning, about the passing of my school years and about the blanked-out future ahead of me? Nostalgic, fearful, and unready; curious, but always, always afraid.

# 5  MIND NUMBING CONVERSATIONS

## THE DOCTOR

'It's certainly more thorough than the ones I'm usually asked to do. What's it for?'

He was pleasant enough, and obviously interested, so I was OK – if a little nervous – about telling him. 'Cryonics,' I replied.

'Hmm... It won't work you know. Far too complex.'

Not quite the reaction I had expected. I must admit that I felt a little disappointed at his blasé response. But I should have realised – I had heard before that he was into sci-fi, and the *Analog* magazines on the waiting-room table had been confirmation of that.

I began to mount a faltering defence of my position but was put off my growing stride by his sudden change of topic to the scientific errors in the movie *Sunshine*. I was forced to agree that it hadn't been a proper 'hard' sci-fi movie.

The samples went into a padded bag, ready for sending to the insurance company's appointed lab. I don't know how I'd describe my feelings of that time – a mix of elation, resolve, and incredulity, perhaps. But, as ever, with the good feelings tempered by the tight knot in my stomach.

This doctor has since 'come out'. Not in the sexual-orientation sense, but in the sense of revealing previously-hidden layers: A video for one of his recently-released songs has him growling his angry lyrics, in flickering candlelight, from beneath the cowl of a monkish robe.

## THE LAWYER

The meeting with my solicitor, a month earlier, had been rather more mundane. Just a few minutes saw the process of dealing with the Alcor legal documentation complete. But it had at least raised a circumspect, professional eyebrow; it wasn't every day that this kind of American legalese passed through the office of a Highland solicitors' practice.

'So you're into all this kind of stuff, are you?' he asked, eyes down, scribbling furiously. The tone was even, but a hint of quizzicality had crept in.

I thought it best, under the circumstances, to employ my stock down-play response. 'I don't see it as being much different to any other kind of anatomical donation.'

'Oh, right.'

He looked up and commented about some strange Will request he had once dealt with. I can't remember what it was. I was too busy thinking about the paperwork, hoping that he wasn't going to tell me it was invalid and he wouldn't be able to notarise it.

He called one of his admin people through to witness the signatures, and it was done. With a quick twitch at the corners if his mouth, which I interpreted as a smile, he handed the signed documents back to me. I had no reason to suspect that the twitch was not genuine – he had always been friendly and polite in all our dealings. It was just a characteristic of the man: all coiled energy, and evidently so busy that he didn't even have time for complete smiles.

## THE MOTHER

'But you'll be out *there* – in America.'

'No, mum, I won't be *anywhere*. I'll be dead.'

'I know but... it won't be the same.'

## THE FATHER

...

## THE MOTHER-IN-LAW

'You'd better tell her,' said my wife. 'It'll be in the papers tomorrow, and they won't know what to think if you don't call them.'

The thought did not appeal to me. I had imagined interesting and heated discussions with direct family, but the 'in-laws' were a different matter. Conservative and religious in their views, and virtually unaware of my scientific awakening, I guessed that they would balk at the announcement.

'Right... and this is going to be in which paper?' It wasn't difficult to gauge the level here, even without the expressive feedback a face-to-face conversation would have provided. This was going to be all about the publicity, and the way it might reflect on *her*. Or maybe it was about prestige? Even a *really strange* pronouncement might confer *some* kudos if published in a respectable newspaper.

Again, I had my downplay at the ready. 'It's really just a different way of donating my body to science.'

'Well... Uncle Andrew did that. Everybody thought it was odd when he decided to do that, but it was what he wanted.'

This was encouraging. I started to explain further but immediately found myself in difficult territory. The issue of decapitation was obviously going to be beyond the pale. I decided, spinelessly [*sic*], to postpone her discovery of that little detail until the publication in the press.

'Well, I'm sure you've thought it all through.'

*No, actually, I am doing this on a whim. I am an impulsive (though oddly long-termist) insane person, who has just decided for no particular reason to arrange for cryonic preservation upon his death.* The currency of sarcasm would not be handled by this woman. I bit my lip. Hard.

After all, the conversation was going better than I had expected. She had not, apparently, decided that I was evil, twisted, or mad. I had done my duty, and decided to wrap it up before it became really awkward.

We visited them a few days after the story was published. An in-law-to-be turned the conversation round to cryonics, and it was quickly made clear by my mother-in-law that she did not wish to hear about the subject again. Her bubble closed around her, the surface featureless and intact.

## GUITAR

(Some weeks later).

'I'm going to the Hallowe'en party as the cryonically-frozen head of Walt Disney.'

'I think that's a bit of an urban myth – the thing about Walt Disney's head being frozen. There's no proof it actually happened,' I ventured, aware that this was a bit of a spoiler.

'It would be funny if it were true, though,' enthused Guitar, unperturbed.

He sent me a photo. True to artistic form, he had produced another masterpiece of grotesquerie. He looked really *cold* in there. It was on a par with his Ahab outfit, as part of which he had sported a grievously-wounded, sad-eyed white whale – lovingly crafted by his wife from an old pillow – roped onto his long back.

Despite his interest in human anatomy, he found the short cryonics documentary I sent him 'a bit grim'.

## SMITH

'*Whhaat?*'

Smith's stifled-laughter routine was always amusing to witness. His face went bright red and the veins stood out on his high forehead. Tears showed at the corners of his eyes prior to the eruption of the deep-mined chuckling guffaw.

'You... are... a... *nutter.*'

## THE TEENAGER

Maybe I'd find it easier to talk to kids if I had some of my own. 'Easier' is perhaps the wrong word. It is not difficult. But I can't do 'child talk'. I just speak to them as if they were adults and hope that I don't come across as too weird. That's not true either. I *do* want to come across as weird. I want to be the weird uncle, and I want them to care about what their weird uncle says. Boring too. The weird and sensible boring uncle. There is little specific that they will learn from me in the snapshots of time that we spend together, but, just maybe, they'll see the dark glee bubbling beneath the surface. That would resonate with me if I were still a child.

'Like in Futurama?' said Alpha's son, his toothy smirk poorly suppressed.

'Not exactly,' I replied, smiling. There was no ridicule in his question. More, I felt, a worldly in-joke acknowledgement of my funny-serious admission. And the image did flash crazily across my mind: my head in a domed glass jar, a similarly-housed Nixon for company.

'My Dad told me about it,' said Alpha's son, now smiling broadly and warmly. He showed remarkable self-consistency for one so young.

I took this rare cue and babbled my disjointed scientific rationalisation. But there was no need. There was no disgust or condemnation in his demeanour, only openness. The detailed explanation could come in time, if he ever wished to hear it. And, for now, perhaps better to just leave him with that image: His weird uncle's cartoon head, at home in a colourful and anarchic future world, preaching enthusiastically to the inside of a jar.

# 6 CONTINUITY

∞

Forever is composed of Nows

—EMILY DICKINSON, 'Forever is composed of Nows'

The subject of death, and the idea of some form of continuity after death, has occupied the minds of philosophers over thousands of years. The subject has, of course, occupied most minds at some point in history – especially those of theologians – but I am, here, talking about the kind of forensic, argumentative, analytical dissection of such issues that was traditionally undertaken only by philosophers. It's not that most of us can't do this type of thinking, it's just that most of us seem to have little use for it outside of those special situations, such as grieving, that may force us to consider these difficult questions. And, even in those circumstances, many of us have ways of 'dealing with' the associated complex and troubling emotions that don't involve precision philosophical dissection of our own reasoning.

I need to do some dissecting. Because we are getting deeper into the subjects of 'self' and personal identity, we need a compass to indicate the direction of *my* understanding of what a 'self' is. My viewpoint is not, by any means, unique. It has been distilled from my own thoughts and from the thoughts of others whose books I have read and/or whose words I have heard. Again, I only give my own understanding of the subject significance *within the context of this book*, as it is, in essence, an examination of my reasons for making a 'radical' choice partly galvanised by that understanding.

Now that I have done my best to wriggle off that particular hook, we can start cutting into some juicy selves.

## THE SAUCE OF REDUCTIONISM

In asking you to think about what constitutes a 'self', I will first ask you to think about which of these camps you fall into:

(a) You think that *there is* some element of what 'you' are that is separate from your physical brain and body. You may or may not think of this 'separate element' as a 'soul' or other *religious* concept, but you *do*, at least, believe that there is *something* more than 'just' your physical brain and body that constitutes *you*. You think that this element is *more* than or *other* than *consciousness* brought about by your physical brain and body. You may think 'soul' is a good word for describing this other thing, and you may believe that this other thing will survive your death.

(b) You *do not* think that there is some element of what 'you' are that is separate from your physical brain and body. You do not think that you have a 'soul' or other element that is *separate* from your physical brain and body. As a result of this, you do not think that 'you' will continue in any 'spiritual' way after your death. You may or may not find the term 'soul' useful shorthand for describing otherwise-tricky concepts such as *consciousness*.

There are, of course, umpteen shades of grey in the spectrum of these beliefs/understandings. (And others may claim that there is a 'third way' that does not fall within this spectrum.[1]) However, I think that there is a clear enough ontological gap between them to treat them as two quite different – in fact, opposing – modes of thought. It is clear which camp I fall into, but I will have to explain at least something of why we have every reason to conclude that one is correct and that the other is not.

My claim that one of these modes is incorrect contradicts the beliefs of billions of people. This sheer weight of numbers (as even a basic grasp of logical fallacy informs us) is irrelevant to the argument. Despite the fact that such believers do not usually require technical corroboration of their beliefs (faiths), they tend (as individuals) to assume that *any* attempt at technical dissection of those beliefs amounts to *ad hominem* assault. My definitions of the two opposing standpoints may not be as clear as they could be, but it is not my intention to hone them for use as a means of attack. While I am always prepared to attack beliefs that are not based upon sound evidence, I am, here, stating only that I will be giving reasons why camp (b) is correct, whereas I would not be able to give reasons, even if I wanted to, why camp (a) would be correct.

In terms of the current scientific paradigm, we need have no more 'technical' knowledge than that – as I stated at the beginning of this book – everything is made of atoms, to understand why (b) is correct. As I mentioned there, there are 'subdivisions' below the level of atoms, but I will argue that even those are not required for the (b) case to be correct. Atoms will suffice. If we allow the introduction of quarks, neutrinos, and other quantum particles into this argument, we end up in the realms of a mishmash, usually strongly pseudo-scientific, of ideas about 'quantum souls' and other such 'New Age' flannel.

I would be happy to allow 'quantum souls', and other ideas of this type, into this argument if they were *required*. By 'required', I mean that it would need to be the case that the generation of 'selves' or 'persons' would not be possible without them. It's worth stating, however, that any newly-discovered requirement for those ingredients would probably defeat the intentions of those seeking to have them included. That's because petitioners for those ideas are vastly more likely to come from camp (a), and they would quickly find that people from camp (b), seeing the requirement, would have then to conclude that something *below* the level of atoms was a *physical* requirement of selfhood. While that might change the 'functional unit' required for (b) to be correct from atoms to, say, quarks, it would not make it any less true.

In the above paragraphs, I have massively (and perhaps poorly) paraphrased some of the ideas of brilliant thinkers such as Parfit and Hofstadter. But I think I have, in the descriptions of camp (a) and camp (b), synthesized my understanding of what is meant by *reductionist* (b) and *non-reductionist* (a) views of what a 'self' is, or is not. I don't particularly like these terms, for they are often misinterpreted: (b) sounds pejorative, appearing to imply that those people merely squash difficult concepts to fit a narrow scientific world-view; (a) sounds falsely libertarian, appearing to suggest that such people exhibit greater freedom of thought than do reductionists. What these terms actually mean, in this context, is that (a) people have a *highly complex* (so complex that neither they nor anybody else can explain it) belief about what constitutes a self, whereas (b) people observe the outwardly-complex phenomenon of self but think that they can explain it in simpler terms – crucially, in terms that other people can understand and, potentially, test.

There are other forms of reductionism and non-reductionism relating to 'selves'. There are, for example, 'ultra-reductionists' who wish to speak of what constitutes a self in terms of units or levels that are not, in the opinions of 'standard' reductionists, required. These ultra-reductionists might, for example, contend that selves, while being purely *physical* phenomena, depend on *superstrings* (down at the Planck scale) for their existence. This general way of thinking has been called 'greedy reductionism' by Daniel Dennett, because it seems to 'gobble up' levels that have not been scientifically proven to be demanded.[2]

Penrose and Hameroff's *orchestrated objective reduction* (Orch-OR)[3] is a well-known example of a 'quantum consciousness' hypothesis. Here, the central claim is that consciousness is not possible without exotic quantum effects such as *Bose-Einstein condensates*. Having been introduced as the result of collaboration between a famous mathematical physicist and a professor of anaesthesiology, it has received more attention than most such hypotheses. It has also been widely debunked. Other physicists, such as Max Tegmark, have pointed out that the quantum states required would be unlikely to arise and that, if they did, they would not last long enough to play any role in consciousness.[4]

In reductionism terms, Orch-OR is gluttonous in its enterprise. It is hardly surprising that its purported scope should induce confusion about whether it is an ultra-reductionist view or a non-reductionist one. Its apparent weight of scientific backing seems to lend it a credibility that run-of-the-mill 'quantum immortality' -style claims do not have. Quantum effects do appear to play some role in biology, and there is growing evidence that quantum fluctuations may be present in *microtubules* within neurons.[5] Nevertheless, it is a huge leap to go from accepting the presence of quantum fluctuations in the brain to asserting that consciousness *depends* on quantum effects. It is an even greater – and totally unjustified – leap to assert that they imply some sort of 'quantum soul'.

In most cases, ultra-reductionism is just a way of 'scientising' what are really non-reductionist points of view. In my view, some non-reductionists, keenly aware that their ideas are open to scientific scrutiny, try to find scientific-*sounding* loopholes that allow them to argue in more technical-*sounding* terms in favour of a view that is, in reality, no more than a New Age notion of what constitutes a spiritual soul.

It's not that there wouldn't be places for these 'quantum entities' to inhabit. The levels smaller than the scale of quantum theory particles, but larger than the *quantum foam* of the Planck scale, are yawningly 'vast' but apparently empty.[6] Could there be entities hiding there, like the 'Lions and Tigers and Bears' of Dan Simmons' *Hyperion* and *Endymion* sci-fi novels?[7] Though immensely entertaining, speculation of this type does nothing to demonstrate a requirement for the existence of such entities. And actually *believing* in such concepts would simply indicate a propensity for religious/magical thinking on the part of the believer.

## WHAT IS MORAL PHILOSOPHY?

The issues I have touched upon above can be classed as part of a vast, venerable, and ancient field of study called *moral philosophy* or, sometimes, *ethics*. It is relevant here because of what it says about the nature of, and indeed about the *existence* of, persons.

Greek philosophers as early as Aristotle (384–322 BCE) and Socrates (469–399 BCE) claimed that self-knowledge was an inherently good and worthwhile type of knowledge to seek. The *Stoics* taught that virtue, personal serenity, and strength of will were the gateways to contentment and understanding. The term *monism* (as opposed to *dualism* or *pluralism*) encompasses a range of early philosophical ideas (some of which have survived into the present day) to do with 'oneness'. I have found it difficult to understand the tenets of monism because of the huge variety in the different forms of this philosophy and in the methods by which the *one* thing, either 'material' or 'mental', is measured. The school of philosophy known as *hedonism* is, perhaps, rather easier for present-day people to get to grips with.

During this early period, Eastern philosophers were also concerning themselves with formulating doctrines for moral standards. *Mohism* encouraged a belief in spirits, specifically for the utilitarian purpose of instilling the belief that people would be rewarded by good spirits for good deeds, and punished by bad spirits for bad. But the utilitarianism of Mohists mostly dealt with more worldly concerns such as thrift, inclusiveness, and promoting 'mentoring' governance. *Confucianist* doctrine was also heavily weighted towards the idea that leaders should set good examples for their people. In accordance with this, it taught the importance of self-cultivation and even self-transformation, so that the results of holding high moral standards would be visible and accessible to all.

Buddhism – in its originally-intended form – also belongs in the field of moral philosophy and not in the realm of religion.[8] There is a common misconception that non-reductionist beliefs, such as reincarnation (for this read 'quantum soul'), are central to Buddhism. This is not, in fact, the case. A central tenet of Buddhism is *anatta*, or *anatman*, which translates as 'no self';[9] a related concept – *mu* (literally 'not' but also implying nonbeing, nothingness, or nonexistence) – can be found in Zen Buddhism. How could a Buddhist truly reconcile anatta with the idea of reincarnation? There can be various tortured responses to this question, but the short answer is that a 'follower' who kept strictly to the original teachings of the Buddha would be unable to reconcile anatta with most *modern* impressions of what 'reincarnation' means. To make them intelligible to the people, many of them poor and illiterate, of that era, early Buddhist teachings were given in parables – a format with unfortunately large scope for ambiguity.

So, through centuries of creeping mythification, the meaning of reincarnation has become obscured to the extent of coming to mean – in the minds of many – the polar opposite of that intended. *Actions* are central to Buddhism, and the real meaning was that actions have consequences, even though the 'actor' does not exist. The consequences of those actions become incorporated into the lives of other entities who then carry the results of those actions with them. In this view the 'entity' need not be human, because consequences of actions can affect all creatures, and indeed, all things. There is strong overlap here with another Buddhist tenet – *karma*. Karma is not, as many people believe it to be, anything to do with fate or pre-determination. It is to do with the ongoing repercussions, good and bad, of one's *actions* in life.[10]

René Descartes (1596–1650) has been called the 'Father of Modern Philosophy'. His *Cartesian* (from the Latinised form of his name) doctrine is still central to many modern philosophical views. For the purposes of this book, I will be focussing on Descartes' *dualism* and, specifically, on his idea of a 'pure ego' that is separate from the physical brain and body. I refer to him as a *dualist* because he believed in *two* distinct parts that go to make up a person: the physical brain and body, and the *ego*. You can quickly see that Descartes was, like the vast majority of people of his time, a non-reductionist. He was a brilliant and important mathematician, and was certainly looking for scientific ways to describe the self, but he could not find a way to depart from the apprehension that there must exist *something else* that makes a self possible. This 'something else', as postulated by Descartes, some philosophers now call a *Cartesian Pure Ego* (CPE).[11]

Unlike most of his predecessors, Descartes did think of the human body as a *mechanism*: He appreciated that certain stimuli, such as those involving pain, seemed to cause an 'automatic' response. But the limitations of the investigative 'tools' available to him – deep introspection only, in the case of his investigations into the nature and 'location' of the self –

dictated that he would be unable to circumvent the notion of a 'seat' of selfhood somewhere in the brain. He decided that the *pineal gland* (actually a hormone-producing gland located near the centre of the brain) was the seat in question.[12] In coming to this conclusion, he was positing a 'central arena' of unification in the brain – others have come to call it the *Cartesian theatre* – where all the 'understanding of' and 'being conscious of' plays out.

Certain later thinkers, such as Georg Ernst Stahl (1660–1734), rejected even Descartes' 'localised ego' hypothesis, instead falling back to an anti-mechanistic position. Stahl, along with other 'vitalists', still believed in an animating-life-force-type soul (*anima*).[13] In that it is was presumed immortal but was also thought to run the more mundane and integrated task of 'powering' the body and mind, this kind of soul sounds less like a CPE and more like an Ancient Egyptian *ka*. (Over a century later, influenced by the scientific discourse of the day and by descriptions of Luigi Galvani's earlier electrical experiments on dead frogs, Mary Shelley's very literal interpretation of such 'powering' would give birth to Frankenstein's monster.) The important philosopher/mathematician Gottfried Wilhelm Leibniz (1646–1716), a contemporary of Stahl, was critical of Stahl's vitalism but also of mechanistic doctrines. Despite his use of the term *vis viva* (living force) in an early, limited formulation of what we now call *kinetic energy*, Leibniz did not accept that perception could ever be explained as any kind of mechanistic process; instead, he favoured a view of the self as a *monad* – not dualistic, mechanistic *or* animistic, just *indivisible*.[14]

I will not go into detail about the ideas of the other important modern moral philosophers who took up the baton from Descartes, but I will mention briefly some of their views on death, afterlife, and the self, as relevant to this discussion.

Blaise Pascal (1623–1662), though perhaps better known as an inventor, physicist, and mathematician, became immersed in Christian philosophy later in his short life. In part of his *Pensées* (Thoughts), published after his death, he laid out his 'rational' and 'pragmatic' arguments for believing in God. The gist of his reasoning was that even if the chance of heaven existing is slim, one would have little to lose and, potentially, eternal life (i.e. *everything*) to gain, by taking measures to get there. (The flip-side of this being the minimal gain but potentially 'infinite' loss of missing out on heaven because of a life lived without faith.) This argument has come to be known as Pascal's wager.[15]

Though couched in convoluted philosophical language, Immanuel Kant's (1724–1804) views on the existence of God appear quite similar to Pascal's. In his *Metaphysics of Morals*, *Critique of Pure Reason*, and other works, Kant argued that religion is a *duty* in life, but also that any relationship between human being and God falls outside the remit of moral philosophy. Lying outside of this scope, his reasoning goes, it is not possible to know whether any kind of afterlife awaits after death, and so it is not, therefore, unreasonable for people to believe in one. As he stated in his *Metaphysics*, 'only the relations of *human beings to human beings* are comprehensible by us.'[16] Kant was driven primarily by a need to understand and lay out the 'moral metaphysics' of knowable relationships, rather than by a desire to examine the unknowable ethics of unknowable ones.

David Hume (1711–1776) has been called a 'bundle theorist' in his attitude towards personal identity and the self. This term applies to theorists who claim that the self is nothing more than a 'bundle' of perceptions or tropes. (This term has also been applied to modern-day moral philosophers such as Parfit.) Hume, an important figure of the 18th century period in Scotland's history that has come to be known as the Scottish Enlightenment, did not believe in a 'permanent ego'. He found only a flow of fleeting 'perceptions' while attempting to 'enter most intimately into' himself. By removing his ability to experience any perceptions of any kind, death, he reasoned, would render him 'a perfect non-entity'.[17] Hume's stance on gods and afterlives was sometimes ambiguous, but it is clear that he didn't have nearly as much time for consideration of 'the next life' as he did for the conundrums of everyday consciousness and selfhood.

I have a set of *ethics* and, by extension, a *moral philosophy*, but I am not a moral philosopher. It seems to me that being a true, modern moral philosopher involves having a great knowledge of the history of this field and applying an intensely detailed, rigorous, and balanced set of intellectual tools and principles to knotty tasks such as unravelling what *does* and what *should* make people and societies tick. I would only have to be this rigorous if I were trying to convince everyone else to make the same decision that I have made. I am not trying to do this. I *am* keen, however, to encourage people to question the notion of a *spiritual self* that is separate from their physical brain and body, for thinking critically about this may lead to novel and valuable conclusions.

The moral-philosophical debate about *continuity* can be condensed down to three main (and very broad) concepts: physical continuity, psychological continuity, and spiritual continuity. The concept of *physical continuity* is that continued survival (as opposed to death) requires a physical brain (or at least part of one) and, in some views, a physical body (or at least the 'important parts' of one). *Psychological continuity* holds that the continued survival of the thoughts and memories of a self (or at least some part of them taken as a whole) would be enough for the continued survival of a self. *Spiritual continuity* is a religious or quasi-religious belief that the person will continue after their own death *even though* both their physical body and their 'psychological body' (here meaning the inner perceptions that were connected to their physical brain) have ceased to function.

I do not think that there is any such thing as *spiritual* continuity. I have some idea of what religious people (and I am including 'spiritual' people in this category) *think* they mean when they talk about 'the soul', but I have no ideas about how such a concept could ever be explained to me in a way that could make sense, let alone be *proven* to me.

There are, among reductionists, coherent definitions of what might constitute a soul, though not in the spiritual sense. In his book *I am a Strange Loop*, Hofstadter proposes that we could have a numerical scale of 'degrees of souledness'; he dubs these 'hunekers', after the arts critic James Huneker.[18] Though he uses – very loosely – the word 'soul', it is clear that Hofstadter's 'hunekers' are really the units on a scale or spectrum to meas-ure *consciousness*. He playfully suggests that by using the huneker scale we could discuss the levels of 'souledness' of things, animals, and people: How many hunekers of soul does a mosquito have? How many a rabbit? How many a zygote? How many an adult person? I added 'things' to the list because we are even asked by Hofstadter to consider the 'desires' of flush toilet float-valve mechanisms, and of thermostats.[19] Most people will find this bizarre, but I am comfortable with thinking about such 'self-balancing' systems in this way. Your thermostat 'wants' to keep your room at a specific temperature; your toilet's float-valve mechanism 'wants' to maintain the correct level of water in the cistern. Each will 'adjust its behaviour' to 'realise' its goal. 'Wants', in this context, does not imply intelligence or consciousness. I will discuss this subject again later in the book.

After excising *spiritual* continuity and, by extension, non-reductionism, we are left with *physical* continuity and *psychological* continuity. Despite all I have said, you may still think that removing non-reductionism from the running is far too big a leap. To this charge, I can only say that my personal moral philosophy is a practical one. I have come to my conclusions from several different directions, and while I find the ethics of the subject fascinating, I am neither torn nor balanced between the reductionist and non-reductionist poles of the moral-philosophical debate. I have *no* spiritual or non-reductionist beliefs. And such beliefs would go against my personal philosophy, not only because I would find myself unable to *believe* in them, but also, if I did believe in them, because I would find myself frustrated by the inability of *anyone* to realise any kind of practical evidence or outcomes related to those beliefs. I am, then, in the terms of this particular discussion, a *practical* or *pragmatic* reductionist – a hungry one perhaps, but not greedy.

If you are the sort of person (and I hope that you are) who thinks about these sorts of subjects, you should perhaps spend some time considering your own stance. In my opinion, psychological life is less rich when one maintains a non-reductionist view. Such a view may manifest as only a *vague*, unexamined, *spiritualistic* notion; or at the other end of the spectrum, it may show up as a *fixed* (formally religious) belief in a numinous, deity-inhabited realm – a full-blown paracosm. A *vague spiritual* view may act as a 'buffer', allowing one to claim a kind of magnanimous agnosticism, or to place importance on 'spirituality' without the inconvenience of the further implications (as the 'soul food' some seek, this may prove stubbornly runny and unsatisfying). A *fixed spiritual* view can close the mind to any possibility of finding 'meaning' in life without a requirement for the existence of spiritual entities. Gods become our drugs.

## THE SHIP AND THE SHOVEL

The dockworker sweated as he hefted another shovel-full of sand into the cement mixer. The March sun, though low in the sky, beat down on him with unmasked intensity.

The naval architect didn't usually stop to chat, but he was in good spirits today; the project was nearing completion.

'You must get tired of shovelling all day, especially in this weather,' said the architect.

'I get tired, but I don't get bored,' replied the dockworker. 'I get into the rhythm of the job, and the time just passes without me really noticing.'

'My job has *a lot* to do with time,' sighed the naval architect. 'Always deadlines.'

'I just need my shovel for *my* job,' said the dockworker. 'No complications. I never liked complications. A broken shovel is about the worst complication I ever get in my job.'

'You must have gone through a few in *your* working life,' said the naval architect.

'No, just the one,' said the dockworker. 'Five new handles and two new blades, but always the same one.'

The naval architect laughed. 'I've heard that one before,' he said.

'It wasn't a joke,' said the dockworker, leaning on his shovel.

'But every part of it has been replaced, so you *can't* possibly believe that it's the same one,' scoffed the architect. 'It's a completely different spade to the one you started with.'

'What about that old ship your people have been working on for years?' said the dockworker. 'The carpenters told me they've replaced every plank on her, after that fire. The masts and rigging were all re-done twenty years ago, and even the ribs were rotten when it first arrived here, so they had to replace those too.'

'HMS *Continuous* is a ship of *historic* significance,' the architect retorted. 'We have restored her to her *original* form.'

'She's not the same ship, then,' said the dockworker. 'She just has the same name.'

'Of course she's the same ship,' said the architect, face reddening. 'The process of maintaining and restoring her has been, just like her name, *continuous.*'

'I hereby name this shovel *Shovel*,' said the dockworker. 'Happy now?'

The story above illustrates a point about the semantic difficulty we sometimes experience in making claims that a thing is, or is not, *the same* even though (or because) it doesn't have any of its original component parts. We could make one of several different claims here:

    (a) Neither the ship nor the shovel is the *same one* as its respective original, because of all the replacement parts.

    (b) Each is the *same one* as its respective original, despite all the replacement parts.

    (c) Only the ship is the same one.

    (d) Only the shovel is the same one.

As the joke implies, most people would reject as ridiculous the claim that the shovel is the *same one* as the original. But they may find rather less ridiculous the claim that the ship, despite the fact that it has *none* of its original parts, *is* the same one. However, if we grant continuity to the ship but not to the shovel, we have a paradox. Is it that the ship, in contrast with the shovel, has a *name* and, by extension, an *identity?* Or is it that the ship earns continuity by being far more *complex* than the shovel? We *could,* I think fairly, argue that the idea of the *physical continuity* of the ship is justified by the *number* and *extent* of the changes over time. Under normal circumstances, the ship's carpenters might have changed one or two planks at a time (although they would, over time, have made *many* such small repairs); at most, each change would only constitute an overall change of one or two percent of the entire ship. In contrast, the shovel has had only *a few* changes over time, but each of those changes altered some fifty percent of the total shovel. It is odd, though, that the comparatively *fewer* changes to the shovel seem to have denied it physical continuity.

The point I am making is an obvious but important one: Whenever we try to attribute a *continuous identity* to a physical thing, we end up having a hard time defining what that actually means. This problem with defining sameness, known as Theseus's paradox or the ship of Theseus (a later variant – 'grandfather's axe' – is closer to my shovel analogy), has been troubling philosophers for millennia.[20] The point holds just as true for selves as it does for ships and shovels.

We accept that human brains and bodies are not fixed in exactly the same atomic configurations throughout a lifetime. Cells are constantly dying, being replaced, exchanging chemicals, and so *our* 'component parts' are always in flux. Though often overstated, it is true that throughout the course of our lives many types of cells in our bodies will be completely replaced, sometimes several times over; there is even some *neurogenesis* (growth of new brain cells), though this appears to be quite limited.[21] Our bodies are, then, in this sense, like ships with names: We appear to have physical continuity despite gradual (though not total) replacement.

We may find it easier to attribute continuity of identity to at thing when it has an individual name, but there are plenty of exceptions to this: We ascribe a form of continuity of identity to things like fires and mountain streams, even though they may have no individual name and are changing from moment to moment before our eyes.

But human beings can have something that ships, shovels, and streams seemingly cannot – *psychological* continuity. Unless you have amnesia, you have an identity that is (ostensibly) continuous throughout your lifetime. That identity is there when you wake up each morning, even though you have been unconscious for some hours. That identity still holds even when you are old and perhaps suffering from some terrible form of dementia. You will, I expect, have to think about that last statement before making up your mind. That's because we are getting into matters of *degree*. The identity issue has here become more nuanced: The dementia sufferer has changed; he behaves differently and forgets things. His internal *psychological* identity has changed, yet we still call him by the same name. We might use phrases like, 'he's not the same' or, 'she's losing it' about such people, to verbalise this perceived change in psychological identity.

Buddhism maintains that, despite our strong sense of having an anchored personal identity, our psychological continuity is both more tenuous and more dynamic than we tend to think. Bioethicist James Hughes puts it this way:

> Buddhist psychology argues that the continuity of self is like a flame passed from one candle to another; the two flames are causally connected, but cannot be said to be the same flame.[22]

We can learn, then, to view our day to day and moment to moment continuity as such a continuously transferred psychological 'flame', and so loosen our attachment to the idea of a fixed self. Acceptance of such a view, however, may not be sufficient to relieve us of the propensity to dwell on the issue of what will happen when our stock of candles runs out.

## VERSIONING ISSUES

It can appear smugly easy for the reductionist, in that she could simply claim, 'I am my brain', thus attributing all continuity of personal identity to the continued physical existence of that organ. I accept that this claim is a reasonable shorthand, and that in most, if not all, *normal* circumstances, this explanation will suffice. But the fact that it will suffice under normal circumstances does not make the claim true (any more than the utility of Newton's laws of motion for measurement at *ordinary* non-relativistic speeds could compel those same laws to hold applicable and true under *all possible* circumstances). The circumstances under which the claim would not hold true would be, I admit, rather unusual. But, crucially, they would not *all* be *impossible*.

In order to show this, I am about to use an example that is *technically impossible*. Parfit illustrates the extreme end of this point with an allegory about a man who is 'teletransported' from Earth to Mars.[23] Imagine this teleportation device working (at least to the casual observer) just like the one on *Star Trek*. The device first scans the man, then beams the captured *information* about the position of every atom in his body to Mars. There, another device creates an *exact copy* of him out of local matter (any old atoms would do). The 'old' version of him on Earth is simultaneously destroyed, so that there is now only the 'copy' on Mars.

You might think the above illustration meaningless: A man has died on Earth, and a copy of him is now living a life on Mars; there has been no overlap of physical continuity between the Earth version and the Mars version, so he is *not the same person*. I admit that this was my reaction the first time I read (out of context) Parfit's example.

Another example of the same idea, also impossible, is to be found in the film *The Prestige*.[24] To perform amazing feats of 'magic', a conjurer named Angier (played by Hugh Jackman) purchases a machine (bizarrely, made by Nikola Tesla played by David Bowie) that can make an exact copy of whatever is placed inside it, at a different location. By entering the machine on stage during the performance (billed as *The Real Transported Man*), the conjurer can 'magically' appear at the back of the auditorium, there to claim the adulation of his audience. Meanwhile, one copy of him (the 'original'?) is left to drown, below stage, in the locked water tank into which he dropped immediately after processing in the copy/move device.

Why was it that I was left feeling puzzled and uneasy about the story? Well, many people who watched the film probably found the idea horrifying: a trail of dead Jackmans left behind by the magician's hideous escapades. But, for me, that wasn't it. Yes, it was horrifying – but fascinatingly, philosophically, existentially so. It was because, I later realised, I couldn't find a word to describe the *state* of the magician while the trick was being performed or immediately after it. I couldn't clearly say that the magician had *died* during the trick, and I couldn't clearly say that he was *still alive* after it. I couldn't even say where 'he' had gone.

I now realise that my inner question about the 'state' of the magician was what Parfit has called 'empty'. There is no *real* answer to this question. As there is no 'something else' beyond his physical brain and body, and his thoughts and memories, *he will still exist* after the trick, if the copy is good. I say 'good' and not 'perfect', because the copy doesn't even have to remember *everything* or have *exactly* the same physical attributes in order to still be the same self, just as you are the same person even if you lose a finger or forget what you were doing on the third of March last year.

Let's now imagine a sequel to *The Prestige*: We'll call it *Prestige II: The Redundant Jackmans Fight Back*. In this film, things start to go wrong, and some of the pre-trick 'originals' survive. Now angered by the wanton disregard, duplicity, and duplication of the enterprise, they decide to rebel. The most recent 'version' is enjoying his fame and fortune and has no wish to have these 'old versions' spoil the fun. Reasoning that they were *supposed* to be dead already, he begins to stalk them and, one by one, kill them off.

That's enough of the plot; now let's look at some of the issues. Shouldn't the old versions be content that the new version is just the same, and so not be concerned about dying? Is the magician obviously a murderer? If he is *now* a murderer, then was he also a murderer when the trick worked properly? When he decides, prior to each trick, that he will copy himself, is he actually committing suicide, or is he, instead, a victim of *himself*?

We *could* think of the Jackman as a file on a computer. In the first film, our Jackman 'file' is copied to a new location, and the 'original' is promptly destroyed. The information in the remaining file is completely intact. In our imaginary sequel, the Jackman file is copied to the new location, but the original file persists. The longer the original Jackman file persists the more time it will have to be altered (by time and experiences in the case of a person, by adding new information in the case of a file). The most recent Jackman file also changes as new information is added.

As you can see, the files are beginning to drift out of sync. Unless we have kept careful track, we might find it hard to say which Jackman file is which, as they both contain new information. You could say that we have a 'version conflict' or 'versioning issue'. The more copies we have the more the problem compounds. Eventually, we would no longer be able to say which file was the original or 'real' file and which were copies.

If you found that you had a versioning issue on your computer, perhaps where you had been accidentally saving various edits of your book with the same filename but in different locations, it's unlikely that you would immediately start deleting files – there might be valuable, unique information in them. Whereas, if you had noticed the issue straight away, before you had started adding new information to a copy file, it's likely that you would simply have deleted the redundant copy.

If we applied the same logic to the magician and the copy Jackmans, might we be forced to say that it was, in a strange way, *OK* for the copies to die *immediately*? Even if so, most would conclude that it would be wrong to kill the copies after they had changed. The most recent Jackman might accept this, but he might also ponder, in his wickedness, *how much* change would make killing the other versions wrong, (or at least a bad mistake).

There is, of course, a closer-to-home question that arises as a result of thinking about the existential states of copied persons. But because it seems so blindingly answerable, it is a question that most people would find it intolerable to let stand as an empty one: What state are the subsequently transported and teletransported men in *before* the process ever occurs? What state, in other words, are they in when they are just as you are right now?

## THE DIVIDED SELF?

Now for a much less extreme, and not impossible, 'trick'. A skilled neurosurgeon divides your brain, cutting through the main bundle of nerve fibres – known as the *corpus callosum* – that connects the left and right hemispheres.[25] You now have what Professor Michael Gazzaniga called 'two independent spheres of consciousness within a single cranium', each with its own specialised skills and cognitive concerns. Are you now one self or two?

This *split-brain* procedure has been performed as a way (a last-resort one) of treating severe epilepsy. It has some fascinating outcomes, including a physical realisation of the saying, 'the left hand doesn't know what the right hand is doing'. A split-brain patient agrees to take part in a study of the effects of such surgery. The patient puts her arms through a barrier behind which there is a collection of tactile objects. While she stares at a central dot on a screen, an image is flashed up to only the right half of her *visual field*. Asked to pick out the corresponding object – say, an apple – she correctly picks it up with her left hand; asked to name it, she also does so correctly. The experiment is repeated, but this time the test image is flashed to only the left half of her visual field. Again, she is able to pick out the corresponding object, this time with her right hand; however, she is unable to name the object she is grasping.

Because each hemisphere controls the *opposite* side of the body, the hand opposite to the side of visual field that received the flashed image does the correct grasping. However, as speech is normally processed predominantly in the *left* hemisphere, the split-brain patient cannot *say* what she has picked up when the image is presented to only her *right* hemisphere (via her left eye). And the left hemisphere can only guess, as it has not *seen* the image.[26]

Is there a further divergence in the way the two separate hemispheres view the world – a divergence in their *beliefs* about it? In one fascinating example, given by the behavioural-neurologist V.S. Ramachandran, the two sides give the same answers to straightforward factual questions about their current location, etc., but disagree on whether they believe in God.[27] Would it make sense to talk of a single *agnostic* brain, in this case? The two halves are *not* in direct communication with each other. The logical conclusion to draw is, surely, that we are now dealing with two separate persons with strongly differing views: In the example given, one of these persons is a theist and the other an atheist.

These studies, and many other related ones, seem to show that, once the connection is cut, we are no longer talking about two separate halves of *one* self but about *two* distinct selves in one body. How else could one explain the results of the research undertaken with these patients? It is tempting for philosophers and neuroscientists to seize upon the results of these studies as evidence that the concept of self is highly malleable and that 'the mind' is not some central, indivisible controller of the actions of an individual. While that proposition is a legitimate one, split-brain patient cases do not provide the clear-cut [*sic*] proof they seek. As Donald MacKay, among others, has pointed out, the two halves of these brains are still connected, albeit by only the lower brain centres instead of by these *and* by the communicational 'fat pipe' of the corpus callosum.[28] This configuration could allow for some form of communication (albeit perhaps slower and more 'primitive') between the two hemispheres. Until there is evidence from cases where living brains have been *completely* split in two and both halves continue to function (such surgical procedures do not yet exist), displaying distinct selves, the critics will not be silenced.

There are human beings alive and well with only *one* hemisphere intact. In some cases this has been due to congenital defects, and in others, due to traumatic accident; the majority of such cases result from a rare and highly-invasive surgical procedure designed to treat severe seizure disorders. Where *hemispherectomy* (surgical removal of one hemisphere) is performed during childhood, 'recovery' of function can be almost total.[29] If you were to meet one of these people later in their lives, and you had not been told of their condition, you would never guess it. After you found out, would you then insist on calling them 'half persons', or would you simply accept them as complete human beings?

## THE SOFT DIVISION

'There will be no pain,' said the neurosurgeon in an even, comforting tone.

'What about the electrodes?' asked the patient, anxiously.

'We'll give you a local anaesthetic to numb the area of the incisions, but your brain has no pain receptors of its own, so you won't feel any pain even though you'll be awake throughout.'

'But will I still be *me* after the procedure?' asked the patient.

'From everything we have learned, we have every confidence in saying that you *will* still be yourself afterwards. We've done this many times, always with complete success. Patients find it easier when we can transfer them straight to the new body, but in your case, we can't afford to wait any longer.'

Later, the surgeon spoke to the patient as he worked. He checked the readings on the soulbox, and all looked normal. He asked the patient to recount stories from her childhood and teenage years. She began to speak brightly and clearly about summer days swimming in the sea, pony trekking on the moors, and singing in the school musical.

After around thirty minutes, the location of her voice began to change, emerging ever more strongly from the soulbox and ever more softly from her lips. Fifteen minutes later, her lips no longer moved.

'How do you feel?' asked the surgeon.

'Fine,' said the patient. 'How much longer will it take?'

'The procedure's complete, and you've been fully transferred to the soulbox,' he said.

'But I'm still in my body,' she said, sounding a little alarmed.

'Your current body is virtual. We've adjusted your points of reference in order to avoid any kind of ontological shock. We'll softmerge you into your new body when it's ready, in about a week's time.'

'And what about my old one?' she said.

'The electrodes gradually phase out your old brain, at the same time as transferring your pattern to the soulbox. How do you feel about your old brain having been phased out?'

'I feel a little sad about it,' she admitted, 'but I'm looking forward to getting my new body. It could be fun being eighteen again.'

No such procedure exists, but it would be incorrect to say that it must *forever* remain impossible, so we should properly say that it is only, to use Parfit's term, *technically impossible*. The point is not (yet) to try to suggest that such a procedure might work, but to think about how one would feel in the patient's situation. We may find that we need to at least reconsider ideas of what constitutes a person. But the rabbit hole goes deeper: I contend that most of us have *no idea* what 'self' actually means; though there may be no objective definition, we fail even to notice our complete lack of one. We spend our lives in unquestioned Cartesian 'I think, therefore I am' certainty, never stopping to wonder about the strange, circuitous phenomenon that makes that possible.

And, despite what you may think, the 'soulbox' itself is not the *most* technically-impossible element of the scenario in the story. Instinctively, we might feel more comfortable with the idea of the person being 'transferred' directly to the new body. But this instinct would be irrational: Remember that no *physical* part of the woman is being transferred, so what we are actually talking about is transference of a *pattern*. If we think in these terms, we can see that the patient's unique pattern is being gradually copied to the soulbox while the original pattern and pattern location is gradually destroyed. If the patient cannot detect a qualitative difference between transfer of her unique pattern to the soulbox and direct transfer to a new brain and body, then there isn't one. In either scenario, the pattern is being copied and the original destroyed. Given this, we could guess that it might be *less technically impossible* to transfer the pattern to an electronic 'soulbox' than it would be to transfer it to a *new* 'blank' brain, with all the vastly complex connection-building that would be needed in order to shape its structure to 'hold' the incoming pattern.

It would be fair to question the utility of such introverted neuron-gazing. But finding *real* answers to such questions can do much more than bring us a new perspective on life. It can help us to learn how not to die.

# 7  WHERE'S THE FIRE?

∞

'Accepted.' Not a status I often got to apply to myself. Perth – 'The Fair City' – beckoned. My cassette-taped piano piece – the one I had learned by ear, due to my inability to read music properly – had done the trick; I was going to 'The Rock School'. This meant a lot to me: a formal (or at least semi-formal) path towards achieving my goal of serious-but-popular-and-quite-famous musicianhood. It was all good: a summer of work at a Skye hotel to accrue necessary resources (beer money), then onward to personal fulfilment, critical acclaim, and financial abundance. I even had a few gigs lined up with the band, though I hadn't really considered at the time that this might be our swansong.

She arrived along with a mutual friend from the hotel where I worked. I noticed her a few times during the evening, though not in the same way I had noticed other, more flamboyant, girls. She seemed to be enjoying our music.

I had never met a female that liked Killing Joke. She blasted *Brighter Than a Thousand Suns* through the tinny speakers of her car stereo as she battered along the narrow country road, blithely flicking ash from her cigarette onto the growing pile in the foot-well; she climbed trees in her pink dungarees; she introduced me to The Macallan malt. What was not to love about her?

A tide of memory: We leave the party and make our way down to the shoreline; she is cold, so I give her my calculatedly-stretched-and-ragged black jumper to wear. On the salt-flats, among the sea-pinks, we tune into the cadences of each other's words and feelings. The pulse of lapping water; slow waves invisible in the dark; two entangled minds; two lives; two beating hearts. The oxytocin is strong with us.

Ten days later, small holdall of clothes in hand, I moved into her flat. I didn't even know her surname.

She had trusted me, had taken me at my word, had let me read her diary. I had insisted that, despite my youth, I was so serious about her that I wanted us to commit to a permanent relationship. No half-measures – I would stay with her forever. And I meant it. My introspective systems, though capacious, had become incapable of computing a future without her – a future where it didn't work out and where we became strangers.

We had the summer, but there was no escaping the fact that I would be going away to college in the autumn. While we felt sure that our relation-ship would be strong enough to withstand the demands placed on it by this situation, we wondered how we would manage it in practice. We didn't. She came down to visit me nearly every weekend, driving the four-hour journey back to Skye late on Sunday nights; arriving home, bleary-eyed, in time for a few hours sleep before work on Monday mornings. In between times, I lived a melancholic, music-laced existence fuelled by black coffee, beer, and the cigarettes and chocolate she always brought to me.

The 'rock school' was a sad joke. The ceiling of one of the rooms in the poky building that housed it had collapsed during the summer recess, and the whole place had been condemned. One of our tutors sheepishly mentioned this on the day we arrived for our induction at Perth College proper. They had secured some rooms in the main college building and at another site several miles distant, but the logistics of this arrangement were a shambles, and it was clear from the outset that our 'education' would suffer as a result.

I persevered, trying to quell my rampant homesickness and a growing sense that nothing worthwhile would come of my time at rock school. My disappointment was compounded by the fact of the dearth of other song-writers there. The interactive, creative buzz I had expected failed to mater-ialise. Songwriting was one of the skills that each of the artificially-assembled bands was expected to demonstrate, and I was, at first, flattered when members of other bands came to me looking for material. This feeling quickly faded when I experienced the hackneyed cod-rock mangling of my hard-won riffs and lyrics.

The parting from my father had gone badly, so it was with some reluctance that I called him. My resources were dwindling, and I saw no alternative. He had (strongly) objected to my moving in with my partner, and had taken the attitude that I was no longer his problem. In hindsight, I can see the disrespect inherent in my nonchalant trampling of his traditional values. I should have talked to him and tried to explain. Instead, I had foolishly taken his acceptance of my partner as a person, and his apparent resignation to the various living arrangements of my older siblings, as signals that he had mellowed in his views. To my surprise, and modest embarrassment, he apologised for his earlier harsh words and for his withdrawal of his initial offer to assist in funding me through college.

Neither the easing of my financial situation nor the progress I had begun to make in reading music made any difference – I was going home to my partner: my Terra. Given that I had made few friends during my short time in Perth, the goodbyes were easy. It's unlikely that they much missed this lovesick oddball with his black-and-white-striped jeans and thinly-cryptic lyrics; less competition but, on the downside, one less person from whom to scrounge Marlboros and Dairy Milk.

The joy (and relief) of homecoming was dizzying.

We moved out of her small attic-flat and in with some friends in a cottage a few doors along the road. Terra worked while I wrote songs and recorded them on the 4-track 'portastudio' she had bought for me. It was a strange, passionate, and happy time. Friends came to visit – Thor, Terra's college room-mate, a complicated and boisterous school-friend who took an instant shine to Terra and demonstrated this by licking her face.

Zeta, only eight years old at the time, also came to visit. Terra had quickly won his trust and, by extension, the trust of my mother. We would drive him to the cottage, a box of toys in the boot of the car (or, once, left on the roof of it), then entertain him in the long lean-to conservatory or in the small garden outside. The rapid integration of Terra into the lives of my family-members brought me great contentment.

*My* integration into *her* family did not go smoothly. Her mother's Victorian traditionalism extended to all aspects of life; strangely-dressed, penniless, eighteen-year-old musicians did not rank high on her list of candidate-types for potential sons-in-law. I had little to say for myself at their formal table. Her mother's chitchat about local happenings was directed over my head; my lack of knowledge about cars, sport, and gardening made opening gambits with her father stilted and faltering. Her younger sister was also formal, but with a quick and dry wit that allowed some room for interactive banter. I was daunted by the obvious intellect – and brute normality – of this family.

We moved house again: this time to an attic-flat in a remote community several miles from Terra's workplace. I spent my days in complete isolation, attempting, in my songwriting, to fuse my new-found happiness with the dark undertow of my recurring themes.

We learned each other there, but something was changing. Some evenings she was uncharacteristically tired and tearful. The thought of having to pack up all our things and move house yet again, seemed to weigh inordinately heavily on her mind.

The next place we stayed was, at least, a longer-term arrangement: a six-month winter let of a traditional 'whitehouse' cottage near to my family's home. Zeta came to visit nearly every weekend, determinedly hauling his bicycle cross-country, or up the steep hill between his home and ours.

Terra's health deteriorated, and she was diagnosed, eventually, with M.E. She negotiated time off work, and I began to nurse her. Drawing on the strands of the culinary knowledge I had picked up from my mother, I learned to cook. Bowls of hot soup, short winter days, and, for Terra, long, long hours of sleep, marked the passage of time. There was still room for music in this twilight existence; and with visits from Epsilon, Guitar, and Smith, room for noisy and drunken Thunderbird-wine revelry.

My friends and I still drew greatly upon our high-school interests for our points of reference. The music and other creative influences we had cared about were the central focus of our chatter. But I was not blind to the fact that they had begun to move on. While I was occupied with the concerns of a deep relationship, and the responsibilities attached to that, they were rapidly developing their social and creative skills in their new lives at college.

The old school ghosts, still heavy with the damp chill of our cramped practice room and heady with teenage literary obsessions, haunted my music. Iain Banks wrote back to me after I sent him a cassette recording of a song I had written under the full influence of *The Wasp Factory* – a recording featuring the echoing laughter of a young Zeta. The arrival of the letter is still clear in my mind: the crisp winter morning, the laid vellum envelope, the ink-stamped wasp on the seal. Quivering hands. A benediction.

By springtime, our previous luck in finding places to live seemed to have run out. My parents came to the rescue, offering to rent us the self-contained apartment at the end of their house. The flat had previously housed an elderly relative – a cousin of my father – with a bi-polar disorder. Lithium had kept her condition in check (although I remember from my childhood some of her outbursts), but her other care needs were rapidly becoming too great for my parents to meet. She was transferred to a care-home a few miles distant, where my parents visited her regularly.

The place was ideal for us, and we made it our own, forming a music/office space downstairs and a curtained-off sleeping area upstairs. I lit the temperamental 'range' stove each morning, and spent my days trying, with Terra now tentatively recovered and back at work, to build up a small business hiring out audio equipment. At weekends, we welcomed in friends and siblings to relax, play music, and party in the comfortable open-plan environment of our living space. Epsilon lounged there frequently: smoking cigarettes with Terra, recording effortless basslines into my portastudio, and gently taking the mickey out of my numerous quirks and foibles. We saw Zeta nearly every day, and we were happy to have him around.

Alpha became a father – to beautiful twin daughters. He was living with his then partner in Edinburgh, but it would not be long before they returned, two almost-identical bundles of life the heavier, to Skye.

My business made no money. Thor drove the gigantic, fuel-guzzling van to remote venues around the island, where we set up cumbersome P.A. equipment and I tried, often in vain, to balance the sound. We would arrive back at my flat in the early hours of the morning, tired, dirty, and with precious little to show for our efforts. I could not drive at the time, and in hindsight, I am glad that, as a result, the calm-tempered Thor ended up with the job of wrestling with the succession of dangerous vehicles I had managed to acquire.

Reality began to bite: Guitar was busy at art-college and could only visit during holidays; Thor had started a mountain-guiding business and we saw less of him; Epsilon was headhunted by Alpha's band. Some of the people I had depended on, even *used*, were no longer available whenever I needed them. I was more practised at being unhappy for myself than at being happy about the good-fortune and personal-development of others. I became disenchanted and (I now think) depressed.

It showed in my songs. 'House of Hands', which I wrote at the time, carries a somewhat uplifting message of generational continuity in the lyrics of the vocal, but is damped by a startlingly sparse drum-machine and organ accompaniment. Only and all me: no Epsilon, no Guitar.

My propensity for damaging introspection made me angry. Anger was a motivator that I understood. Guilt, too, featured in my decision to find other ways to make money. Terra paid for everything, and I felt that this emphasised the gap – in age and in life-experience – between us.

A feeling of contented exhaustion, a handful of tenners, and a steaming plateful of mini pork-pies became my regular reward for hefting the heavy and hard-won sacks of whelks back along the shore, ready for collection.

With my aversion to 'real work' now broken, I got a job in a local shop selling over-priced designer jumpers to American tourists with far more money than discernment. My colleagues, all female, showed consummate skill in persuading the over-weight, middle-aged clientèle that the garish knitted creations were 'perfect' on them. I didn't mind the work itself, and I ended up as production manager, co-ordinating the various poorly-paid home-workers who made some of the garments. But our boss – wealthy owner of the shop, and follower of a whacked-out narcissism cult – displayed a manipulative streak to which I took strong exception: Regularly reducing certain other members of staff to tears was not, in my view, acceptable.

Beta married again in 1991, this time to a kind and confident Englishman with a ready sense of humour. We already knew him well by the time of the wedding, which took place in grand style at a castle tucked away in a forested part of rural Stirlingshire. We were happy to see Beta so radiantly in love. I drank too much, and made myself green smoking an enormous cigar.

Terra was excited and called me at work. She had seen a 'For Sale' sign in the window of a bungalow just along the road from my parents' house. Her father offered to pay the deposit, and we decided that, despite our meagre finances, we could afford the mortgage on this first home of our own.

1993 was a year of welcome change for us. Even the fact that I was 'made redundant' only a year or so later failed to affect our mood of contentment and optimism. Terra revelled in having a garden to tend, while I learned to deal with the various 'man jobs' that go along with being a homeowner. It also gave me a way to connect with the senior male influences in my life. Terra's father came to help with the DIY. Seeing my reasonable proficiency at dealing with repair tasks around my new home, he was then happy to enlist my help with jobs around his. My own father, too, became a regular visitor – offering advice, tools, and an ever-critical eye.

Zeta continued to spend time with us when we moved to our new home. We felt protective of him. When *he* became ill with M.E., our opinions about how best to deal with his condition sometimes clashed with those of my mother. But, for the most part, we got along well and took care, for each other's sakes, to avoid being judgemental.

Alpha's twin daughters often stayed with their grandmother, and consequently they spent much time with Terra and me. Clever and funny, if often rather grumpy, they brought further warmth to our home with their cute totterings and entertaining tit-for-tat prattle. Unlike me, Terra was never too preoccupied to play with them. Her ability to let go of all adult affect and simply join the children as an equal in their balloon-bright world was heart-melting to watch.

My commitment to music had slipped, and I wished to renew it. I bought new equipment that would allow me to sequence my own songs, and to load in cover versions to play at gigs.

My eclectic mix of styles did not go down well in the exacting environments of Skye pubs on Friday nights. I was not asked back for repeat performances.

I found a form of semi-integration into the local community after taking a job at the village shop. The premises came up for lease, and I prepared to take on the shop as a going concern. But it all fell apart at the last minute. The local feudal overlord, owner of much land and property in the area, decided to insert unworkable clauses into the lease. I blew up at him in unbusinesslike fashion, and he responded with the extraordinary intransigence and blankness characteristic of people far removed from the realities faced by people like me: people trying to eke out a living, and some sense of self-respect, from even the slightest and most humdrum of opportunities.

Confidence and optimism began, yet again, to elude me. Where was my *direction?* What was I *supposed* to be doing? I was trying to dig out some nugget of transcendent beauty in my music but was finding only repetition and cliché. And yet more darkness. I needed to make some money but was afraid of sacrificing...what? My creativity? Evidence for the existence of any such thing seemed sparse. I questioned myself constantly: *I am happy, my home life is happy, so what is wrong with me? Why am I so self-obsessed? What am I trying to find?*

Leaving for Dundee, though only for a few days to stay with Guitar, was much harder than I had anticipated. Terra guessed that I was struggling with something, while I knew that I was struggling with everything. And none of it was her fault. She was the lifebelt and I the drowning man, but I kept swimming in the opposite direction.

This was a mistake. Guitar and I got blind drunk and traded paranormal and metaphysical insanities. I flipped out; thought I had been *possessed* by something. So little self-knowledge; so little knowledge of *anything*. This was a breakdown. This was a cruel mind-trick that I had played upon myself.

It was wonderful to return home, to normality. I tried, haltingly, to explain my dreadful experience, but found that the previously-solid shape of it in my mind was dissipating fast.

I set up another small business, this time selling army-surplus clothing, which I ordered in great rummage-boxes from the north of England. Thor's younger brother was in need of cash, so I took him along on each of the ventures, in my bright-yellow ex-British-Telecom van, to musty little village-halls around the North-West Highlands. In those venues, we sold our olive-drab and camouflaged wares to dour crofters and militaria-obsessed kids. My father liked the boots.

The business was not lucrative. I took a job at the local hotel – making beds, cleaning bathrooms, and, in the evenings, waiting tables. My awkwardness around strangers, and my tendency to panic when things went wrong, decreed that I would fail to shine in this part-time profession.

For a while, I even sold Amway products, seduced into the cultish business by the promise of great wealth, or at very least, financial freedom.

What was I reading at the time? Remembering that might give me a better handle on what must have been my state of mind. I can't remember. Books did not yet feature strongly in my post-school life. But I can guess: ufology, fantasy novels, Iain M. Banks, some other sci-fi, music-related, self-help.

Alpha married his new American partner, and so, once again, we came together as a family, for the tartan-bedecked, lakeside celebration of their happiness. In the evening, we danced to ceilidh-music, then to Alpha's band's unique brand of rock, before staggering back to the chalet where we were staying along with Gamma and his partner.

I loved to be together with my whole family: the noise, the revelry, the banter, the smiles. The gathering-in.

The local Gaelic college, where we had lived (and where my father had worked) when we first moved to Skye, had a vacancy for a night-warden. The thought of having to be away from Terra overnights troubled me, but the money on offer was reasonable, and it would allow me free time, during the day, to work on my songs.

There were computers everywhere in the college. There was also a blazing-fast broadband connection by virtue of an expensive microwave link to the mainland. My room had Ethernet sockets in the walls. I learned about the internet from those lecturers and students who hung around late into the night web-surfing or playing, respectively, Unix-workstation *Tetris* and multiplayer *Doom*. A small, bespectacled boy, son of the college Director, taught me how to code web pages. Information, intoxicating information, was now within easy reach. I downloaded music software, engaged in online political debate, learned to touch-type, experimented with computer-based graphic design.

It wasn't enough just to *write* songs. I wished to package them and get them 'out there'. Most of the resources I needed were at hand in the college, except for the music-recording facilities and expertise; I needed the garrulous Glaswegian and his 'shack'. He was kind enough to put up with me again, for several weekends of lyrical deconstruction, double-tracking, and *double-entendre*.

The resulting self-produced CD-album was not a success, but I was proud of it. Nestling amidst the cheap-synth melodies and too-thin drum sounds, was something of *me*. The self-designed inlay included the lyrics; I relished the exhilarating, if frightening, sense of laying my feelings out, in glossy and immediate type, for all to see. What those feelings were is hard to say; the lyrics seem an expression of that unschooled, dark euphoria. A search for meaning in a graveyard of purpose? Perhaps I just wrote what my ears wanted to hear.

Whilst on holiday in Paris, atop a cloud-enshrouded Eiffel Tower, I asked Terra to marry me. She, in quick succession, thought I was joking, cried, and then accepted. The setting for the proposal, while something of a cliché, was not *my* usual kind of cliché; she genuinely had not expected it. We drank champagne, courtesy of Beta and her husband, in the restaurant within the Tower.

Our wedding celebration took place within the Gaelic college campus. We arrived late, to warm applause, after my father-in-law had taken us along to see an old friend of his, intent on showing us off in our wedding finery. Epsilon's best-man speech was mischievously confident; my father-in-law and I each struggled with ours, choked by nerves and emotion.

I flick back through the photograph album: she and I so young, true-love smiles in every frame; parents and siblings, delighted for us; my father's family, most of them now gone, relishing a proper Highland wedding; real friends (the ones who never forget this); acquaintances (the ones who do). Alpha and his wife have left their baby son in the care of family in America; Gamma's son, also still a breast-feeding infant, is present at the celebration.

After the romance of the first dance, followed by the semi-formal hilarity of the ceilidh-dancing, I joined my brothers on stage to scare the 'oldies' off to bed. We celebrated into the early hours of the morning, Terra and I having no intention of leaving our own party while there was still fun to be had.

Marriage suited me, suited *us*. Along with the confidence born of settled home life came a renewed determination to find direction in my life more generally. I dropped the cover-versions and organised gigs in venues that I thought might be more receptive to my kind of music. With the 'dance music' craze then in full-flow, I found that the repetitive electronic tunes went down better than my thoughtful electro-rock. Gamma joined me at these gigs, his charismatic cool mapping perfectly onto his new role as a late-night DJ.

The night-work at the college continued. Terra sometimes came down to stay with me in my small but comfortable room there. The staff knew us well, many of them having been involved in our wedding arrangements and celebration. My mother worked there, part-time, in the kitchens.

Collision. White van hits parents' white Audi. Father is dazed. Mother hauls heavy-bodied aunt from car, just as it catches fire. Father comes-to and helps uncle.

'We're OK,' they reassure us, later in hospital. They certainly don't *look* too bad, considering. Aunt and uncle look worse, but they are older, and it is to be expected, the doctor says, that their recoveries will take longer. But Mum is not 'OK'. Her bowel has been ruptured. The doctor and nurses fail to notice this, for days, while her body poisons itself. One of the nurses takes me to one side and tells me that she thinks my mother is depressed, and asks if we can try to 'cheer her up a bit'.

We visit her in the evenings, after Terra has finished work. The doctors have discovered, at last, the peritonitis-causing rupture, and she has begun to heal.

Screaming. Violent knocking on the door of my room. There has been a car-crash, they tell us, just a few hundred metres down the road – a young guy who works in the kitchen. We know him.

The four-wheel-drive has ploughed into a tree. Two bodies lie in the road. The driver stands weeping in the weakly-pulsing lights. A brash American fool-of-a-man from the college is panicking the few bystanders, shouting about 'fixed and dilated'.

I go to the familiar body. *So this is what death feels like to the touch – a smooth, cold neck, in fading-orange glow.*

By the time we see her, the next day at the hospital, my mother already knows. She heard the commotion during the night, and asked who they were bringing in. Two boys – one of them the loveable rogue she had worked with, a person I had known most of my life – now, irretrievably, dead.

A pattern. Things are breaking. Destruction. Man questions futile imper-manence of life. No, not really. I had long been godless; what questions (let alone answers) would I find buried in *this* wreckage? Meaninglessness and impermanence I knew to be the norm. I found the musings of certain others, that there might be some 'meaning' in such tragic events, not just odd but profoundly *creepy*. There appeared to be something terribly *wrong* with these people.

And reasons? Everything, as they say, does *indeed* happen for a reason: The events that preceded the event – *all of them without exception* – are the causes of the event. An indeterminate number of mostly-indeterminate causes have determined this outcome. The result of a coin flip could be predicted if one knew the precise settings of *all* the variables. But the infin-itesimal details of these variables and settings *cannot be known*; better, if still incorrect, to call this randomness than to brutalise the laws of physics by ascribing 'purpose'.

A song, 'Dead Lights', was stillborn. Its haunting melody and lyrics distilled something of my ragged emotions, but it lacked rage. It needed work, and tearing-up with discordant guitar. I couldn't bring myself to finish the empty thing.

A strained conversation with a locum doctor brought no respite from a physical sensation I was experiencing of numbness and delay in my percep-tion of the world around me.

Encounters with dead bodies, especially with those of people we know, inevitably change us. I felt hollow and liminal; in danger of slipping into acrid solipsism, but ready for change.

My father had been a supporter of the Scottish National Party since his university days. As a teenager, I had imagined myself a communist; the fact of the pathetic dearth of ideological/political labels had not occurred to me (nor to many others) in the 1980s. With the referendum on Scot-tish devolution imminent, I became concerned about what would happen to my small nation should the majority of Scots choose to vote 'No'. The public-faces ranged against devolution comprised mainly Tories, banking fat-cats, and other spiteful, vested-interest naysayers. Their mealy-mouthed imperialism did not appeal to me. The 'Yes' camp, in contrast – particularly the SNP – displayed the kind of positivity, pragmatism, and openness to change that I felt were a welcome tonic in Scottish politics.

With no experience of political campaigning, I wasn't sure where to start. After contacting the 'Yes' campaign head-office, I received an assortment of leaflets, badges, and stickers, which I handed out round the college. I took to the 'streets' of my own small community and knocked doors. My 'sales pitch' came surprisingly easily: This was an issue I cared about and seemed to understand quite well.

A car pulled up beside me as I patrolled in the rain, soggy leaflets in hand, on the day of the vote. The occupants – twin brothers who I had seen in the pub but had previously had little reason to speak to – congratulated me on my efforts and invited me to their house to watch the results come in.

Post devolution-settlement Scotland was different – at least for me. Time spent at SNP meetings, on newsletter-design, and out campaigning, ate into my evenings and weekends. The accommodating but irascible twins became something of a fixture.

Suit and tie, salary, Apple Mac – my new job came with various unfamiliar add-ons. I don't doubt that the confidence I had gained from political debate helped me during the interview. Computer-literacy was another requirement – perhaps my only skill that would have held up, bluster-free, under intense scrutiny. It wasn't that I couldn't carry off the sales-pitching demanded by my role as 'design development manager', it was just that it caused me excruciating embarrassment. My initial confidence veiled this, but as pressure to deliver intensified, sweating and redness-of-face began to give me away. That pressure to deliver was mostly coming from inside of me, rather than from my boss. Shyness, anger, and rampant inferiority-complexes are not propitious in the bruising arena of sales.

The computers and network in the design office were out of date and could not properly communicate with each other or with the various printers. I relished the task of reconciling them, and spent as much of my time as possible doing that rather than performing my intended role of hawking our questionable services to uninterested potential-clients.

The green-cast hue of *The Matrix* persisted for a time after I left the cinema. It was 1999. Silly pre-millennium tension was evident everywhere in the popular culture, so my expectations for the film had not been high. What it did, in fact, was to flex the walls of my reality. The black-clad hacker cool of it was, I suppose, bound to appeal to my past 'goth' sensibilities, but my attachment to the movie had little to do with subcultural identification. It was to do with my growing realisation that I, like the character Neo, had been living in a dream-world. Mine did not (fortunately) consist of a computer-generated reality designed by malevolent AIs to enslave the human race; it was, however, one built around a tightly-wound core of inaccurate assumptions.

There are many ways of thinking about the mind, but the key is to begin thinking about it. The 'splinter' in *my* mind had always been there, but I had sometimes assumed that it came from my failure to integrate fully into the accepted schema. I had, unwittingly, internalised an essentially *conservative* view of reality – one impoverished by a dearth of levels. The portrayals of alternative realities I had found in science fiction books had always interested me, but even the virtual-reality 'cyberpunk' novels of William Gibson had failed to make me *feel* them. The idea of multiply-nested realities, each seeming just as 'real' as the levels 'above', *should* feel unsettling and exhilarating. *The Matrix* delivered on these missing sensations.

In his *Parable of the Cave,*[1] the philosopher Plato (427 BCE–347 BCE) provided a lasting allegory – one intended to aid comprehension of the fact that our version of reality can turn out to be a severely restricted one. He envisioned a group of prisoners, shackled, for their entire lives, facing the wall of a cave. Their shackles prevent them from turning away from it. A fire burns behind them, and people sometimes walk in front of it, casting shadows onto the otherwise-blank wall. These shadow-forms, their interactions, and their context are – to the prisoners – a complete, integrated *reality*. Plato's intent with this allegory was to suggest that without the knowledge of the real world outside of the 'cave' that could be provided only (at that time) via the study of philosophy, an individual would always be like an ultra-blinkered prisoner in the cave. *The Matrix* (and various other films and books including, to some extent, *Flatland*) is a play on this old theme – the theme that self-knowledge is the key that unlocks the truth of reality.

Plato's mistake, however, lay in his belief (faith) that there exists a real, *ideal* reality at the top of a some hierarchy of realities. A coherent response to the idea of layers of reality is to accept instead what philosopher Nick Bostrom calls 'indexical uncertainty'[2] – we have no idea whether we live in the real, so we don't know which existential level we occupy in any hierarchical index of realities. The indexical uncertainty we face is total.

'Know thyself' has, in some quarters, mutated to become a bumper sticker for having confidence in one's own abilities and actions. This, to me, is quite wrong. How can you possibly begin to know yourself without first knowing something of your own architecture? I was certainly ready to learn, but the idea of learning about the workings of the human brain had not yet occurred to me. The brain was interesting but somehow *mundane*. A heart is a pump for blood; a brain is an organ for thinking with. Right? Even a child can decide to sweep away the metaphysics, but what are we left with by way of a conceptual framework? Only lumps of 'stuff' doing discrete 'jobs'. This is not, despite a certain *tabula rasa* appeal, a very inspiring view. It is only when we plug that 'organ for thinking with' back into itself, then dive into the yawning spaces between the words, that we begin to get somewhere. We begin to emerge, blinking, from the mouth of the cave. This will indeed feel, as Morpheus says, 'a little weird'.

Sheep in the office, burning electrics, their brotherly brawls on the computer desks; being in business with the irascible twins was eventful. They were disorganised, so I manifested in a confident 'new broom' incarnation and began sorting through the formal detritus of account and bad debt. They were the creatives in this environment. I decided that my role was to take care of the management aspects of the business while they exercised their creativity in designing the buildings. The twins were ambitious, their cosseted principal-architect merely abrasive. The business prospered, but ever-present tensions built – block by grey block, day by eventful day.

My friends noticed the changes but commented on them little. They could see what I was leaving behind in pursuing financial stability over creative fulfilment. I thought myself more flexible than was actually the case: *I would get back to writing songs, I would pursue other ventures outside of the business.* In truth, the business, and the world of new situations and acquaintances it opened up for me, soon came to dominate my time.

Considering how recent a period of my life I am here writing about, it surprises me how difficult it is to remember it clearly. It seems boxed and bounded somehow – ten years of *other*. The trap of convincing myself that I was an *almost-entirely* different person then, is gaping wide. Why a trap? Because it makes no sense to cut myself off from any part of my past; to parcel up huge dripping chunks of it and put them to one side, perhaps to fester. Anger that some memories of my personal life at that time have been smeared out by the dull minutiae of business should be allowed vent, perhaps chasing those memories back to the surface. How easily we can come to set large sections of our lives within hard parenthesis.

Control. My indecision sometimes disgusted me, so making rigid decisions – *control* gestures – came to feel like the best way to break that tendency. So it had been with my decision to make money, later forsaking many of my other interests. I sought greater control of my body, too. When I first heard, during a noisy pub conversation, of low-carbohydrate diets, I was incredulous. Within a matter of days, I had decided to try one. My usually-restive intestines seemed to appreciate the gesture, I lost weight, my brain happily burned ketones rather than glucose. There seemed now a progression to things – each day a little stronger, thinner, better. I ran, but on my own terms: back from the pub in the pitch dark, sensing my way from memory and from the sound of my footfalls on the narrow road. Breathless, exhilarated, and sometimes quite drunk, I would sneak into the carpeted silence of our small bungalow and snuggle into Terra.

Those around me at work were university educated. I talked the talk of indifference about my lack of further education, but, in reality, I felt inferior. I took a first-year Open University module on the social sciences. Terra bought me an expensive and powerful bottle of cask-strength Caol Ila – one dram to be taken after each successfully-completed assignment. The essays were, to me, simply exercises in manipulating words to sound like believable concepts. This I could do. I drew upon my personal 'agency', applied it to the 'structure' at hand (it was all structure and agency), and passed the course with ease. Then I ran out of steam. There would be no degree for me.

Much had changed in the lives of siblings and friends.

Alpha and his wife had returned from America, with their young son, after a time living in Atlanta. They built a house on a croft that had belonged to my father, and to his father before him, in a remote township on the mainland north of Skye. There, now with two new daughters as well as their son, they settled into a new environment and lifestyle. Alpha's twin daughters from his previous relationship stayed with their mother.

Beta and her husband were settled in France, after several years spent living in oil-rich countries, reaping the fruits and tribulations of the ex-pat lifestyle. Although we are very different, I found that Beta and I could now relax in each other's company. The princess, while still in her, had become practical and open.

Gamma's relationship with the mother of his son had broken down: She and the boy eventually moved away to Newcastle. For a time, Gamma looked to Terra and me for support. We were happy to provide it, and although I was glad when he found his way again, I was sorry to see him go.

After a time living in a draughty old Victorian mansion in Argyll, Epsilon had returned to Skye with his partner. They had moved north, where Epsilon used the talents he had learned at college in Glasgow to set up a small business as a luthier. The relationship with his partner was fading. They had always seemed ill-matched. To my relief, they split up, and Epsilon began, gradually, to sound like himself again.

Zeta, now in his early twenties, had recovered from the M.E. that had blighted his teenage years. He had received almost the entirety of his secondary schooling at home. A bright and non-judgemental young man emerged, smiling, from the fog of long-term illness.

Guitar had married at around the time I had started with the business. It was a snowy winter wedding in Portree – Guitar in leather kilt, his wife in cowled cape like an imposing gothic-fairytale queen. She became ill a couple of years later – also diagnosed with M.E. They bought a house out on the coast north-east of Dundee, where they had both studied at art college.

Smith had broken up with his long-term partner, and was now living in London pursuing a career as a music journalist. He had, some years before, suffered from depression. I remember now how much this had frightened me. His prodigious talent, stubborn self-consistency, and sheer drive had always seemed to carry him through. I had wondered how *I* would manage to keep it together when *he* had plummeted, for a time at least, to the depths of despair. Of the three of us, he had always seemed the least likely to struggle in this way. But, I suppose, everything had been wrong for him back then: wrong place, wrong job, wrong relationship.

Without the correspondence from Guitar and Smith, I would struggle to piece together my eroded memories of those years. I have a file, now open here beside me, containing those letters from them – handwritten and typed, illustrated and plain, vellum and stark white; letters that brim with camaraderie, and with shared interest in art, music, and revelry. Correspondence from my family is, by contrast, thin.

Supernovas, neutron stars, holes in space with physics-perturbing singularities at their black hearts: These things called to me from television programmes and books, carrying with them fragile memories resonant with the butterfly-inducing lilt of Carl Sagan. They pushed into my dreams, sometimes making grand, terrifying space-operas of them. My crushed and spaghettified subconscious recovered each time with surprising rapidity, and in any case, those dreams were always preferable to the prosaic financial/security worry variety.

I wasn't deaf to what my instincts were telling me, but I didn't know what to do with that information, and so, most of the time, I simply ignored it and got on with my day job. Isn't that what you are supposed to do? Isn't that what everyone does?

A precarious compromise, then. I kept taking the blue pills, and Wonderland kept a respectable distance.

Now, open *wide*.

# 8 MAKING IT TO THE BRIDGE

∞

*Omnes vulnerant, ultima necat.*
(Each hour wounds; the last one kills.)[1]

—Traditional sundial motto

'It looks like mud,' Terra says, throwing back the clear-gelatine-encased *resveratrol* capsule along with a gulp of lukewarm tea.

'Red-wine and Japanese-knotweed extract,' I respond. 'Bridging medicine.'

I, like most of us, have sometimes fallen prey to 'health supplement' hype. In this case, I read a bit about it, found out that there was a paper or two suggesting that it might have a role in the activation of a protein called *SIRT1* – a protein shown to be activated during severely calorie-restricted diets – and decided that it was interesting enough to be worth experimenting with.

And if it really *did* have those beneficial properties (and it now looks very much like it does not), what difference would it make anyway? A few days? A couple of months? It's hardly radical life-extension.

'Healthy living' and I have an uneasy relationship. I do things that are (supposedly) bad for me: I drink, I don't exercise much, I eat lots of fat and red meat. Surely my priority should be to sort out my 'unhealthy' habits, rather than wasting money on unproven supplements. That is true, but I, like most of us, am rather lazy about my health: I want the benefits but without all the effort, so I try to take shortcuts.

Many of us assume that the health advice we receive, often via the popular media, is correct. We sometimes find it confusing, but we assume that information that seems to fit with a pattern we recognise is *probably* correct. This advice is coming from *experts* – they would know, wouldn't they? Alternatively, many of us reject the entirety of the standard text, choosing instead 'medicines' and 'therapies' that we believe to be more 'in harmony' with our bodies: acupuncture, aromatherapy, even homeopathy.

What we *can* choose to do, rather than glugging snake oil or sponging up every drop of popular commentary on the subject, is to educate ourselves properly about what may or may not be healthy for us. We can then make our own informed decisions, but, crucially, *if* we choose this route we must also be prepared to take the consequences when we get it wrong. Is that a level of risk we can live with? Should we always stick to the advice provided by our doctors? Should we just use our 'common sense', and if so, where does that come from?

The conventional version of the 'fat hypothesis', for example, sounds correct, because it seems obvious (even common sense) that *eating* lots of fat will always *make* us fat. However, this hypothesis is flawed. Eating fat can make us fat under certain circumstances, but it can make us lose weight under others.[2] This is perturbing because it is not simple. Complex relationships between things do not play well in the snappy-soundbite context of popular television and internet sources. The vitally important difference between *correlation* and *causation* is smeared out by sloppy language.

The wide field of 'nutrition' steps up to some of the requirements of genuine scientific endeavour, but it falls down spectacularly on others. This is to be expected: It is a new field of study that has accumulated relatively little of the hard data required to build solid theories. Its protagonists are still a mishmash: doctors, scientists, statisticians, well-meaning and well-qualified dieticians (but often with outdated knowledge), enthusiastic amateurs, and downright charlatans. Broad scientific consensus in this field has yet to be established, so there is still a great deal of elbow room for plausible-sounding opportunists.

I will not belabour the point, but it is worth noting that any detrimental effects of self-education about (and experimentation with) diet and lifestyle are usually limited to the individual. If a critic were to insist that I was a 'fat-hypothesis denier', I would be forced, despite his/her ugly use of language, to agree. Deniers of scientific *consensus*, on the other hand, belong in a different category to those of us who are merely questioning the conventional wisdom of ideas that have yet to withstand the full rigour of the scientific method.

Because I don't have the information and tools I might require, my current day-to-day approach to extending my lifespan is comparatively half-hearted. I do not have the self-discipline for full-blown *caloric restriction*, though I can usually manage a couple of calorie-restricted days per week. My general philosophy on health is that most of the 'settings' for our longevity are 'dialled in' before we leave the womb; there is little I can do with currently available techniques and medicines that will have an impact on my longevity proportionally greater than my genetic and gestational set-up will dictate.

And so, for now at least, I just do the easy things: a few supplements, no sugar, cold showers, a little fasting.

## BRIDGING VEHICLES

Does *our* bridge already await us? Perhaps we are the first of the bi-centenarians. Or maybe we are the last of the dying-breed. The radical-life-extension populariser Aubrey de Grey, head of the SENS (Strategies for Engineered Negligible Senescence) Research Foundation, thinks that the first one-thousand-year-old may have already been born.[3] This sounds like an implausible, even ridiculous, claim – until, that is, you listen to what he is actually saying. Living to be a thousand does not require us to have bodies with an *initial* thousand years of resilience *built in*, any more than keeping a classic car running for a hundred years requires that it have an initial hundred years of running built in. Why not think of human bodies the way we think of cars? Give them a full service every few years – a course of 'preventative maintenance' – and then put them back on the road.[4]

I cannot fault de Grey's logic. Our bodies wear out over time; the fact that they are harder to repair and 'service' than cars should not deter us from finding new ways to keep them rolling over the bridges and into the future. Scientists can incorporate each new life-extending medical technology into the 'service plan' to keep us alive for, incrementally (even exponentially), longer and longer. With even the first stage of this approach – what some call 'Bridge One' – in place, we would begin to find it difficult to determine the upper limits of human lifespan. Certainly Bridge One will be the hardest to cross, but it may well prove to be the only one bearing significant obstacles.

The arguments against such an approach usually centre on why we should not do it rather than on reasons why we will not be able to do it; most are ethical, rather than technical, objections. The 'population explosion' argument is a commendably practical one, but it tends to ignore the obviousness of the changes in human birth patterns that would accompany such a *paradigm shift*.[5] Population growth and shortages of resources to sustain a booming – and, potentially, long-lingering – population are, of course, essential considerations. However, those voicing such utilitarian Malthusian concerns should be prepared to respond to the growing body of evidence for the *slowing* of population growth.[6] Changing attitudes, improving productivity, better utilisation of available resources, and other mitigating factors will also increasingly come into play; the planet may well be able to sustain a larger population, of longer-living people, with better life-quality – even without the kind of Golden Age of super-abundance envisioned by some futurist thinkers. Many life-extensionists think that super-abundance may be readily attainable; in the meantime, however, the sensitive question of priorities still hovers: Should they lie with sustaining the people who already exist or with making and sustaining new ones?

We are all, in at least the 'weak' sense, advocates of life extension. We expect doctors to treat us when we are sick, even when our chances of survival begin to look slim. The formal withholding of sustenance and medical treatment is reserved for only the most hopeless of 'terminal' cases.

We should have, then, little (or no) objection to the approach advocated by de Grey and others. They are, after all, only talking about the routine application of certain medical techniques intended to give participating 'patients' another *few years* of life. The moral-philosophical tension seems to be born of gut reaction against the idea of using such techniques as an exponential 'crane' to build the participants a platform from which to reach out for *very much more* than a few extra years. Others see the idea as merely a fanciful 'skyhook' (an impossibility, a miracle).[7]

But what about the cost? Wouldn't such techniques be the preserve [sic] of only the super-wealthy? This argument is a non-starter. For such an argument to have merit, we would also have to decide that it might be right, because of their high (and sometimes *prohibitively* high) cost, not to produce *any* new drugs. I am not an advocate of the current 'trickle-down' system of drug licensing where large companies make vast fortunes from expensive new drugs until such time as the licences run out and they become cheap *generics*. However, as Hughes states, 'the solution is not to ban corporate technologies but to strengthen democratic funding of scientific research...'.[8] Many people already die needlessly, awaiting drugs denied to them purely on the grounds of cost. If I *was* an advocate for this system, I could argue that it would work in just the same way for life-extension drugs as it does for other drug-types. But the accompanying paradigm shift in thinking and behaviour would affect *all* aspects of our current conventions of pharmaceutical development and production – pushing them rapidly in the direction of democratic funding and control – rendering many of the orthodox points of this particular debate moot.

You can reasonably assume that whenever I *do* advocate for a particular technology or technique I am also advocating for its being made universally available. An appreciation of the equality of all human beings is a central hub that connects various spokes of my philosophy. Robert Burns (1759–1796), among other progressive Enlightenment thinkers, spoke up for this equality over two hundred years ago. For example, while obviously written in the male-dominant parlance of his day, his poem 'A Man's A Man For A' That' looks to a future when the *worth* of *all* people is recognised.[9] (There are, of course, *technical* ways in which we are not truly 'equal' – our brains are different, for example. However, I think it would be wrong to lose within the technical details the moral utility of the idea of a type of equality that recognises and embraces difference. Because it is fundamental to an understanding of the value of human life, I will return to this subject in a later chapter.)

The complexity of the human body, indeed of any mammalian body, is baffling to those of us with little knowledge of biology. And the entire biological perspective itself is merely one of many methods we use to try to fathom the deep complexity of living organisms. Does it matter if many (perhaps billions) of us spend our entire lives in the dark about the very processes that cause us to be? Life doesn't care: It happens whether we understand it or not. Traits selected for since the beginning of life on Earth produced the form that you now take. The bones of the hands with which you hold this book are comparable in form to the essential bone structure of the fins/limbs of your 'fishapod' ancestors, such as *Tiktaalik*, which lived some 375 million years ago.[10] You are at the tip of a twig on the evolutionary 'tree of life'. Ninety-six percent of your genes are identical to those found in the genetic code of chimpanzees.[11] You are related to all the life forms that have ever existed on this planet.

A more immediate kind of natural selection – the specifics of the particular ovum that was available that month, and the ultimate winner of the race of 200 million or so sperm to that egg – resulted in *you*. The mix of *DNA* (Deoxyribonucleic acid) contained in the *chromosomes* of these *gametes* determined your unique biological makeup. The division (via *mitosis*) of the fertilised *zygote* resulted in *somatic* (body) cells that each, apart from those that go on to become gametes, contain copies of all the genetic information that was in that original zygote. This is the reason that cloning works; a copy of an entire life form can be created from a single cell. *Embryonic stem cells* are early-stage cells that have yet to 'differentiate' to become specific types of somatic cells – heart-muscle cells, *neurons* (brain cells), blood cells – and so they are prized in medical research for their 'plastic' capacity to be coaxed into becoming whatever cell-type is required for the purposes of the research or treatment.

Already, even in this simple description, certain people will infer challenges to their beliefs about their own uniqueness. Some persons do not wish to be the result of happenstance: They wish to believe that they were *always going to be* composed of a *particular* ovum and a *particular* sperm; they believe that this was divinely pre-determined. They might also object to any medical use of embryonic stem cells, believing that each of those cells is, in effect, such a pre-determined entity with a right to life or 'normal' death as decided by some external supernatural agent.

It is not so very long ago that, with the emerging use of the newly-invented microscope as an aid to scientific enquiry, some people came to believe they could see *homunculi* (little men) inside individual sperm. Antonie van Leeuwenhoek (1632–1723) was one of the first to observe and record what he called 'animalcules' – what we now call *microorganisms* – via his improved-lens version of this revolutionary apparatus. He believed that sperm, as fundamental precursors to people, must already have souls. This type of notion, in this case later transmogrified into the idea that they must already contain tiny *people*, came to be known as *preformationism*.

The medicalisation of the soul was less problematic back in the days when there were so many places in the body still to look, and so many levels of magnification still to be achieved. It is hardly surprising that, at that time, fervent believers in the soul should expect to stumble across it in the brain, in the heart, or in a human animalcule under a microscope. The immateriality of the souls of today might disappoint those early investigators. Were they to be somehow transported to the present, they might side with those who *still* believe in them, formulating dualist hypotheses based upon as-yet-undiscovered physics. However, I prefer to think that those enquiring minds would quickly come to reject any such notions in the light of the wealth of new scientific facts available to them.

The fantastical avenues by which there could have been, or could *still* be, scientific evidence for the existence of souls, wind on. A darkly amusing section in Jostein Gaarder's *The Ringmaster's Daughter* posits a terrible condition of soullessness – 'Lack of Soul Disease' – that emerges when the population of the Earth hits 12 billion. Babies born after this tipping-point grow up uncontrollable, licentious, and murderously violent, caring nothing for themselves or for anyone else. As soon as the population drops back below the tipping point, 'LSD' children stop being born. Humanity quickly comes to realise that only 12 billion souls are available to populate human bodies; babies are born LSD because the stock of souls has been used up. Many come to view the soulless as no more than raw materials, suggesting that they should be kept alive and healthy but separate from normal persons, so that their body parts can be harvested for transplant.[12]

Parfit points out, in his heroically restrained style, that proof of *reincarnation* might have been the soul-clincher. In his refutational example, a modern-day Japanese woman believes that she is the reincarnation of a Bronze Age Celtic warrior. She knows a great deal about this warrior, and archaeological evidence emerges that provides irrefutable proof that she is telling the truth: In some way, she must have *been* that Celtic warrior. Many similar cases, with similarly strong evidential backing, come to light. Scientists are left, therefore, with no choice but to conclude that some essential (and separate) conscious element of long-dead persons can transfer into *living* persons. Parfit rounds off his rebuttal with characteristic dryness: 'We do not in fact have the kind of evidence described above. ... Nor do we have evidence to believe that a person is any other kind of separately existing entity.'[13]

My brief excursion back onto the territory of souls is mainly intended to show the belief context within which the emerging biological facts of human life were viewed. I thought it would also be useful to show that the same kinds of misapprehensions still exist in modern-day structures of religious belief, and that they exist *in spite of* the fact that, unlike in the past, an understanding of biological processes of life – albeit an imperfect and incomplete one – is now available to all who care to pursue it. My uses of the refutational example by Parfit and the satirical one by Gaarder are intended to illustrate the folly of making grand claims for the existence of immaterial entities – claims for which there is *no* evidence and of which, if the evidence *did* exist, the truth would be blindingly obvious to everyone.

The desire to live much longer – to cross the bridges to the future – makes no sense if we believe that we will continue on, in spirit, after our deaths. While I understand the attempts of cryonics organisations to find forms of words for what they do that will not immediately exclude the religious, I think this approach will fail. The faithful believe that they already have their means of survival; why would they need cryonics?

Though we might sometimes try to cast ourselves as mavericks, those of us interested in life extension are keenly aware that we do not really travel alone. Our decisions to 'hack' our lives – to turn them into a kind of *experiment* – are not made in isolation. It *can* be, for example, a straightforward matter to stick to a particular diet or exercise regime. Until, that is, the influence of friends or family is brought to bear. What had been clear-cut independent decisions are countermanded or, perhaps more often, simply diluted out by 'real life', fading will power, and general apathy.

For most of us, our inbuilt need to find common purpose with others about whom we care provides a constant pressure to seek approval, or at very least tacit acceptance, of our decisions. Our vanities feel less vain when others in our group display similar ones. We want to be 'better' (whatever that may mean), but we do not want to be all alone in individual cocoons of self-metamorphosis.

Most people are interested in 'self-improvement' of one kind or another: Many of us wish to improve our skills and knowledge; some people wish to change their bodies (these days, the fashion is to make them thinner, more muscular, more sun-bronzed); some regard self-improvement as having more to do with meeting 'the right people' and climbing some perceived 'social ladder'. I tell myself that my philosophy is different: *I* am engaged in an experiment; I am not so much trying to improve myself as seeking to find out *what will happen*. My tentative steps up to the edge of this ocean of chaos are as nothing compared to the total immersion of some.

Cryonics is a last resort.

This is something that is often misunderstood. Neither I (nor anyone else I know involved in cryonics) ever wants to find that their preservation has become necessary. We do not want to die; it really is that simple. We are not the sort of people who will 'go gentle' into Dylan Thomas' 'good night', nor, for that matter, anywhere else that the other little-questioned dogmas of this particular era – this time that we just *happen* to live in – tell us we must go. We are radicals, but only as viewed within this myopic context. Yes, we rage – perhaps impotently, often alone – 'against the dying of the light', but we also stand with all the other light-bearers raging at the margins of a wider scientific dawning.

The *ideal* situation, for those who agree that living longer is a good thing, would be for human lifespan to increase at a rate that would allow us to attain the type of longevity pace-keeping state that de Grey calls 'longevity escape velocity' (LEV). Though de Grey and others have begun to flesh out the details, the idea has old origins. Over two centuries ago, in hiding and facing imminent arrest and execution, a French philosopher and revolutionary called Antoine-Nicolas de Condorcet (1743–1794) was pondering it:

> Would it be absurd now to suppose that the improvement of the human race should be regarded as capable of unlimited progress? That a time will come when death would result only from extraordinary accidents or the more and more gradual wearing out of vitality, and that, finally, the duration of the average interval between birth and wearing out has itself no specific limit whatsoever? No doubt man will not become immortal, but cannot the span constantly increase between the moment he begins to live and the time when naturally, without illness or accident, he finds life a burden?[14]

Condorcet killed himself rather than face the guillotine. But the Enlightenment of which he was part, and the ideas penned in his final *Esquisse* (*Sketch*), lived on. I do not know how or when we will reach the point that Condorcet envisioned, but unlike in his age, there is today no shortage of developing scientific insight into medical interventions that might become part of our LEV service-schedule.

There is evidence, for example, that *telomeres* – the DNA (once thought of as 'junk DNA') at the ends of chromosomes – may act as 'life clocks' for our cells. Each time a cell divides, the repetitive telomere *domains* (portions) of the chromosomes get shorter. When the telomere tails have been depleted, the cell can no longer be copied without catastrophic replication errors, leading inexorably to arrest of the cell cycle and/or *apoptosis* (programmed cell death). The absence of telomere material has effectively called time on the continued division of that cell. We can see the benefits of such a set-up from the point of view of natural selection: Potentially 'abnormal' cells cannot usually divide out of control, achieving immortality at the expense of other vital cellular material. And indeed, scientific studies have shown a correlation between the functioning of the enzyme *telomerase* and the incidence of certain cancers in laboratory mice; it may be the case that certain cells do not fall prey to the telomeric 'countdown' process described above, allowing them free reign to blossom, accruing replication errors and spreading unclocked and unchecked.[15] Immortality of cells does not seem to have been a trait routinely selected for in nature.

In his 1997 documentary *The Way of All Flesh*, film-maker Adam Curtis documents the extraordinary case of Henrietta Lacks – a woman whose cancerous cells became 'immortal'.[16] Lacks' cell-line, now known as *HeLa*, continues to grow in laboratory cultures all over the world, despite the fact that she has been dead for over sixty years. The abnormally robust proliferation of these cells *in vitro* makes them ideal for use in a wide range of experimental research, including that on telomeres. In her book *The Immortal Life of Henrietta Lacks*, Rebecca Skloot cites an estimate by Professor Leonard Hayflick that the weight of HeLa cells ever grown may total more than 50 million metric tons.[17] Implausible as this may seem, it is certainly true that an enormous quantity of HeLa has been grown, and that a huge cumulative stock of this cell-line exists today in labs across the globe. And, though neither Lacks nor her family gave permission for the tissue (originally from a cancerous tumour in her cervix) to be harvested, it is certainly the case that many people are alive today because of research performed on HeLa cells.

I earlier mentioned a protein called SIRT1. Caloric restriction – where a person reduces his daily calorie intake to around two-thirds or less of the normally-recommended level – has been shown to activate this protein.[18] This is important to scientists because SIRT1 seems to be involved in cell durability and fat loss. It may seem obvious that eating fewer calories will make a person thinner, but it is not obvious that it should help their cells to replicate for longer before degrading. (The reason may be a stressor-induced *hormesis* mechanism[19] that kicks the body into food-crisis survival mode.) It is in this context that we can see where the hype about resveratrol comes from: A pill that would activate SIRT1 could bring all the benefits of caloric restriction without the unpalatable gastronomic abstinence.

It is tempting to think that some kind of telomerase therapy or SIRT1 tweaking might become the 'magic bullet' that knocks out the ageing and self-destructing proclivities of the cells in our bodies. However, as de Grey and other *gerontologists* have pointed out, ageing is a multifaceted problem (hence the regular 'preventative maintenance' approach). And, as always in medicine, the precautionary principle will apply. Caution against introducing therapies that might increase cancer risk will slow the development of telomerase therapy and other anti-ageing treatments.

It is easy to mock the faltering efforts of those who seek to improve their health and vitality with the latest pills, potions, and treatments. But the drive is clearly strong in us not merely to *survive* but to *thrive*. Every time you pick up a tub of blueberries with the conviction that they are a health-promoting 'super food', or buy the latest 'probiotic' yoghurt expecting that it will improve your intestinal well-being; when you choose 'organic' foods, or kid yourself that homeopathic 'medicine' has therapeutic merit; you are sending out a message that you want to *live well* and to *be well*. It says something about your view of your life: that it is worth trying to stay as healthy as possible for as long as possible in order to enjoy it. Perhaps it also says something about your view of your death: that it can be pushed back – little by little, blueberry by blueberry – ever further into the future, ever further out of sight. If this is your view, then you may be at least partially right. Blueberries and yoghurt, however, are not the interventions required.

## KIDDING YOURSELF

Some view their *children* as compact bridging vehicles. The demise of old social structures that once firmly demarcated one generation from the next has meant that parents can now identify ever more strongly with their children. We now see mothers who dress and style their hair and makeup like that of their teenage daughters; they spend much time together and share similar tastes and interests. Children are often dressed in the type of designer-label clothes that their parents either wear or aspire to wearing – the type of clothes that were probably not available to the parents when they were, themselves, children. And many doting parents display as their own 'profile image' on social media websites photographs of their *children's* faces.

Obviously, there is nothing new about the idea of continuity through our children. At the genetic level, nature has (thus far) dictated that this is about the only type of physical continuity available to humans. Natural selection has, in effect, programmed parents to care about their offspring, and it has programmed offspring to be (at least initially) appealing to parents. Strong families show love and respect for one another; the deep origins of those feelings of mutual care are evolutionary ones.

A parent's feelings of satisfaction about *being* a parent are inextricably linked, then, to the drive to send his or her genetic payload forward in time by means of reproduction. I think, however, that this inbuilt sentiment sometimes becomes stretched to a point where the parent begins to see the child almost as a *reincarnation* of him or herself. Such feelings could provide great comfort to an ageing parent: The body of the offspring is still strong and beautiful when their own has begun to wrinkle and weaken; death does not seem far away for the parent but is still decades distant for the offspring; the offspring can still reproduce whereas, perhaps, the parent cannot. There may come a point, though, when the pressure from the parent to conform to a narrow set of expectations pushes the child away: The child is now his *own* vehicle and no longer wishes to carry the identikit camouflage of a fading other.

I do not wish to condemn parents who seek strong overlap with their children. I seek only to point out that the drive to *continue* – to avoid the prospect of *utter* extinction – is present in the mundane fabric of parenthood just as it is present in the dreams of life-extensionists. The standard argument – that one is 'natural' and the other is not – ignores the hidden motivations of the parent in choosing to have children; motivations sometimes brought to mind, but more often, buried so deep in the code of their genetic inheritance that they do not recognise them. The difference, perhaps, is that we can readily understand how our species benefits (e.g. in terms of adaptive capacity) from *new* individuals, whereas we have yet to see any benefits – 'natural' or otherwise – from greatly-extended lives of existing ones.

Inevitably, my attitude towards parenthood will have been influenced by the fact of my own childlessness. I have caused no one to exist.[20] There may come a time when I regret our decision not to have children. I would have made a reasonable father, and Terra a wonderful mother. The drive towards parenthood was, however, when it actually counted, weak in us. It just didn't seem that important. Were we too *selfish* to become parents? Perhaps, yes. But there can be selfishness, too, in having children – it's just that it is selfishness of a different, less obvious, kind.

## THE MAN IN THE WHITE SUIT

Recently, while on a Mediterranean holiday, I noticed a wiry old man at the table next to us tucking into an enormous quantity of fresh fruit. He smiled a lot (between mouthfuls of orange) and often reached out to touch the hand of the woman – presumably his wife – who sat with him. His skeletally fit appearance, and his choice of food, rang a bell; I commented to Terra (and to my parents who were with us) that I thought he must be on a caloric-restriction diet. A few days later, we saw him in the bar doing an impromptu tap-dance routine, to the delight of onlookers. And again, on our last night there, he stood out from the crowd, leading his lady in a skilful waltz on the otherwise-empty dance-floor. He wore a spotless white suit, his balding head peeping, tortoise-like, out from his too-wide bow-tied shirt collar. Here was a wizened and special man, not merely clinging to life but *living* it, and with a verve and panache far greater than that of many of those with much more of it left than he.

(I reimagine him now, this time more *Turritopsis* than tortoise: Shrivelling with age, he eventually disappears into his shirt-collar, emerging as a newborn polyp-baby from his pile of crumpled clothes.)

For some people, the man in the white suit represents a problem: He is the white tip of a hard blue iceberg of life extension. At what point does he go from endearing, sprightly old gentleman to self-centred life-grabber? How many oranges are too many? Which step in his ongoing vital dance is a step too far?

Is it acceptable for us to employ all available medical techniques to stay on the road and make it to the bridge? And, if it is not acceptable, then where lies the dividing line between reasonable and necessary medical intervention and stubbornly vain refusal to admit defeat?

Then there are the vexing questions of 'enhancement' and 'augmentation'. Is it acceptable to take drugs or to have implants fitted that make our bodies and brains function 'better' than the baseline? Few credible voices would argue that *cochlear implants*, for example, are an affront to nature – probably because most people see these as enabling a function that *should* have been there in the first place. Adding *novel* functionality is, though, another matter. People like Professor Kevin Warwick who experiment with implanted technological 'enhancements' – in Warwick's case a miniature radio transmitter and then an electrode – strongly divide opinion. Warwick does himself few favours (his claim to be 'Cyborg 1.0' was, I think, a bit silly) but at least, unlike most other academics in his field of study, he has a strongly practical take on raising the complex issues involved.[21]

Progressive medics, too, recognise that their historical role as fixers (and excisers) of damaged body-parts is changing. As Roy Porter points out in his book *Blood & Guts: A Short History of Modern Medicine*, 'The twenty-first century will go beyond replacement, deep into the realms of transformative and other elective surgery.'[22]

The dividing lines between life-preserving, life 'normalising', and elective treatments will become increasingly blurred. And few people yet realise the paradox of the life-*preserving* kind, or *bridging technologies* as I call them: Scientists constantly strive to create better and better replacement parts for human bodies – hearts, livers, kidneys, and so on – to allow critical patients to live for longer and longer, but *perfection* of such devices will never be required (after all, the parts you were born with are far from perfect, and the initial replacement doesn't need to be nearly as resilient as those). As long as such patients can survive to receive the next 'upgrade', then near enough will be good enough. Accepting this principle allows such parts to become more affordable, more quickly, for more patients: Why waste money and risk lives trying to develop *perfect* bridging tech when it's only going to be swapped-out for something better as soon as it becomes available? What is important is that life-threatening gaps always be bridged, so that patients no longer die awaiting medical technology that is available but not yet affordable.

Unsurprisingly, I am not perturbed by the prospect of human enhancement. But I am, I suppose, still somewhat conservative about what goes into my body. That mostly has to do with the fact that the vast majority of the supposed enhancements have been banal, or primitive, or both. Few could yet be described as transformative. However, that situation is changing rapidly, and the temperature of the ethical debate is rising. My slight conservatism on this issue is also due to the fact that I struggle to properly utilise the powerful tools I *already* have at my disposal: my brain, my senses, my hands, my creativity. I am not ready – yet. Maybe this applies to the majority of people, but I have no wish to prejudge the capabilities and readiness of others; why deny anyone the *morphological freedom*[23] they seek?

The are many reasons why the answers are not clear-cut. But, perhaps more importantly, the nature of the *questions* is coloured by the current context – a context of confusion about what persons are and why we should even *care* about their continued existence. A revolution is required. A struggle remains to be joined. It is a struggle that begins, with a slow detonation, inside our own minds. The effect of the shock wave is unpredictable, but it would be fair to say that it turns some persons inside-out.

# 9 SHADOWS, REFLECTIONS, AND TRANSPARENCIES

∞

Memories were clenching their fists
And quartz pulsed the time on their wrists

—ROOTLOVE, 'Bellyache'

## PRESENT AND STATELESS

No clock-tick. Breathing. Rumble from heating system. Men working on the new house on the hill. Breath. Sea-lap. Cars go. Intestinal roiling. Breathe. Bird-chatter. Ankle-click. Exhale.

There should be more, surely. But this is all. This is the sum total of my input. No, not really. If I were doing this properly then maybe this would be all. There's too much feedback. The thoughts keep pushing up with anxious intent, and the mental hall of mirrors keeps aloft the panoply of reflections; a magnificent feat, yes, but one from which I seek relief. Be still my beating mind.

Another strategy. And, in this stateless state, strategies become nothing but fluttering notions: subtle nuances of micro-focus on breath, eyelid, hand-relaxation. There should *be* no strategies because there are no *goals*. There is nothing. I am a dream of myself.

My mind finds mindfulness alien and tries to push it away. The reflections have no intention of suspending their light mid-flow, letting me pause in the moment. So I will trick myself. I will focus on the biological processes occurring in my brain and body, and tell myself a practical story of how I can change myself: adjust the balance of some neurotransmitter here, loosen the connections of some tight neuron-grouping there. Would this help even if I could map the intricacies of my own neural architecture? I doubt it. I would be focussing on *things* again, and this is not a thing.

The religious use the 'thing' trick. Believing that they are communing with some other thing – some other *entity* – they find a focus for the reflections and bring the light to a point of *stupefying* brilliance. Instant epiphany – just add the always-missing secret ingredient. Without the thing, there is nothing to hang onto; no limpet-studded rock in the numbing ocean as the water in your boots drags you under.

Over-dramatising now. Depressive grandiosity creeping in. There are just as many reasons why this should be easy as difficult. Without the plughole of Cartesian ego, I should float: kick off my boots, kick off my body, and simply hang between reflections. But I have no experience of such a state, no memory of it. It might make sense to say that I am not even *evolved* for it. What would be the result of humankind living in the moment, without the constant thorn of *the future*? Would anything get done? Would there be any progress?

Better to think of mindfulness as time off. My intention is not to stay in this state indefinitely, only to be able to find it when I look for it. But I want the tranquillity of it to pervade my life, like that feeling of clarity that sometimes lingers after a special holiday.

Light leaks through my lids, forcing me to stand and draw down the blinds. Such distractions should not matter, but I am a beginner, and focus is fleeting. I adjust my breathing and try to find my previous state. It has gone, and needs to be replaced by a subtly different one. There are levels here. No, just the *illusion* of levels. No valence. No satori? There is no good reason why I should struggle to find my previous focus, but my brain seems to require this flailing interregnum. There must be due process in the processes due. Logical even in this, where a logical state of mind holds no sway.

The stubborn pressure in my chest remains. The breath catches there. I can't seem to get enough air into my body to clear this blockage. Peristaltic diaphragm. Stubborn *enteric nervous system.*[1] When I probe the sensation further, I recognise it for what it is: only tension. Holding myself in this taut-string configuration, true relaxation will always evade me. I try to let myself droop into limp-muscled torpor, but I still try to keep my back straight – strong back, soft front. Greater bodily comfort might help. I would probably find other distractions.

I will focus on the breath at the tip of my nose. The idea here is to refine the focus down to a sharp point; my nose should be ideal for this. Put myself *there* in the simple loop of inhale and exhale. This sometimes works for a brief moment, before my heaving chest or leaden gut drag me back to my wet interior. So kick back up into cortex, avoiding the weed-wrap pull of intestinal entanglement. Now restless *occipital lobe* wants to dominate – starved of vision 'it' tries to stare through my closed lids.

I give up, and find that twenty minutes have passed. This is curious. And something does remain. I sit, contented, for a few moments, palms to the reassuring roughness of carpet tufts, before rising to resume my day of input.

Is any of this familiar? If you have tried meditation, you may recognise something here. If not, you may see nothing more than stream-of-consciousness babble. But we all make choices to alter our brain states: We take a drink or pop a pill; we 'pull ourselves together' or wallow in despair; we pursue activities that make us feel happy or resign ourselves to tiresome duties. Our choices determine much of the input that our senses receive, and that sensory input is processed by the brain in myriad different ways. We are, intuitively, used to the fact that our brains do not tend to respond immediately to 'instructions' such as 'be more optimistic', but we still assert such instructions to ourselves in attempts to gain some measure of control.

Some choose to assert these self-instructions in mantra-like ways. They repeat the relevant phrase multiple times, perhaps in front of a mirror. They use such rituals as tools to try to program themselves to behave in ways that they think will be beneficial to them and/or to others about whom they care deeply. Most of us do not do this so overtly; our self-assertions are internalised, more subtle, and less like mantras, but they may still be powerful and highly repetitive.

We often use these kinds of self-instructions to drown out 'negative' thoughts. We pursue the notion that some types of self-referential thoughts are good for us, while others are bad for us. We find that particular qualities are valued in our everyday lives: Our partners may want 'calmness' and 'commitment'; our employers may seek 'efficiency' and 'enthusiasm'. There seems to be no valued place in society for the kinds of vexatious thoughts and feelings that may push to our conscious attention when we awake at four in the morning, or when we rise in the winter blackness to prepare ourselves for work.

## THE TOOLBOX

We all self-medicate. Alcohol is one obvious method. We may say that we are drinking only to be sociable; nevertheless, we have made a choice. We have chosen to imbibe a substance that we know will alter our state of mind: We will become less inhibited and, at least for a while, less bothered by our day-to-day concerns. The next day we find a new situation: tension, anxiety, perhaps illness. We do not 'feel ourselves' – our *self* has been, quite literally, fragmented by the drug. It *feels* like our 'true' self must be in there somewhere, subdued by the discomfort and distress. However, until our workaday collection of self-governing precepts coagulate into something recognisable, we are truly more rabble than regimen. Some choose other drugs: caffeine, cocaine, cannabis, LSD, Valium, chocolate, heroin. The highs vary from mild stimulation to soaring euphoria, the comedowns from deflation to physical and psychological devastation.

Thinking, too, can be self-medication. And not in some roundabout philosophical sense. Our brains are drug-squirting machines: Everything we experience and every thought we have will alter, in large or tiny ways, the chemical balance of neurotransmitters in our brains. Taking a rollercoaster ride may give you a rush of *adrenalin* or *cortisol*; making love may flood your brain with *endorphins* and *oxytocin*. Even *thinking* about these things affects your chemical balance, though not to as great an extent as does experiencing them directly.

In my view, there is a disconnect between *this* understanding and the role of psychiatric *medicine* and psychiatry. A psychotherapist might try to dig out the 'true' cause of a patient's stress and anxiety by asking him a series of methodical and carefully-circumscribed questions. The therapist's aim is to make his patient feel better by getting him to understand *why* he feels a certain way and, in so doing, give him a path to reconciling troubling emotions within his mind. A psychiatrist may conclude that the patient requires medication to help him along this path. But does the psychiatrist always explain the role of the patient's *own thoughts* in the chemical mix? I do not think so. The patient may be left wondering how the 'real' and 'direct' course of taking psychiatric medicine relates to his inner world of irksome thoughts and feelings. It might help a great deal if the patient understood, from the outset, that both the medication *and* his own thoughts are acting on the *same* biological substrate in the *same* sorts of ways: Tell the patient *what he is*, and then involve him in the decision about what type of treatment might be appropriate.

I am sure there are excellent practitioners in the field who do exactly this, but I am equally sure that there are others who prescribe medication as a matter of course. I realise that it must be difficult for some psychiatric practitioners to admit to their patients that 'mental illness' is just a figure of speech and that, in reality, all illness has a physical (sometimes called *organic*) basis.

Despite great progress, some forms of psychotherapy are still mired in simplistic Freudian notions. There can be an implicit assumption that negative thoughts are rooted in a build up of unexpressed inner torment, usually posited to be about sex and/or death (*eros* and *thanatos*[2]). This 'pressure' must be *vented*, lest the patient have some kind of metaphorical mental explosion. And we tend to use this kind of terminology when talking about people whose mental balance seems to have gone dramatically awry: We may say they have had a 'breakdown' or even 'meltdown'.

Perhaps the realisation that Freud's hypotheses sprang from quite primitive ideas about the brain would temper our language.

The early anatomist Galen (131–201 CE) thought the solid matter of the brain unimportant, assuming instead that the fluid-filled *ventricles* performed the functions that we now know to be handled by networks of neurons. And those ideas came from earlier Greek thinkers such as Hippocrates (*c.* 460–370 BCE), who believed that the workings of the mind and body were controlled by the ebb and flow of 'the four humours': black bile, yellow bile, blood, and phlegm. Maintenance of the proper balance of these fluids was considered the key to good health.[3] We still use some of the related terminology today, when we say that a person is 'phlegmatic' or 'sanguine'. The philosophical baggage of such fanciful concepts survived into the twentieth century and was bundled into Freud's nascent hypotheses. There is a strong correlation between his ideas about verbal venting of deep-rooted neuroses and those much earlier ones about literal venting (by means of incision) of 'bad humours'.

Our modern knowledge of the brain, built upon since the brilliant work of Andreas Vesalius (1514–1564) correctly identified it as the main source and destination of nerve fibres,[4] means that we can put earlier mistaken concepts behind us. Though it is true that complex chemical *secretions* relay information within our brains, we are not under the simplistic, hydraulic control of 'humours'.

However, we should also be suspicious of the still-pervasive influence of certain more-recent hypotheses, such as those of Freudian psychology; in my view, it is a poor, blunt instrument with which our psyches were once probed for their non-existent 'id' and 'death drive'. Should we feel compelled to delve into the deepest recesses of our worst nightmares and blurt them into the cold light of day, in order to gain a feeling of 'mental balance'? Forgetting *might* be good for us. Painful childhood memories can recede into the inaccessible past; bad dreams can begin to fade, unexamined, to insignificance from the moment we cry out in the dark.

And we can change. We can think ourselves anew. I can alter my*self* using nothing but the contents of the 'toolbox' I have created. Like Epsilon the luthier, we must first make our own tools in order to craft our unique instruments. Like Alpha the stonemason, we can hew the rough shapes and place them, close-fitting and even-faced, in the walls of our windswept towers of *I*.

Mindfulness can feature in such a toolbox, but, for me, its effectiveness would be limited by undertaking it in a prescriptive way. I cannot follow a meditation script, because I would not *believe* it. Tranquil background music would remind me that tranquillity is fleeting and goal-centric. Though this may sound like control-freakery, I feel that I must take 'ownership' of my neural machinery. But *who* could possibly be this 'owner'? And how does he work his way round this kind of impasse? Only time and gentle cajoling seem to have any effect.

Mindfulness is both meditation and medication. I have *chosen* to alter my state of mind via the expedient of this particular type of quiet self-reflection. This practice has altered some neurochemical balance and reconfigured some groups of neurons in my brain. This is slow, powerful medicine.

In the context of this book, *my* reasons for wishing to try mindfulness meditation are relevant. Part of my motivation was a simple desire for self-experimentation: I wanted to find out what would happen. But, as we have already seen, it would be wrong to conclude from this that I indulge in wholesale self-experimentation. I drink alcohol and coffee, I happily smoke a 'joint' on the rare occasions when one is passed to me, but I do not take other recreational drugs. I am not altogether sure why this has happened. Perhaps I just didn't move in those circles, but opportunities have arisen and I have chosen not to join in. In truth, I am wary of the potential loss of self-control. I would, I think, be happy to experiment with drugs in a safe, controlled environment. It would also be true to say that I haven't found the idea of taking drugs very *interesting*. Recreational-drug-takers – far from being glamorous – can be excruciatingly dull people to be around.

It is extremely difficult to create drugs with highly-specific effects, because of the nature of the complex chemical balance on which they must act.

*Cocaine* affects the release and reuptake of the neurotransmitters *dopamine* and *noradrenaline*, causing feelings of exhilaration and euphoria;[5] *LSD* affects perceptions, often causing hallucinations, by mimicking *serotonin*; in effect taking its place at *synapses*,[6] thus blocking its normal effects and redirecting signals to 'older' parts of the brain. Cocaine is addictive, whereas LSD is not. *Most* widely-available and -affordable recreational drugs have *imprecise* and *diffuse* effects and side effects. The attraction of wild, chemical bludgeonment into a different state of mind is, for me, limited.

*Nootropics* – 'smart drugs' – are a different matter. As with all other drug-taking, there are ethical issues connected to the taking of smart drugs, but it would be illogical for me to condemn their use. Students regularly use drugs originally designed for other purposes – such as *modafinil* and *Ritalin* – to improve attention span and to increase alertness while cramming for exams. However, there are new drugs on the way, such as *ampakines*, designed primarily for use as cognitive enhancers. Would I take them? Probably. We sometimes hit insurmountable 'roadblocks' in our learning – problems that appear so abstract to us that they are simply beyond our baseline ability to compute them. I see nootropics not as drugs for everyday use but perhaps as a method of breaking through such blockages to reach important new avenues of enlightened comprehension beyond.

Enlightenment is a fine thing to aspire to, but the majority of drug-takers seek only a kind of balance in their lives: a balance found in relief from pain, abeyance of the spread of cancerous cells, maintenance of normal sleeping patterns, suppression of intrusive thoughts, and so on. Their drugs are provided by medical professionals via the normalised structures of commerce and society. Gigantic pharmaceutical companies seek to develop new drugs, but they prefer to tweak, re-brand, and re-licence existing ones because, without the huge additional costs of research and development, this is vastly more profitable for them. Drug-types such as *SSRIs* (*Selective Serotonin Reuptake Inhibitors*) – often prescribed to treat depression – have appeared in various forms over the last few decades. *Fluoxetine*, for example, is available under the brand-names *Prozac*,

*Seromex, Fluctin, Sarafem,* and various others. In case you hadn't guessed yet, the method of operation of SSRIs is not so different from that of certain 'recreational' drugs – it's all about synapses and neurotransmitters again. The normal reuptake of serotonin into the *axon terminals* curtails the amount of that feeling-of-wellbeing-inducing neurotransmitter available to 'slosh around' at synapses. *Selectively inhibiting* the *reuptake* of serotonin means that it stays in the *synaptic cleft* for longer, making those feelings of wellbeing more persistent.

Are doctors too quick to reach for SSRIs and tranquilisers when patients arrive at their doors feeling 'down'? Do we have a right to chemical numbing whenever we are in emotional pain? As is so often the case, it *should* be all a matter of degree – clinical depression is not remotely the same as just 'feeling down'. But we should recognise that there are players in the system with *agendas*. The agendas of drug companies are obvious, but what about those of doctors, hospitals, and governments? And what about our own? Should we face our emotional pains and really *feel* them, or should we forget them as we bathe calmly in the cleansing flood of our boosted neurotransmitters? A sea of tranquilisers may not wash away our pain, but perhaps – in the absence of any 'actual' tranquillity – they'll get us through.

Although I have not been tempted to take SSRIs or similar drugs, I have felt uncomfortable with myself and wished to change this. I thought mindfulness might help me to feel more content. 'Uncomfortable' is nowhere near being the correct word, but it may be the closest I can get to a straight answer. Sometimes I felt *made* of fear – an animated streak of pure worry adrift in a terrifying, irrational world; like that child in the shadowy dark, clutching his glass of milk. There were recurring loops in my mind that seemed to be self-defeating, and detrimental to my ability to function and interact as a consistent, reasonable, and empathetic person.

In his book *Revelation Space*, sci-fi author Alastair Reynolds introduces us to the 'Pattern Jugglers'[7]: an alien race of marine microorganisms capable of recording and reconfiguring the minds of those who come to swim on their ocean worlds. For now, and in the absence of the outlandish ministrations of alien beings, those of us who seek cognitive change may just have to do the pattern-juggling for ourselves.

Mindfulness may or may not have brought about some of the changes I originally desired. My motivations for undertaking it have changed – perhaps at least partly as a result of my having undertaken it. I still use it to find temporary respite from clamorous thoughts, but I also use it to remind myself that I am a figment of my imagination. This may sound self-defeating, but I find it liberating. I am not a Cartesian ego trapped in a fleshly prison. I am a pattern. I am a reconfigurable, jugglable pattern. And I am a process: part of the moment-to-moment flow of the entropic universe. I am, as Steve Grand has put it, a 'persistent phenomenon'.[8]

And this is true.

It is, at least, truer than the *intuitive* view of selves as discrete (and isolated) vehicles/persons moving through a real world of objects, encountering other such discrete vehicles/persons along the way. Moreover, if you think of yourself as the 'driver' within your brain then you are mistaken. Who, in that case, would be the driver of you the driver? You would be positing what could only be an infinite regress. You would be invoking a *homunculus*: In this context, a little person in your brain who monitors your input on internal screens and controls your responses. He would be the 'audience' in, and 'stage manager' of, the Cartesian theatre. To think of it another way, your head would have to contain one or more characters similar to The Numskulls – the tiny people from the old D.C. Thomson comic strips who steer their hollow-headed man-vehicle into a colourful variety of humorous situations. As Bruce Hood points out, in his book *Supersense: From Superstition to Religion – The Brain Science of Belief*, such dogged homunculi are not easily shaken out:

> We feel like pilots controlling a complicated meat machine. There is only one Numskull in control inside my head, and it is I. But how can a nonphysical me control the physical body? How can a ghost inside my head pull the levers?[9]

There is a sense in which the concept of homunculi – or Numskulls – *can* be useful in understanding the brain. Philosopher Daniel Dennett uses the analogy of hordes of progressively less 'intelligent' homunculi ('agents' or 'demons') operating all the way down through the various 'levels' of consciousness, until we get to the really stupid ones at 'the bottom' who don't *know* anything about anything.[10] This is a sort of 'hive mind' view, a view that is also expounded by *artificial intelligence* (AI) pioneer Marvin Minsky in his book *The Society of Mind*. In his later work, he uses the term 'Critics' to refer to sub-processes in the brain which sound (superficially) homunculus-like.[11]

I can see what they are getting at with such analogies, although I wonder about the usefulness of using terminology, at least in the case of Dennett, which has such a strong connection to the discredited ideas of Cartesian dualism. Despite his claim of neutrality about it,[12] I think Dennett's use of *homunculi* owes something to his mischievous streak. He has cleverly taken an old, redundant concept then altered it so radically that he *forces* it to mean something else entirely. I see this as a clear demonstration that coherent descriptions of consciousness need not be dazzling, jargon-laden *pieces de resistance*; they can, in fact, be stitched together even from old rags of conceptual language. Calling them homunculi may or may not be mischievous, but it is certainly more colourful than, as he states, 'calling them simply . . . units.'[13] Dennett has since refined his view to allow the low-level 'agents' more, well, *agency*. He is now moving towards the idea that *neurons* are the agents in question and that they have, in some sense, agendas of their own.[14]

The idea of hordes of *self-bootstrapping* (I will expand on this term later) low-level Numskulls making a mind holds a certain, if somewhat comic, appeal. And I have no doubt that a coherent explanation *could* be built using even *this* faintly-ridiculous terminology, if only in a spirit of playful provocation. The key reason for using such analogies is to get across the point that the various 'agents' involved in the process of generating the thing we call consciousness need not be self-aware. But if the agents are not self-aware, then where is the self-awar*eness* happening? And what (or who) is having it? The traps, as you will see, are many and varied if we continue, in our descriptions of the mind, to over-privilege and over-use words such as 'consciousness' – words that are, by their very nature, imprecise and ill-defined. Minsky uses the term 'suitcase words' to describe 'words that we use to conceal the complexity of very large ranges of different things whose relationships we don't yet comprehend.'[15] The word 'consciousness' is beginning to look a bit like a well-travelled – but I think rather battered – old suitcase.

Author and occultist Alan Moore, interviewed for print in his graphic novel *A Disease of Language*, says on the subject of consciousness that the 'I' is its own blind spot.[16] In a way, he is correct, in that consciousness is, like his statement, oddly *self-referent* (although I am not sure what he would expect to see *without* such a 'blind spot'). However, being a confessed magical-thinker, he goes on to make it clear that he believes there must be something else going on – a something that sounds suspiciously like a Cartesian ego. He contradicts himself, having just admitted the self-referential nature of selfhood. Uncharacteristically, his inductive reasoning has failed him, leading him to establish in his own mind a high probability that there must be 'something else' based solely upon the 'evidence' that he cannot seem to look into his own consciousness blind spot.

As Hofstadter points out, there are fascinating parallels between the looping self-referential nature of conscious thought and 'liar paradox' statements like 'This statement is false.'[17] Can we ever rely on careful use of language to keep us out of self-referential hot water? Is there anything *wrong* with self-referential language and concepts? On the contrary, I think them essential. Perhaps the true 'disease of language' lies elsewhere: For all our linguistic luxuriance and guile, we repeatedly fall back on discredited ideas about the nature of self out of a simple inability to give ourselves a coherent narrative of what we *really* are. There is a cure for this disease: We can, like Dennett and others, co-opt old words and change their meanings to suit our new ends. Alternatively, we can simply create new words.

Where, then, *is* my mind? If I shift focus during mindfulness I can, like Parfit, sometimes shock myself into remembering that *I* am not my body and brain. I am a *result* of the *structure* of my brain, but I am not my brain. So where am I? In one sense, I currently happen to be, in the words of W.B. Yeats, 'fastened to a dying animal'.[18] This turn of phrase could equally be used by a non-reductionist as a metaphor for the current 'location' of their 'soul'. Nevertheless, it is more powerful when used in a reductionist sense, for *we* are keenly aware that we exist only as unique self-patterns running on substrates that will not preserve them indefinitely; for now, the medium *is* the message,[19] and the medium in question is wet biology. This is a *real* problem, one that I think needs to be tackled with practical measures.

Come the time when my pattern cannot persist upon the substrate of the dying animal, I would wish it transcribed to *some other* (protracting) substrate. But if it happens that my *only* choice is 'vitri-fastening' aboard the feeble raft of my own dead brain, then I choose that. It would be, I admit, a poor (and probably futile) substitute for unbroken continuity, but it would be, after all, only a very last resort.

My implied separation of the pattern and the substrate here should not be confused with any claim that they are already, at root, separate and divisible. I am saying that they are inseparate but not necessarily inseparable. To quote Yeats once again: 'How can we know the dancer from the dance?'[20] You are – at least for the duration of the *current* performance – both. A *season* of shows, incorporating some changes of lead and step, is not, however, utterly inconceivable.

This 'lightness of being', as brought into inescapable focus by mindful thinking, can sometimes feel unbearable. The application of philosophical reasoning, and a certain 'flexing' of the self, alleviates this. Placing an imaginary doppelgänger on the shore on the opposite side of the stretch of water I face towards, I try to shift my focus to 'him' and to look back from that vantage point towards the bedroom floor upon which I now sit. This may sound strange, even crazy, but it helps to remind me of the truth: that this body is *a* body breathing, this heart is *a* heart beating; this brain is *a* brain thinking thoughts. I don't have to worry about *owning* them or *wearing* them like some suit of armour; they just *are*. I am a person on an island, not an island person. Though ostensibly like those of a bubble in a volume, my boundaries are mutable. I can overlap with others and feel the warm kinship of shared experience and mutual understanding. This is joyful; we are in this together.

## TICK...FOLLOWED...TOCK

And what of 'the moment'? Can mindfulness help us to live in the present, freed from past mistakes and from worries about the future? Glimmers of that state of mind do arise. Years of practice might bring it to the fore. I do not know. As I try to let my mindstream flow,[21] I sometimes find that I cannot help but think about the flip-book continuity of my moments: They must be quantised within my brain, and they must have a 'granularity' or resolution. Physics states that there is such a measure as the shortest time possible – the *Planck time*.[22] The universe may follow this infinitesimal signature, but my deathly-slow awareness off its workings will always smear out any possibility of my experiencing a 'singular moment' of reality. This technical explanation is not, of course, what is meant by 'being in the moment'. What, then – apart from learning to ignore invasive thoughts of past and future – *could* it mean? An appreciation of the simultaneity and equality of the processes of attention; an awareness that what we happen to notice is not the same as what is happening; an attempt to re-weight cognitive priorities in recognition that many of our modes of perception have been culturally instilled.

There is a conundrum here; and a set of choices so abstracted from our learned ways of *deciding* that, without the proper groundwork, they seem utterly intangible.

It seems good for me to spend more of my time 'in the moment', but what am I *really* doing here? Sensing the world in my time-delayed fashion while actually living in a Planck-time-perfect movie of reality? *Temporal divergence* within the brain makes a nonsense of 'the moment'.[23] Neurons receive the original sensory input and reflections of that input *asynchronously*, at widely (in millisecond terms[24]) divergent moments. Our brains scramble linearity; our woven reality is more skein than fabric. Our version of the present is always old, knitted hat.

Immanuel Kant suggested that we should refer to the world of our senses, and of our sense of the passage of time, as *phenomenal*, and the 'real' one – outside of time and not accessible to our senses – as the *noumenal*.[25] (There is a certain resemblance here to Plato's theory of ideal Forms, where he suggested that manifestations of 'forms' and 'ideas' in the material world were pale imitations of their true forms in some higher reality.) Kant used this as a philosophical device to explain how actions that might seem right in the phenomenal world could still be, *in actual fact*, wrong in the noumenal one. If he had heard of the Planck time, he would have had to place his noumenal world even beyond this. In some ways, Kant seems to be discussing a realm of *subjectivity* accessible to every *one* (every individual). Seen in terms of 'ideal Forms', however, the noumenal appears to be a realm of absolutes accessible to absolutely no-one.

While I find Kant's noumenal/phenomenal formulation problematic, I accept that we should neither underestimate the complexity of 'subjective' experience nor overestimate what *is* accessible to us. Minsky refers to the notion that we are seeing things exactly as they *really* are in *the present moment* as 'the Immanence Illusion.' He concedes that this sense, although an illusion, 'may be indispensable in everyday life.'[26]

And the present moment illusion is far from being the only one to which we are in perpetual thrall. *All* that we experience via our senses is highly filtered. We are only able to view objects of a certain scale with our naked eyes; we only see a minute sliver of the electromagnetic spectrum, in the form of visible light (if we could see *infrared*, for example, we might talk of 'turning up the whiteness' rather than 'turning up the heating' in our homes); we cannot hear things above or below a limited band of audio frequencies; our sense of touch is confined to material things within current reach of our bodies; and so on. Dawkins refers to this filtered view as 'the mother of all burkas',[27] in that, as with a burka, we have only a small 'slot' through which to view the world; all else is screened out. In a similar vein, neuroscientist and author David Eagleman talks of the *umwelt*.[28] Each organism can (as I have discussed above for humans) detect only a 'thin slice' of input from the totality of what we might call the *holoverse*; Eagleman – drawing again on terms originally devised by Jakob von Uexküll in the late nineteenth and early twentieth centuries – calls this the *umgebung*. Organisms inhabit, according to Eagleman, 'micro-realities' imposed by the limitations of what is detectable to their senses. Umwelts are, at least superficially, a little like Eli Pariser's 'filter bubbles',[29] but here the filters are mediated by biological evolution rather than by what an advertising-company-designed computer algorithm has 'decided' to show us. The umwelt concept is also similar to the Buddhist idea of 'conditioned existence'; Buddhism stresses that *all* existence is conditioned.[30]

The umwelt sounds, to me, prison-like – a larger version of Plato's cave. But, fortunately for us, humankind has methods if not of *escaping* our current umwelt then at least of peering beyond it. We do this with our intelligence and with our technology; in many ways, our brains are a type of umwelt-busting technology, albeit one that came about by accident rather than by design. We use our inventions, such as telescopes, micro-scopes, and infrared cameras, to view a greater slice of the holoverse than is possible using only our evolutionarily-dictated eyes. In such ways, we *augment* the reality of our umwelt. We use our brains to ponder, some-times using advanced and abstract mathematical concepts, what the grand structure might look like 'from the outside'. Other organisms do not do this; they are confined, as a result, to the umwelt of their evolutionarily-adapted sensory input.

We should revel in our ability to peer beyond *our* original umwelt. It could be argued that the technologically-enhanced augmentation of our sensory input only serves to fool us: that we believe we have greatly *expanded* our umwelt while we are merely staring out between the prison bars or through the burka slot. This viewpoint is valid in the sense that our ability to have 'direct experience' of anything beyond it is severely, if not totally, curtailed. But it ignores the potential for our technological advancement: that we may, in time, develop methods that allow us to gain such direct experience. And of course, there are always those with a door-locking agenda – the 'prison warders' of the human umwelt. They seek demarcation of the one we currently inhabit, in the mistaken belief that the decision to step outside it is not ours to make.

*Changing* the umwelt is far from straightforward.

Would we always know if it had happened? What if we were to take our 'moments' of reality and spread them out over a far greater period of time? To put it in computer terminology, we would be slowing our *clock rate*, so we would receive and compute the input of our senses less quickly. This might not matter, provided it had also happened to everybody else. This would become our unnoticed new reality.

If we were able to transfer self-patterns onto computer-like systems, we might have a genuine divergence in the clock rate of perceived reality. If we assume, as in Greg Egan's book *Permutation City*,[31] that the clock rate of putative computer-based persons is substantially slower than that of biological-based persons, we end up with a situation where every day of 'virtual' lifetime is equal to two weeks or so back in the 'physical' world. To the outside observer (if there were a way of observing it), the life of the virtual person might look like a stuttering movie. The virtual person, on the other hand, would notice *no* gaps. Egan's book ponders the issue of *how slow* the computation rate could go before the virtual persons started to notice gaps – one computation per second, one per minute, one per century? As his story concludes, there is no reason they should *ever* notice gaps, whatever the computation rate. Their lives would still be, to them, a continuous flow of sensory experience. Whole civilisations in the 'outside world' could arise and crumble in a perceived moment of time in the virtual world.

Similarly, physicist Freeman Dyson has suggested that such a slowing of *computation* – and, therefore, *thought* – could be a long-term survival strategy for future human-descended entities. As (due to entropy) progressively less *ordered* energy is available in the universe for thought, and as the entities expand to utilise the remaining *useful* energy, the rate slows until there are gaps of hundreds or thousands of years between computations.[32] Due to the gradual thermodynamic *heat death* of the universe, all surviving entities would be in the same cognitive boat, hence there might be no other entities left capable of perceiving any 'gaps'.

We have a strong sense of being in the world just *as it is*, with time unfurling along with us just *as it does*. The inclination to question this sense does not seem to come naturally to us; nonetheless, it is gapingly wide open to question. In one of his early science fiction stories, Robert A. Heinlein wrote: 'Duration is an attribute of consciousness and not of the plenum. It has no *Ding an Sich*.'[33] As his time-traveller character struggles with his thesis, he employs Stoical and Kantian terms – 'plenum' (space filled with matter; the opposite of vacuum) and 'Ding an sich' (the thing-in-itself; the *noumenon*) – to convey this complete absence of temporal reference. This suggests that 'duration' is just as elusive as 'gaps': Gaps in what? Gaps with reference to what?

Erich Harth uses the analogy of *daguerreotypes* to illustrate the ways in which 'gaps' affect our perception of reality. The exposure of this type of early photograph took so long (many minutes) that the people in street-scenes simply disappeared: Their appearances in the scenes were too brief to register on the photographic plates, so only the roads and buildings remained.[34] Your visual perception system is no slouch, but in comparison to those of some other creatures[35] (let alone to the rate of the Planck time), it's worse than a daguerreotype.

It's all too easy – even for neuroscientists – to get hung up on gaps, sequence, and processing speeds. But Dennett regards talk of 'filling in' processes happening in the brain to be loose and wrong. What, he asks proponents of such ideas, would all these 'gaps' be 'filled in' *with*? 'Figment?'[36]

This concept also gives us a hint of the way in which the 'soft division' I mentioned earlier could be a trick – a trick on two levels. On one level, it's a thought-experiment 'intuition pump'[37] to get you thinking in the direction of seamless transfer. On the other, it's the patient who is being tricked: Her 'transfer' to the soulbox (and to the new body) could actually be quick, slow, soft, or hard. It would make no difference, for it wouldn't really be a transfer at all. There is no 'phasing out' (and no 'filling in'), there is only replication and destruction. As long as the doctors involved later gave the patient a memory and sensation of a seamless transference (some input but no 'figment' required), there could actually be long delays in the procedure. *Where* and in *what state* would the patient *be* during those gaps? Alive? Suspended? Dead? Where are *you* during yours?

'The moment' is a slippery concept, in any umwelt.

The past and the future seem to provide context and purpose, giving us something to hold onto. Are they not also real? The straight answer is *no*. Our smeared-out version of present reality is the best one we are currently able to perceive. The past exists only in artefact and memory; the future has not yet happened. The only context is now. The only *purpose* is that which we make for ourselves. Attributing purposiveness to anything that existed before conscious brains is pure *teleology*: According to Cox and Forshaw's definition of the word, 'that is to say things appear to happen in order to achieve a pre-specified outcome'.[38] The philosophical concept of teleology is of limited use in science, but it finds a comfortable home in faith-based doctrines, where it sprouts wings to become a belief in supernatural 'intelligent design'. Purposiveness requires thought, and thought did not exist before conscious brains.[39] If we are the only intelligent beings in the universe then we are the *only* entities that perceive this specific type of linking together of the discrete, granular moments of our reality into an intelligible flow that has a 'past' and a 'future' state.[40] We join it up, smooth it out, and not concerning ourselves with questions of our clock rate, call *this* bit 'now'.

We may as well live in the present because that's all there is, folks. It would be wrong to infer from this we should all adopt a selfish, careless, hedonistic lifestyle. Remember the others with whom you overlap. Remember the entropic effects of your actions. Early Buddhist teachings emphasised the *reality* of actions contrasted with the effective/relative *nonexistence* of persons. In other words, we – these stories-of-selves – initiate actions that reverberate into the future and affect other future and contemporaneous stories-of-selves. That future does not yet exist, but we should have some regard for its inhabitants. We manage to hold a fair degree of such regard for the stories-of-selves that will occupy *our own* older bodies, but we struggle to extend it to persons we do not know or who do not yet exist.

In Adam Curtis' *The Century of the Self*, the theme that emerges (one that also arises in other examples of his work) is the cynical manipulation of the masses by those in power. In this film, he highlights the breaking of society and stripping-away of real freedoms by means of the top-down promotion of selfishness. There are many overlapping strands to his argument, but one of them is the role that Freudian psychology and the quest to 'find oneself' have played. He believes that the real agenda of governments and ruling classes in offering us more 'choice' and 'personal freedom' has been to take away the choices and freedoms that should really matter to humankind. In his view, they have manipulated us to make us focus inward, on changing ourselves, instead of outward, on changing society and politics.

Having wholeheartedly agreed with Curtis when I first watched *The Century of the Self*, particularly on the subject of Freudian psychology, I now think that, while broadly correct, he has bundled too much into his hypothesis. That is because I now think that a great deal of good can be done by allowing people to find out *what they really are*. This is not something one can readily demonstrate to others; they need to find out for themselves. Introspection is not the enemy. If, by means of some quiet time spent focussing on her true nature, a person can come to realise that what matters is not *the self per se* but rather *the actions of the self*, then little but good will come of the exercise. By learning along the way that she is an end in herself and that she must, logically, apply this same rationale to others (instead of treating them as means to her own ends),[41] she may, at last, find everyone else. True introspection of this kind will also force persons to examine the appropriateness of any *beliefs* they may hold dear; religious dogmas and extremist views do not tend to stand up to the realisation of *self* as 'mere' pattern.

*Belief* does not feature strongly in my vocabulary of useful concepts. And, throughout this book, I will always try to enunciate what I *think*, rather than what I *believe*. You may argue that I am splitting hairs; in that case, you should have no particular objection to my choice of word. *Belief* (and the strongly connected concept of *faith*) has been given a privileged position in our vocabulary that sets it somewhere above *thought* and *opinion*. This position is not justifiable. When a person says that he 'believes in' something, he is, in fact, revealing only what he *thinks* or *reckons*. But, although they actually equate to the same thing, 'I *reckon* God exists' doesn't seem to have quite the semantic gravitas of 'I *believe* in God'. I also find *threat* implicit (or overt) in certain declarations of belief: a thinly-veiled 'back off because I *believe* this'. I make an effort not to be ideologically wedded to my own current ideas or to the ideas of others that I use in support of my arguments. The emergence of better theories, with greater weight of scientific evidence behind them, should always force me to change my stance.

The subjects of introspection, belief, and changing stance bring to mind Egan's radical example. The virtual persons in *Permutation City* are able to alter utterly their opinions, interests, *beliefs*, and obsessions by simply reaching into their own virtual neural-architecture and making the 'required' changes. At one stage in the book, his character Peer changes himself in this way so that he becomes content to do nothing but turn table legs on a lathe, for over fifty years. These table legs will never be used. Earlier in the book, we find Peer walking – for recreation – the side of an infinitely-tall skyscraper, in blazing sunshine. Again, he never tires of this. Why would he? His interest is constantly refreshed by his 'programming'; time is irrelevant. Such ideas raise deep questions about what it would mean to be human in a world where we could instantly and easily change our personalities, altering, as a result, our views on anything or everything – those we currently love and cherish included.

And philosophers deconstruct the morality of such ideas. To illustrate his *thin theory of the good*, Rawls envisions a man who has decided that the best use of his life, as it is his 'only pleasure', is to count the blades of grass in lawns, and who then proceeds to do just that.[42] Parfit, being an objectivist about reasons, disputes Rawls' claim that such a choice can be described as good or moral.[43] However, *Egan* is not raising questions of morality, only suggesting that, in theory, such radical self-programming *could* be done. Blithely marrying Egan's sci-fi hypothesis with Parfit's objectivism, I claim that it might be good *and* moral for us to re-program ourselves, but only as a means of making us more inclined to pursue objectively-rational, moral ends.

In *What Sort of People Should There Be?* Jonathan Glover takes a similar line, in considering which specific aspects of ourselves we should keep:

> It is not just *any* aspect of present human nature that is worth preserving. Rather it is especially those features which contribute to self-development and self-expression, to certain kinds of relationships, and to the development of our consciousness and understanding. And some of these features may be extended rather than threatened by technology.[44]

We all self-program, but as I mentioned earlier, some people make the changes more consciously than others, using the 'technology' of learned technique. Meditation is a real-life example of a way in which certain people can seem almost absurdly content to perform the same task for countless hours (though meditating may not even qualify as a *task*). But if we allow the possibilities that participants' interest in such introspection is limitlessly refreshable because of their ability to 'program' themselves to *want* to do it, and that the nature of the 'task' makes them lose their sense of the passage of time, then their contentment seems less absurd. We could still argue that having an *explanation* for why they want to do it does not make it any less pointless; like producing countless table legs that will never carry tabletops graced by meals enjoyed by happy 'normal' families.

Such concerns about the self-centric nature of what I am 'attempting' bubble up when I meditate. Thoughts of oblivion also surface. I have an almost-mantra for this situation, but I am only able to use it because it is true. If this almost-mantra were a physical tool, it would probably be a torque-wrench: I remind myself that my death will only mean that, in Parfit's words, 'there will be no one living who will be me'.[45] The meaning of this phrase may appear, superficially, to be the same as 'when I am dead I will be dead', but *Parfit's* intended meaning avoids that shattering tautology. He means that because our *self* exists only as a pattern, and because our continuity as selves consists only in a particular relationship (which he calls 'Relation R'[46]) between our past and present selves, we can say no more about the situation to which the word 'dead' pertains than the quote I have given. It is appropriate for me to feel sad about this situation, but I should *not* feel (just because my instincts tell me that it will mean more than this and because there is a terribly ominous word for it) that it *will* mean more than this. I have found that the nut on this dead bolt keeps coming loose, hence the wrench.

We use the words 'dead' and 'death' in ways that increasingly seem odd to me. I have heard and pondered over phrases such as, 'He's been dead ten years now'. What does this *mean*? The person in question *died* ten years ago, but, in the ten years since his death, *he* has not *been* in any kind of 'state of being'. Saying instead, 'He hasn't *been* for ten years now' might lead to confusion, but it might also help us to expunge the notion that 'dead' is a *state* or that 'death' is a thing. We don't talk about not-yet-conceived people in this way; indeed, we don't even have a word for them or their 'state', because they *are not* and, as a result, do not have one. We have a propensity to fill such voids with words to the point where we end up in Gordian knots, in the mistaken belief that we are saying something about a situation that is not, in fact, a *situation* at all. I admit that simply rephrasing to 'He died ten years ago' readily slices through this particular knot.

Here, then, am I, using mindfulness techniques to try to be 'in the moment', and attempting to neutralise concerns about my own future death (and about the future deaths of others); all the while aware that I have taken steps to avoid the death that I am now trying to tell myself does not matter as much as I thought it would. This is not, much as it may appear so from the outside, a conflicted situation. It does not require resolution into linear dogma. Wishing to live *now* and appreciate the moments of my life, and the lives of others, is not at odds with the realisation that death, while to be avoided, is a subject that can viewed through a different, if convoluted, lens.

I don't doubt that mindfulness is changing my brain, although this may be for others to judge, as *I* cannot now remember how I was before. I have clear memories of how I *acted* before, but I am not now entirely familiar with the person who was connected to those past actions. The nightmares still come, and perhaps they always will. Not so much nightmares as *change dreams*, as they are always strongest at times in my life when I am making significant changes. They are, however, distressing, and I have decided because of a recent and particularly traumatic one that I need a device to release myself when I feel trapped in them. Not, in this case, a torque-wrench, but a hairpin.

## THE ANATOMY OF A NIGHTMARE

*Walled prison camp strewn with dead insects.*

The North Korean prison camp mentioned in a news item the night before. A young boy lived there, among thousands of other prisoners, starved and living on whatever he could find – grass, mice, insects.

*Bruce Willis (the protagonist).*

The half-mad character Cole played by Bruce Willis in *Twelve Monkeys*, one of my favourite films.

*The gigantic millipede that crawls overhead along the prison wall, emitting a dry rattling sound and dropping down scythe-like hooks in its attempts to snag Bruce and me. The millipede seems to take several minutes to pass. We cower by the wall, stupefied with terror.*

Millipedes crawl into my house at this time of year to bask in the dry-warmth and sunlight. I often find dead ones trapped in the metal door-threshold. The hooks came from an image of the microscopic barbs found on gecko feet, from a recent documentary I watched about Andre Geim, the inventor of *gecko tape* and discoverer of *graphene*.[47]

*The 'inner sanctum' building to where I follow Bruce, in the hope of finding another way out.*

'The Tower of Rassilon' from a childhood episode of *Doctor Who*. In the plot of this particular episode, the tower is a place where, according to legend, the secret of immortality can be obtained. It eventually turns out that the type of immortality on offer is no prize at all but, instead, eternity trapped in stone as a motionless face.

*(With the plot of the dream now taking this turn for the pastiche, I realise that I must be in a movie alongside Bruce.) The movie is real, and the plot now*

*involves a man becoming entombed forever in some kind of vault. My panic grows.*

The self-mummifying 'Sokushinbutsu' zealot monks,[48] from a conversation I had with Guitar a few weeks before.

*Bruce insists that the elevator in the sanctum will take us to another level, from where we can escape. But, after getting into the elevator, we find ourselves trapped between levels and now in another sub-prison that looks, from the outside, two-dimensional.*

The 'elevator' and 'levels' from a song I wrote at high school. The 'two-dimensional prison' from a combination of the flat prison that contains the *Superman* supervillain General Zod, and the Andre Geim documentary where graphene is described as 'the first two-dimensional material.'

*A happy family appears in our prison. Bruce then proceeds to butcher them, explaining all the while that nothing here is real. The dismembered family continue to shamble around, smiling and laughing.*

A mixture of worries about family and reality. But also references the torture of the North Korean boy in the news story – torture that followed the execution of his own parents on the basis of incriminating information that *he* had willingly given to the guards. There is also, in the post-slaughter motion of the 'happy family', an aspect of the 'automatone' 'Kourai Khryseai' (the 'Golden Maidens' of Greek mythology),[49] which I had been reading about the day before.

*The rules of the movie/reality do not allow me to escape, although I am aware that there are software protocols for such situations. Emergency.*

The 'bale-out' option in Egan's *Permutation City*, disabled by the protagonist so that the virtual copy of himself (which later turns out not to be a 'copy') cannot escape/die.

I awoke then, hyperventilating in the dark, the rattling of the blinds immediately explaining part of the dream. Terra, realising I had had a nightmare, snuggled up to comfort me.

'I couldn't get out,' I explained. 'There were *rules* stopping me. I need a way... a trigger to get me out.'

'A hairpin,' she said, calmly. 'It always works in the movies.'

'Yes,' I agreed, 'good idea.'

(I have not yet had the opportunity to test the efficacy of my hairpin,[50] but I have been making an effort to shape and sharpen it each night as I lapse into unconsciousness, just in case.)

I have contradicted myself, having stated earlier that there may be *no need* to delve into nightmares. I have included this one because it happened while I was writing this book, because it contains themes that are relevant, and because it demonstrates the banality or semi-banality of night-terrors that *feel* like they should mean something more. We also see, from such examples, how the magpie tendencies of our brains gather and reflect some of the glittering (and light-sucking) fragments of our daily information-diets. Often, we can conclude of our nightmares that much of this information was bad enough the first time round and that we really don't need to see it again, during sleep, with excruciating new twists.

In *The Creative Loop*, Erich Harth compares the sometimes hallucination-inducing effects of sensory deprivation to what happens when we sleep and dream:

> No *real* sensory images from the retina are there to interfere with the patterns generated by the positive feedback from above, which thus is free to generate full-blown patterns that can mimic sensory input. Just like the mapmakers in the days of Columbus the brain places the monsters where the map is otherwise blank.

Perhaps we should not worry so much that nightmares are trying to tell us something deeper about ourselves. There may indeed be 'messages from the subconscious' hidden in the content of our dreams, but, in many cases, most or all of it is just noise. 'Subconscious' messages are also part of our waking lives; maybe they are easier to ignore because they are not quite so aggressive or colourful once blended with the mass of wakeful-state input.

I seem unable to gain back the clearly-labelled, playful, imaginative humour of my childhood 'nightmares': *the savoury shuffle-plod of the*

*Cheeseback Turtle*, the spidery *giant-pompon-and-pipe-cleaner architecture of the Mittelmice*; though I can certainly live without the *indescribable furniture-scraping terror of The Mobenhost*.

Labels. The unnameable thing is always worse.

Why must I now do this to myself: the cold sweats, the crushing dreams that I have lost everybody, or that everyone else is safe and it is I who am lost? Give me back the rapture – the black and purple symphony where I walk with Terra through a forest conducting the trees to raise beautiful music for her as we go.

The only lesson I feel that *I* can take from my current nightmares is that I need to find a reliable way of stopping them in their tracks. The idea of the 'hairpin device' does beg the question of why, given that fact that neither it nor the dream content are real, I would need such a tool rather than just *deciding* that I will not have bad dreams any more. I should simply morph the bad dreams into good ones. Perhaps the efficacy of 'the hairpin' will be demonstrated by the fact that it never emerges, because I never again end up in a dream situation where I need it: an imaginary tool that did its job despite the fact that it never existed, even in the realm of dream-imagination for which it was designed.

The long, dark nights of winter are now a fading memory. Spring light dapples the mountains as I look out across the stretch of rippled steel-grey sea towards the mainland. Cheery yellow gorse flowers further brighten the scene. I know their coconut-sweet scent, and will pick some from their thorn-laden bushes when next I step outside. I am relaxed and happy.

*Becoming* takes its toll, but it seems that I find little contentment unless I can convince myself that I am becoming something else. In writing this book I am becoming, I hope, a writer. That drive to *become* also makes me keenly aware that I am nowhere near to mastering 'the moment'. There is a constant dynamic tension between my appreciation of the beauty and wonder in my life, and my desire for change and growth. It *feels* to me that the dynamic tension is important, but, in reality, I have no way of knowing whether that is true or whether it is just a useful trick my brain plays on itself in order to make me feel that I have purpose.

'You think too much,' is an 'accusation' that has been levelled at me from time to time. I try to bring to mind now Rodin's *The Thinker*,[51] wrestling in sober isolation with some weighty intellectual puzzle. I imagine the man depicted by Rodin thinking with cool clarity, his thought like a river in bronze evening light. He is in a state of mind that psychologists call *flow*.[52] My thoughts do not feel like this. Nevertheless, I am able, using my *imagination*, to experience, 'second-hand', a taste of how the thought-flow of *The Thinker* might feel. Pure *thought* is that reflective process where we 'bring to mind' images, words, fuzzy/symbolic interpretations in the absence of direct sources of stimuli: Our minds talk to themselves. In truth there are *no* times when our brains are completely starved of input – there is always something going on. Even during wakeful rest, the *default mode network* (DMN)[53] of our brains is active and thrumming with neuronal oscillations. These cognitive ripples and reflections are highly complex, and to all intents and purposes, impossible to predict. And, as in my thoughts about the thoughts of *The Thinker*, they can be multiply nested, like infinite Russian *matryoshka* dolls.

The issue of the *initiation* of the type of thoughts that 'come to us' in isolation causes consternation for some. I remember a conversation with my father about thoughts and ideas where he was insistent that they must 'come from somewhere', the implication being that he thought they must come from somewhere *else*. A state of 'flow' requires that we have an immersive task in hand or, at least, an immersive thought to ponder. But where is the source of the stream of ideas? Whence comes the rain that replenishes it? I find it odd that some people are not satisfied with the answer that, in the absence of normal external stimuli, these thoughts are generated, sometimes randomly, within their own minds. I say only *sometimes* randomly because it may be the case that *completely* random thoughts are rare; we usually have some current set of concerns that lead our minds off to fish for reflection in particular directions. However, even in the absence of the stimulation of pressing current concerns or interests, there is always *noise* in our default mode: the constant, low-level 'background radiation' of cognitive activity that acts as a seemingly-bottomless pool of reflective upwellings.

Feelings of disappointment are not appropriate to the realisation that there is no 'divine inspiration'. Rather, take delight in the rich, nested dimensions of imagination available *to* your own mind *from* your own mind. Because we are only tenuously connected to the 'outside world' – there are many more interconnections within our brains than between our brains and our sensory organs[54] – it should not surprise us that our 'inner worlds' can be this way. A dogged determination to roll out and explore these nested landscapes has been a characteristic of great thinkers down the ages. (Unfortunately, however, many of us just seem to get an amplified version of the noise.)

The topographies of these 'mindscapes' need bear no relation to anything in the real world. Distance and dimension lose meaning. We have left the Spaceland of our sensorium, and entered Thoughtland.[55] The laws of physics do not apply here. All time is open to us: past, present, and future accessible without recourse to fabulous inventions (although fabulous inventions are commonplace here). And they are all different; each is special and unique. A mathematically-minded *synaesthete*[56] might stride forth in a weird landscape of shapes and colours that represent numbers; a poet might bathe in a sea of buoyant words; an astrophysicist might weave his way between red giants and white dwarfs regardless of the real-world limitations imposed by the speed of light. This is our workspace, our playground. And it is limitless.

And here we loop back to the beginning of this chapter – to the subject of finding a balance between intense thought and *not* thinking. Both mindfulness and deep introspection have a deserved place in our self-exploratory (and self-explanatory) toolbox, but they are limited tools. Alone, they are not enough to open our minds to our true nature and potential. Each of us has a brain in our head – 'this teetering bulb of dread and dream' referred to by the poet Russell Edson;[57] why, then, do we consider the study and understanding of it the preserve of only a small band of adepts? Neuroscientists can be brilliant, yes, but they are neither magicians nor high priests and priestesses of the mind and brain. The days of that sort of blind reverence are gone.

Armchair psychologists are ten a penny. But where are the armchair neuroscientists?

We are all capable of gaining some understanding of, at very least, the gross biology and biological processes of the brain. Why shy away from gaining this knowledge when we could take it on board in the same way that we do any other subset of general knowledge? Perhaps such reluctance is, at least in part, to do with ingrained attitudes toward the 'hard problem' of *consciousness*. When we hear even *experts* calling it 'the hard problem', we are disinclined to think that there can be much accessible knowledge of this field comprehensible to *the rest of us*? But why assume that a neuroscientist must be more capable of nailing down consciousness than you would be? They suffer from the same kinds of problems in describing it that you would encounter: They end up drawing over-rigid boxes and flowcharts; they get confused by feedback processes; they struggle to find the correct analogies (for example, despite his otherwise perfectly-clear ones, Eagleman can't resist using a questionable 'CEO' analogy in describing conscious thought[58]).

I am tired of hearing it called 'the hard problem' and now think I have valid reasons, based on what I have so far learned, to question whether we should continue to describe consciousness as a *problem* at all.

## DRIVINGNESS EXPLAINED?

The little girl giggled as the wind from the open car-window ruffled her curly hair and blew it, ticklishly, across her face.

'I like driving along. I like watching the other cars driving too. And all the pretty colours they are,' she said. 'Where's the drivingness in cars?'

Her father smiled at her from the driving seat, 'I'm glad you're having a nice time, but I'm not sure what you mean by "drivingness"?'

'You know,' she replied, 'the drivingness as it goes along.'

'Oh, you mean the *fuel*, like we just put in back at the petrol pump.'

'No, not *fuel*, the *drivingness*,' she said, a little exasperated.

'Well,' said her father, 'cars have *engines* in them. The fuel gets lit up in lots of little explosions that move things called *pistons*, and the movement of the pistons gets changed so that they can move a thing called a *drive-shaft*, and that makes the wheels turn.'

'I don't *feel* little explosions. I just feel nice *drivingness*,' said the girl, now clearly dissatisfied with her father's responses.

'Maybe you just mean *driving*,' he said, a gentle frown creasing his forehead. 'The engine makes the car *go*, but *I* am driving the car.'

'But nobody drives *Lightning McQueen*,' she said. 'He *goes* by himself.'

'Yes love, but you know he's just a cartoon character. He's not real,' said her father.

'So the drivingness must be *in you*,' she said, happy again. 'That's why it's nice!'

It was harder to concentrate on the busy road while answering his daughter's naïve questions, but he believed in trying to explain things to her, rather than giving her 'that's just the way it is' responses. 'I have a *brain*,' he said, 'and that makes me smart enough to drive the car along the road.'

'The drivingness is in your brain, like *Numskulls*,' she asserted. 'My friend Tom says his dad was telling him about Numskulls about how there's little men in your brain that make you be able to do stuff and stuff. They can make you be able to drive too, cos they have drivingness in them too.'

'No, love,' he said. 'There are no Numskulls in your brain; they're just cartoon characters from when I was a little boy. And there's *no such thing* as "drivingness". I drive the car using *my brain*, and the *engine* makes it *go*.'

The little girl started to cry.

'What's *wrong?*' asked her father, utterly baffled.

'I don't like it now cos you said there's no drivingness and I'm scared cos nothing is driving the car.'

Children ask some odd questions. Sometimes the questions are playful or just plain silly, but at other times they are based upon a genuine desire to understand phenomena they may be coming across for the first time. The little girl in this story has identified a phenomenon that she thinks requires an explanation. The fact that it is not a phenomenon that we have ever heard of, and that we don't even think *exists*, is irrelevant to her. It is real *for her*, she has given it a name, and she now wishes to have an explanation to complement the name. Her father's confusion is understandable: He doesn't even know to what she is referring. Is it a noun? Is it a verb? He does a reasonable job of giving a *mechanistic* explanation of how the car's engine works, even though that is not what the little girl is talking about; but he fails to give any kind of explanation – mechanistic or otherwise – for what his brain is doing. In this context, the little girl's explanation is better than his own. She at least posits something *inside* his brain that is providing the 'drivingness'. He, on the other hand, has left his brain as an unexplained 'black box': It just does what it does.

Despite his efforts, the father has given his daughter a 'that's just the way it is' response.

# 10  BITS OF SELVES

∞

Everything should be made as simple as possible, but not simpler.[1]

—Usually attributed to ALBERT EINSTEIN

The virtual hypercolumn is complete: a fully-functioning cortical module of ten thousand exquisitely-interlinked neurons, all held aloft by the RAM and multiple processors of an IBM Blue Gene supercomputer. The visual interface taps into the generated output, and the scientists watch the beautiful dance of colour-coded morphing connections on the screen. Patterns. Intricate, delicate, self-referent patterns.

This is not science fiction. This goal was realised in 2007 by scientists working on the Human Brain Project (formerly called the Blue Brain Project), at the École Polytechnique Fédérale de Lausanne, Switzerland.[2] The *hypercolumn* mentioned in the passage above simulates a full hypercolumn (also known as a *cortical column*) of a rat brain. This successful stage in their work was important because it was the first time anyone had succeeded in doing this: building a fully-functional, cell-resolution, *virtual* version of a biological hypercolumn. This stage was a powerful demonstration of proof-of-concept, along the path to achieving the same goal for a human hypercolumn.

The project is progressing well, and the team plan to complete a full rat-brain simulation soon, before pressing on towards a full *human* brain simulation by around 2023.[3] I use the word 'simulation' with caution, and only because the Blue Brain scientists seem content to use it, for the eventual goal here is to produce a *virtual* version that operates just like a physical one.

The science behind the project is complex, but the method and principles are easier to understand. The scientists hypothesised, then made predictions (based upon their own knowledge and upon that gleaned by other scientists from decades of detailed study of brains) about what *functional unit* they thought might be required to achieve successful simulation. Based upon knowledge and experiment, they decided that accurate neurons were the key to accurate modelling; they used observations of the behaviour of biological neurons to model virtual neurons that behaved in a similar way. They were then in a position to test their predictions: by building a virtual hypercolumn – an interconnected 'stack' of working virtual neurons – to see if it behaved, as a whole module, in the way that a physical hypercolumn behaves.

I have already mentioned the term 'functional unit' several times in this book, and I understand that it may seem like I keep changing my mind about what it should be: Is it atoms? Molecules? Neurons? Hypercolumns? The answer is that it depends upon the context. If one can build an accurate and fully functional virtual neuron without going down to the molecular or atomic level, then that is technically easier and will suffice for the task in hand. The Blue Brain scientists found that they could produce a brain-cell simulation at *cellular* resolution. It is important to note, however, that their cellular-level simulation cannot tell them what is going on at the molecular level. And because molecular-level hypercolumn simulations would be extremely useful for purposes such as studying *gene expression*, they are also working on simulating that resolution.

The virtual neuron employs *algorithms* to mimic the behaviour of a physical neuron. An algorithm is a set of computational rules or procedures for solving a mathematical problem in a number of steps. So, in this context, we can imagine the *input* from some virtual neuron (or neurons) triggering the algorithms in another neuron to cause it to behave in a particular way, depending on a number of *variables* such as the number of incoming connections, signal strength, signal speed, and so on. In the standard view of the *physical* neuron, the resultant output is 'all-or-none': Either it fires or it doesn't, so, in effect, its output is *binary* (on/off, one/zero).[4] There is, however, debate on this point: While it might be accurate to say that the *main* output could be represented as binary, other neuronal behaviour, such as the 'accumulation' and 'grading' of past signals,[5] appears to point to non-binary, *analogue* behaviour. Faithful *virtual* representations of neurons would display these same types of behaviour. (It would also be possible, however, to code virtual 'neurons' – given names such as *perceptrons*[6] in the field of AI – that behave in different, or even more 'sophisticated', ways than physical ones. The *output* of such a triggered perceptron could carry a whole new *set* of variables, or even whole subprograms, produced from some combined effect of the various inputs and the products of its own algorithm.)

As well as raising the ire of other neuroscientists over the issue of the huge amount of funding it has received,[7] the Human Brain Project's leader, Professor Henry Markram, has been criticised for focussing too much upon what they see as the tiny details of neuron structure – such as *ion channels* – and not enough on the wider issues of neuron interconnectivity.[8] And it is true that even if their virtual neurons act as faithful simulations in isolation, the Blue Brain scientists cannot be sure whether they have endowed them with the ability to connect up in the right ways. We don't yet know the rules by which a neuron chooses one synaptic connection over another. Maybe, for example, the randomness inherent in their *Peters' rule*[9] wiring scheme is too random; perhaps their synaptic *weighting* and *reweighting*[10] algorithms are flawed.

Watching the visual representation of the hypercolumn 'running' on the Blue Gene allows these scientists to get some idea of how the neurons are interacting. I say 'get some idea' rather than 'see', because the patterns produced are referencing *themselves*; therefore, they can – like a sound or video feedback loop – produce strange, unpredictable, and unreadable effects.[11] So the 'visual representation' is just that. But would it be too much of a leap to say that they are, at some level, watching the hypercolumn *think?*

What would it look like if neuroscientists were able to tap into your brain in a similar way? What would they see on their observation screens? Tapping into general output from your brain is actually quite easy. Even simple toys, such as Mattel's *MindFlex*, can include small, cheap *electroencephalogram* (EEG) sensors that read neural signals emanating from your scalp.[12] The signals received by such devices are faint and distorted by the barrier of the skull, but they are, nonetheless, genuine neural *electrical potentials*. Medical EEG devices can provide clearer readings of these signals. They come in a variety of forms, but usually involve an array of small sensors forming a net that is placed over the scalp. Each sensor is pressed to the scalp, coated with a gel that aids signal conduction.

The output from EEG devices is often represented as a wave pattern. This shows spikes in neural activity as peaks on the graph and readings of low activity as troughs. This kind of representation seems easy for us to follow because we have an intuitive understanding that sometimes our thoughts are 'more active', and at other times they are 'calmer'. But there are many ways of representing brain activity. For example, I have an EEG-type headset that shows a picture of a brain with each main area lighting up in a different bright colour depending on how 'active' it is at that moment. This headset is actually designed for gaming: It is possible, with patience and practice, to train oneself to move game elements around by increasing or decreasing particular ranges of frequencies.[13]

The electrical signals, created by the *synchronous activity* of large numbers of neurons, emanate from the outer part of the brain called the *cortex*. The majority of the cortex is known as *neocortex*; the 'neo' part of the word indicates that this is a 'new' part of the brain – it is, in fact, the newest part in evolutionary terms. It sits on top of evolutionarily older parts: the *midbrain*; and underneath that – attached to and merging into the spinal cord – the *brain stem*. This description is a huge generalisation, but it will give you some idea of the 'layer cake' anatomy of the brain. Most of the discussion in this chapter focuses on the neocortex rather than on other parts of the brain, because it is where most of your conscious *thinking* takes place. Brain structures are, by their nature, highly interconnected, but our most complex, conscious, 'higher-order' thoughts can be considered to be processed mainly in the neocortex.[14]

Only mammals have neocortices, although the outer part of bird and reptile brains is sometimes referred to as cortex. The human neocortex is larger in proportion to body size and has more neurons than those of other mammals. Stretched out flat it would be about the size of a large dinner napkin (the fabric ones that you get in posh restaurants).[15] It is around two to four millimetres thick, and the neurons within it are arranged in vertical 'columns' of microcircuits (the 'hypercolumns' I mentioned earlier). Your neocortex is scrumpled up inside your skull so that all of that area of neuron-rich cortical tissue can actually fit. If you look up a brain anatomy book and find images of small mammalian brains – such as rat, rabbit, or cat – you will see that the proportion of the whole brain devoted to neocortex is smaller, and that it is smoother than ours is. The sophistication of those brains is also limited by their neuron size: Being larger than our own neurons, scaling up their number to match our own would result in structurally and energetically unsustainable brain sizes. It is widely known that certain mammals, such as dolphins, have large brains with large areas of neocortex. Again, however, the neurons are larger and fewer in number than our own (also, their brains do not have as many *cortical layers* – humans have six whereas dolphins have five). Only in other primate brains do we see neurons similar in size to ours. Although around three times as large as a chimpanzee's, the human brain is still clearly recognisable as that of a primate.[16]

Getting clear readings from the cortex requires direct interface with it, and that involves getting through the skull. Sensors can then be placed directly onto the cortex, or better still, implanted directly into the relevant *area* at the relevant *depth* of cortical layer. It is even possible, with this kind of technique, to get readings from the firing of *individual* neurons. Neuroscientists involved in such work can then find out which stimuli cause particular neurons, or groups of neurons, to fire. The process can also work 'in reverse': A neurosurgeon can apply an electrical signal to the neurons in order to *make* them fire, and then observe what effect this stimulus has on the patient.

These techniques may sound like inhumane tinkering, but they are, in fact, fundamental to our understanding of the brain and to the treatment of brain disorders. An effective treatment for tremors caused by Parkinson's disease is a form of *deep brain stimulation* (DBS) that involves implanting an electrode into one of three specific sites (*thalamus, globus pallidus, subthalamic nucleus*) and applying an electrical pulse to it from a signal generator implanted in the chest. The pulses help to regulate the brain area responsible, thereby decreasing the frequency and severity of tremors.[17]

Recent advances in brain scanning techniques, using *functional magnetic resonance imaging* (fMRI), are now allowing scientists to 'eavesdrop' on *thought*. In 2011, a team at UC Berkeley announced that they had succeeded in decoding neural signals generated by moving images. The scientists showed each subject a set of movie clips, then used the reconstructed signal – recorded from the subject's brain during the experiment – to decide which clip they had been watching at a particular time.[18] This is not, of course, the same as watching general output from a subject's brain and being able to tell what he or she is thinking about at that moment, but it may be a first step towards that situation. Bear in mind also that these experiments were performed from outside the skull with non-invasive fMRI. *Brain-computer interfaces* (BCIs), which *are* invasive and so interface *directly* with the grey matter of the cortex, will greatly widen the possibilities in this field of study.

If you find the idea of BCIs grotesque or frightening, it's worth remembering that you are already using one or many of them. Bodies interface with brains in myriad ways, and of course, brains interface with *themselves* at many levels. For example, the automatic novel-stimulus-induced *orienting response* we share with other animals is almost like a simple big-button BCI marked 'press here for alarm'.[19] Some evolutionary biologists argue that the brain has evolved in the way it has primarily because of evolutionary pressure towards ever-more-sophisticated *movement*: It allows us to move within our environment and to move our environment around to better suit our ends.[20] The evolved flexibility of the human brain means that it can quickly adapt to interface with its environment even when presented with novel methods of input. With the 'BrainPort', developed by neuroscientist Dr Paul Bach-y-Rita and colleagues, the input is received from a video camera that transmits electrical pulses directly to the tongue. The resulting patterns generated allow a blind person to 'see' and interact with her environment.[21] And do not assume that this means that she learns to *translate* the input explicitly, something like: *X pattern of pulses on my tongue equals Y object in front of me*; it no more means that than does *X wavelength of light hitting my eyes equals Y colour in front of me*, for a sighted person. The brain takes the available input and, after a period of adjustment, begins to use it to interact with its environment – no explicit translation required. Brains do this.

As we are now reaching a stage in neuroscience where it is possible to build virtual hypercolumns, eavesdrop on thought, and interface computers directly with the brain, it becomes increasingly difficult for non-reductionists to claim that there is something *extra* – some kind of 'secret sauce' – at work in the brain when we think. That is an argument for them to have among themselves; for me, it's a redundant discussion. I would guess, though, that a non-reductionist might say something like, 'OK, I accept that we can do *simulations* and get some idea, from brain output, of what a person is looking at. But that doesn't tell us anything about their *soul*, because that's something you can never see.' This is a deeply entrenched view, evidently one that many humans find deeply profound, elegant, and beautiful. To me, however, the elegance and beauty is to be found in the brain's generation of processes so intricate and self-fortifyingly self-referential that they can make a biological entity *feel* that its consciousness is ineffable and eternal.

In Chapter 6, I talked a little about the idea of selves as patterns, and gave the hypothetical example of a patient's *self-pattern* transferred to a device I called a 'soulbox'. I would now like to pull the *idea* of self-patterns together with the actual science (neuro and computational) I have discussed. If you feel somewhat lost at this point it is either because I have not made myself clear or because you sense yourself being dragged further down the rabbit hole. Don't worry, Alice, I'll be with you all the way.

*Patterns* are common to both the speculation and the facts. It would be reasonable, at least in one sense, for a non-reductionist to call himself a *patternist*. I, as a patternist, could state that I did not believe in any *non-physical* element in the phenomenon of self, but I would also be happy to say that the physical elements of self create patterns so complex that it's easy to see why others *would* believe that there was some *extra* non-physical element to it. The odd, looping, self-referential processes that generate selves *are* highly complex, but at the same time, they emerge from *simpler* patterns that *become* progressively more complex. They are self-creating and self-sustaining; to use a term originally applied to 'living machines' such as cells, they are *autopoietic*.[22]

You are looking at an apple tree covered in white blossom. The spring air is fresh and cold. A gust of wind whips some of the blossom from the tree and carries it away. Your eyes follow the floating blossom for a while. This is an example of an *actual experience* you may have had: It is a thing that *actually happened* at some point in your past. What happens when you try to bring it to mind? Can you measure the size of the tree? Can you say how many bits of blossom blew away on the wind? The chances are that your memory of the experience will not be accurate enough to give you this information. There may also be a lot of other information missing, about the context of the experience. It is likely, however, that you will have developed *associations* with the experience: other things that it reminds you of, feelings connected with it, and so on.

It may now be difficult for you to remember the exact details and context, but the memory always brings to mind a certain set of images, sounds, scents, and *feelings*. You would probably admit that the set of feelings you have about the experience *now* is different from the feelings you had *at the time*. You may also admit that the experience *looks, sounds, and smells* somewhat different in your memory to the way it actually was at the time. We accept this as normal: Feelings, associations with, and exact details of memories change over time as our minds mull them over and 'find a place' for them.

If you think of the *event* as the original *input*, then you can say that the input generated a pattern in your mind. Gradually, over the years, your mind has adjusted the pattern by adding *new* thoughts, associations, and feelings to the original input. But it gets more complicated: The modified pattern is constantly *re-input*. The specific results of such a looping feedback process are highly unpredictable. It would be impossible for any 'external observer' to guess what the pattern would look like in the future, after even a few cycles of this process.

Note that this patternist view of feedback processes in the brain does not attempt to argue that these processes are simple. Reductionists are often accused of *over*-simplifying highly complex effects, but if we extend our definition to include the patternist view, we find that the reductionist respects the fact that the *outcomes* of initially simpler processes may *become* highly complex over time. Particular levels of complexity will generate particular outcomes in terms of transient thought or even long-term plans, but we must avoid the trap of assuming that once all this complexity has bubbled up 'into consciousness' it has, in some way, come together there. The 'script' generated by such processes is never 'published' anywhere in a functioning brain; it is better to think of the versions passing through as what Dennett calls 'Multiple Drafts'.[23] Where are *you*? Certainly not complete on the bookshelf, but perhaps in the flow of drafts.

You may still be wondering, even if you have accepted the general premise, what these *mental* events, such as feelings and memories, have to do with your *physical* brain. What I am trying to demonstrate is that mental events *are* physical. In order to take that point further, I will need to talk about the physical *structures* that make 'mental' events possible.

## THE FOREST OF THE MIND

Have you ever wondered what it looks like in there, down in the deep structure of the neocortex where neuron-firings generate thought? The subject fascinates me. I realise that not everyone is interested in such topics, but now that I have learned a little about it, I find it hard to understand why some people don't seem to care. It may just be the case that it's enough for them to be alive and conscious, but I suspect that, for many, it has to do with the notion that the subject would be too complicated for them to understand.

And neuroscience *is* highly complex and specialised. There are medical specialisations such as *neurosurgery* and *neuropathology*, and there are experimental ones like *computational neuroscience, neurochemistry,* and *molecular neurobiology*. Doctors and scientists worldwide strive constantly for better understanding of the brain, and for better techniques to fix it when things go wrong. They study and treat disorders of the nervous system such as Alzheimer's disease, epilepsy, multiple sclerosis, schizophrenia, stroke, and depression. And by *reducing* down complex – even chaotic-looking – neural effects, some of them attempt to explain the fundamental molecular interactions that make us what we are.

Non-reductionist beliefs would be of little use to a molecular neurobiologist. (I am not saying, however, that no specialists in this field hold such beliefs.) Experimental neuroscientists need evidence of *physical* processes in action, in order to prove that their theories about the function of the brain are correct. Over the decades, their studies have taken them deeper and deeper into the microscopic structures of *neurons, axons, dendrites,* and *synapses*. They have discovered amazing things: the fact that the human brain contains 86 billion neurons (around 16 billion of those in the *cerebral cortex*);[24] the electrical *action potentials* (nerve impulses) of brain cells; the release of chemical neurotransmitters at synapses (around a hundred trillion of them), where one neuron communicates with another. This electrochemical communication is taking place in your brain right now, as you read this book. And if you disagree with that statement it is because some group/sequence of neurons has fired, or been *inhibited* from firing,[25] to make you feel that way.

This fact should give you pause for thought. Your thoughts are *caused by* these neuron firings and *are* these firings. Your *brain* would still physically exist without them, but your *mind* would not. Under circumstances of a temporary loss of them, we would normally say that you were unconscious; with a permanent cessation of them, that you were dead.

When I first learned about action potentials and the electrochemical activity in the brain, I quickly developed mental analogies with the way computers work. I pictured a flow of electrical activity generating a mind from the interconnected responses in the 'wiring' of the 'circuits' of the brain. Those analogies were satisfying to me on one level, but on others, I found them crude and frustrating. A large part of the problem was due to my poor understanding of biology and chemistry: I could not imagine how the comparatively slow-acting 'wetware' of the neocortex produced such rapid and complex effects. I found over time that, for those who are used to thinking in that way, computer-related analogies *can* be useful. However, I had a feeling that meaningful insight would also take different, perhaps simpler, analogies.

*Dendrites* are the branching parts of neurons that receive incoming nerve impulses. The term comes from the Greek for 'tree'; if you have seen a picture of a dendrite, you'll see how this term came about. An *axon*, the main filament sending output from a neuron, carries signals to its *axon terminal*, where the neuron's electrical signals are transmitted to the dendrite of another neuron at the *synapse* (actually a tiny, specialised gap rather than a direct connection). Projections from neuron cell bodies (axons and dendrites) are known collectively as *neurites*. The molecular-level details of how signals transfer from axon to dendrite are complicated: The transfer involves the release of *neurotransmitter* chemicals, stored in *synaptic vesicles*, that bind to specific *receptor proteins* prior to conversion back into electrical signals to continue along the receiving (*post-synaptic*) dendrite. In common with many types of manufactured wiring, axons are insulated, and for similar reasons: A sheath of a fatty substance called *myelin* coats most axons, minimising electrical signal degradation and 'crosstalk', which would otherwise greatly limit their signal-carrying speed and efficiency.

Imagine the network of dendrites projecting out from the neuron, awaiting signals from incoming axons. If we zoom in on the dendrite (we would need an electron microscope to do this for real) we see tiny *dendritic spines* projecting out from the surface of the dendritic 'branch'. These spines are essential to the proper formation of neuronal connections because they act as 'targets' for synaptic input; without them (or with fewer of them) it becomes difficult or impossible to form synaptic connections. Fewer synapses means less communication between neurons. Retardation of dendritic-spine growth is evident in conditions such as

Down syndrome and foetal alcohol syndrome. There is also evidence of retarded dendritic-spine growth in children brought up in environments of impoverished mental stimuli (neglected children who have lacked the conditions of environments thought to be conducive to healthy mental development).[26] In contrast, we see an *increase* in the number and health of these spines in specific areas of the brain when a person is exposed to new, then regularly-reinforced, stimulating experiences such as learning to play a musical instrument.[27]

I sometimes picture the mind as a forest. Having long found this kind of analogy useful, it was interesting to learn that Dr Sebastian Seung – a computational neuroscientist pursuing what I think is one of the most powerful theoretical routes towards understanding the mind – uses a similar one. Drawing on the writings of the great Spanish pathologist Santiago Ramón y Cajal (1852–1934), he calls it 'the jungle of the mind'.[28] While this is appropriate, and neatly encapsulates the sense of the vast tangle of complexity that we are discussing, I am trying to give you a way in that doesn't immediately require a machete; for that reason, I will stick with Cajal's 'forest'. (Bear in mind that the derivation of *dendrite* is only partially relevant here, and the analogy breaks down if we focus too much on that particular neuronal structure.)

At first, I tended to picture the type of spruce or pine forest one can see on the mountain slopes of the Highlands of Scotland, where I live. This gets across a sense of the macro-regularity we see before we zoom in. As we get closer, however, we find that any initial appearance of regularity was deceptive: The forest is not only vastly more complex and far less uniform than it appeared from a distance, but also contains other kinds of flora. Think of the forest, as it currently stands, as your brain in the structural configuration of neurons and inter-neuronal synaptic connections it is in right now. Like your brain, the forest will change over time – some trees will die off, while others will grow strongly and branch out. They will branch out in *somewhat* predictable ways – the 'options' available to a tree are limited by its biological 'programming' – but the precise configurations that roots and branches take on will vary greatly from tree to tree. Most of the trees will touch others, and the more they grow out in root and branch the more likely they are to touch.

The above analogy is a very broad one. Unlike tree branches, neuronal connections are not restricted to touching their near neighbours;[29] they can, in fact, stretch far across and deep into other brain areas. Nor do trees 'communicate' with each other in the ways that neurons do. It is *the forest* and its ever-changing configuration that is relevant here. This analogy is, in some ways, as much a moral-philosophical as a scientific one. We would usually consider a forest to be the *same one* through most (if not all) of its changes of configuration. There are cases, such as that where the forest had burned to the ground, where we might not consider the 'new' forest to be the same one as the previous one. Even in those cases, however, the forest may still keep its previous name. This, again, touches on the identity issue.

Nevertheless, dendrites and their spines are, at some level, relevant to the forest analogy. I am encouraging you to think of the brain as a highly-complex and ever-changing *structure* – one that becomes more complex the more you zoom in on the detail. In this context, it may be useful for you to think of dendritic spines as part of the fine detail of the structure, like the individual needles you would see if you were to zoom in on a forest of pine trees.

A more complete analogy would begin to make for a very complicated kind of forest (I'll admit that it is beginning to sound more like a rainforest,[30] or even a jungle): one incorporating groups of trees with partly-specialised functions that can change species depending on which other trees they touch; trees with branches and roots that can stretch out for miles in any direction; branches that can connect to the branches of other trees and communicate electrical signals to them; twigs with needles that change their quantity and complex configurations in response to incoming electrical signals, and, if they result in healthy growth, attract more incoming branches.

This is not as strange as it may sound. Plants may not behave like neurons, but their behaviour is actually far more complex than we sometimes assume. *Thigmotropism*, where a plant responds to touch signals, is but one example of what some botanists refer to as 'plant perception'. I was prompted to find out about this after talking to Terra about the runner beans she often grows, and the way that they wrap themselves tightly around the sticks she erects to support them. Because neurons are found in brains, we sometimes tend to think of them as 'smart'. If they are indeed smart, then so are Terra's beans: They, too, respond to their environment; they use the supporting structures available to them; they 'strive' to achieve their 'goals' through the use of advanced chemical 'machinery'.

We don't tend to attribute the ways in which individual plants, or entire forests, grow, to 'motivations'. You can think of plant growth behaving according to a set of rules: absorb water and nutrients, extend towards light, etc. The various *tropisms* (like the example given above) that plants exhibit are controlled by chains of hormonal molecular reactions. At the genetic level, these sorts of rules are *encoded* into the DNA of the seed from which the plant sprang. They are the plant's algorithm – the instruction set that, when implemented, allows it to persist and reproduce in its environment.

Nor do we attribute the ongoing 'behaviour' of forests to 'external' forces other than the obvious ones: sunlight, rainfall, animal grazing, perhaps some forest management, and so on. Soil conditions may be rich or poor; disease may spread throughout the forest from time to time; winter storms may knock down groups of trees, which will become, in turn, new fodder for insects and fungi; clearings may form. Once we start to think in this way, it can become hard to decide what is internal and what is external to the 'forest system'. Sometimes conditions are conducive to growth, and sometimes they are not. We accept that much randomness is at play in such a system.

Interestingly, though, many people still wish to reject an algorithmic definition of the way *minds* work. We, as persons, *feel* that we have complex, conscious motivations of our own[31] (although we also seem to be guided, buffeted, or at least influenced by the 'external' forces exerted on us by other people, and by the systems of which we and they are part). Because of this, it may not feel natural for us to ascribe the workings of our own brains to what seems like a *basic* set of forest-system-like rules. But I have not suggested, at any point, that we – as selves – behave in basic, robotic, or pre-determined ways. I have only suggested that neurons and their neurites behave according to a set of rules that determines how they will react in response to stimuli. Trees 'want' to extend their canopies towards sunlight and their roots towards water; neurons that 'fire together' 'want' to reach out and 'wire together' more strongly with each other.[32] (And, conversely, connections will be 'pruned back' when relevant stimuli weaken or become absent for prolonged periods.) There is a streamlining process at work here, a natural selection of sorts: a process that some have termed *neural Darwinism.*[33]

I am saying that we have a *comparatively* simple starting point (the neuron), with its comparatively simple 'wants', that in massively interconnected combination with a vast number of similar but not identical *nodes*, generates a highly complex and *very different* phenomenological outcome – a self.

## EMERGING FROM THE TREES

If we were to look *only* at the operation of individual neurons, then no matter how detailed a description we formulated, we would be failing to see the forest for the trees. The exact mode of operation of individual neurons is interesting and important, and I will talk about it again later, but for now, I am discussing outcomes and not details. If you acknowledge the premise – now a widely accepted one – that selves can emerge from this kind of 'ramping up' of comparatively much simpler processes, then you will be accepting that 'self' is an *emergent property.*

*Emergence* is a much-contested field of study. The term is used in a variety of different ways in subjects ranging from philosophy to computer science. I'll use Jeffrey Goldstein's definition here because it's short and rather easier to understand than some of the more long-winded definitions: (Emergence is) 'the arising of novel and coherent structures, patterns and properties during the process of self-organization in complex systems'.[34] Goldstein is neither a neuroscientist nor a philosopher, but an economist. And the study of emergence is highly important in this field: Economists seek to build accurate 'models' of the outcomes of vast numbers of comparatively small financial transactions, although they often find that their models turn out to be poor predictors of actual results.

Physicist Murray Gell-Mann's definition of emergence is even more concise: 'You don't need something more to get something more.'[35] If you have seen him speaking on the subject you may note frustration in his demeanour as he tries to get this point across. Emergence from simplicity, evolving toward ever-greater complexity, is a deep fact of certain systems. It is not possible to understand those systems without accepting the principle of emergence. Once understood, the presence of emergent properties in seemingly-complex systems becomes ever more apparent. Gell-Mann's frustration is triggered by people's tendency to believe that complex effects must always have complex causes.

In his book *Incomplete Nature: How Mind Emerged from Matter*, Professor Terrence W. Deacon takes an unconventional stance on the concept of emergence. Instead of concentrating on complexity emerging from simplicity, he focuses in on the *constraints*: getting something *less* than everything; therefore, something useful. 'Emergent properties are not something added,' he says, 'but rather a reflection of something restricted and hidden via ascent in scale due to constraints propagated from lower-level dynamical processes.'[36] Deacon goes on to re-frame entropy as a measure of constraint: 'An increase in entropy is a decrease in constraint, and vice versa.'[37]

Philip Ball, in his book *Critical Mass*, uses the example of *cellular automata* to illustrate the concept of emergence. The 'automata' in question are computer algorithms (called 'boids' in this case) programmed to behave according to a set of simple rules.[38] The boids are 'set free' to interact with each other, and the emerging results of their interactions are displayed on a screen in pictorial form. The results are seldom the kind of simple configurations that might be predicted by the casual observer. In fact, the resulting configurations are often *so* complex that they could not have been predicted by *anyone* – to the extent that it can appear that the boids have 'taken on a life of their own'.

Cellular automata with complex behaviours were first demonstrated in John Conway's *Game of Life*. His early experiments were based on an extremely simple starting point: a grid with groups of five dots, and four programmed rules. These rules equated approximately to evolutionary pressures and conditions of over-population, under-population, population balance, and reproduction. Conway found that when he ran the program over many *iterations* the groups sometimes behaved in complex ways, evolving a kind of diagonal movement. He called these groups 'gliders'. Other, much larger, complex groups also emerged, such as multi-dot 'spaceships', and 'gospers' that 'fired out' a continuous stream of gliders via a 'glider gun'.[39]

We can usefully employ the computing term *bootstrapping* – derived from the saying 'pulling oneself up by one's own bootstraps'[40] – when discussing the emergence of complexity from simplicity. You may have heard of *booting*, in the sense of a sequence of processes that *run* automatically when your computer *boots up*. Bootstrapping is, broadly speaking, the linking of booting-like (automatic) sequences into themselves, so that they can initiate – or *seed* – other sequences, and thus run in a self-sustaining *recursive* manner. Each successive iteration of this recursion can cause the program to grow in complexity and capability. The term is now used in the field of *machine learning*. Advocates of certain 'strong AI' hypotheses think that machine intelligence could emerge, from a correctly-programmed seed algorithm, via an exponentially-growing bootstrapping sequence.

Though the emergence of complexity from simplicity is a palpable fact of life, it is still a subject seen by many as bristling with teleology: What/who initiates such emergent processes? In the case of a computer-based seed algorithm, an operator is clearly involved in the origination of the resulting recursive process. Positing an originator of neural processes sounds decidedly homuncular; the proximate causes may become progressively better defined, but we still seem to be lacking an ultimate cause. While computer analogies to brain function are just that, we do need to bear in mind that *we* are (in computational terms) *already* 'booted up' and 'running' – we do not need an originator. I will, however, qualify that by admitting that, in some sense, we do come to find one: one that becomes essential to our ability to function within our surroundings – our so-called self.

It's time to bring some of these ideas together into a framework that says something about my view of the way in which selves emerge from the massively interconnected networks of neurons in our brains. The theory of *connectomes* (popularised by Seung,[41] but originally proposed as a term by Sporns and Hagmann[42]) is highly relevant here. We have all heard of the concept of *genomes*, particularly in relation to the Human Genome Project, but the word 'connectome' – and the related scientific studies, including the Human Connectome Project (HCP) – may be quite new to many. The principle, though, is similar. We often hear in the news that possessing a particular *gene* could increase our susceptibility to conditions such as colon cancer, heart disease, or diabetes. Other genes seem to carry glad tidings: greater physical stamina, longer lifespan, full head of hair. Genes are sequences of DNA that *encode* for particular behaviours in the cells in our bodies, and in the bodies of all other biological organisms. A *genome* is, then, the genetic 'instruction set' for building an entire organism. A connectome provides 'instructions' in the form of a connectional map of neuron wiring, but it 'encodes' for something rather different: It encodes for a specific, conscious self.

Imagine that a complete 3D scan of your cortex is taken. For the sake of clarity, we'll say that the scan is at a resolution high enough (around twenty *nanometres*) to show all the neurons, neurites, and synapses. Given that the only connectome fully mapped to date has been that of the tiny nematode worm *C. elegans*, we can see that the mapping of full human connectomes is still in the realm of scientific speculation. Also, bear in mind that detailed connectome mapping can currently be undertaken only on dead brains. A one-millimetre cube of adult human cortex – containing some fifty thousand neurons and up to half a billion synapses[43] – could take one million person-years to map manually and could consume one *petabyte* (over one million gigabytes) of data storage space.[44] But if we think about the way that automated connectome-mapping technologies might well develop *exponentially*, it becomes less difficult to imagine full-human-connectome mapping being achieved within decades rather than centuries.

Scientists might do this by having many teams each mapping a one-millimetre cube of cortex, working out the interconnections within each cube, then joining their mapped cubes together with those of the other teams.

If we scaled up such a (theoretical) full 3D map and turned it into a hologram, we could walk around in it and look at individual neurons and neurites. If we scaled up by a factor of a thousand – the point where the *cell body* of an average neuron would be two centimetres across, and so possible for us to examine without a microscope – then our flattened neocortical map would be some four hundred metres long, four hundred metres wide, and two metres high. Even at this scale, we would struggle to make any sense of connections between neurons because of the vast tangle of axons and dendrites. There are usually *thousands* of incoming connections per neuron. And the axons would still be only around a millimetre thick (about the thickness of 'angel hair' pasta). This vast structure would be a scaled-up map of a unique connectome. No two connectomes are the same.

You can see from this type of illustration how difficult it is to make detailed maps of such a complicated structure. And I haven't even included the smaller-scale components such as dendritic spines, or the nanometre-scale *microtubules* and *neurofilaments* projecting along the axon from inside the cell body of the neuron, let alone the *molecular* components. Given this enormous complexity, we should agree that making a connectomic map of even a one-millimetre cube of cortex would be a significant achievement.

But what if full-connectome scans of living human brains *were* readily available? What would they actually tell us? They would tell us our 'brain state', showing the specific configurations of neurons and their interconnections at the moment of the scan: a 'neural snapshot', if you like. Any software designed to extract *meaning* from this tangled mass would have to be highly sophisticated. It would need to be able to search rapidly through a vast database of other scans to establish what *functions* particular groups of interconnected neurons perform. It would then have to collate all this information, categorise it, correlate it with the various brain areas and *their* functions, and create a *probabilistic map*, based upon which it could give some kind of meaningful readout to the operator.

The readout might be able to tell the operator the precise location of some neuron firing-sequence or grouping that indicates a particular type of recurring depressive thought with which the patient has been struggling. The operator could then access information from the software database about how to modify the grouping in order to alleviate the patient's problem. Various forms of intervention might be deemed appropriate: meditation, highly-targeted cognitive therapy, *transcranial direct-current stimulation* (tDCS), *neurofeedback* training, targeted medication, or even neurosurgery.

The possible applications for treatment of neural disorders and for greater understanding of the brain are almost limitless. But I wish to focus, for the time being, on what such scanning techniques will reveal to us about *selfhood*.

The kind of self that emerges as a result of its connectome depends on many factors. Nature and nurture become blurred to a point where telling one from the other rapidly becomes extremely difficult, if not impossible. Would you, for example, consider your physical mannerisms to be part of what makes you a unique person? Maybe it's a particular way you move your head when you laugh, or a particular gesture you make with your hands when you are having an argument. You may see where I am going with this: In many ways, you consider your mannerisms to be unique to you, *but* you sometimes see the same (or very similar) ones in your siblings, children, or parents. The main reason for this is not to do with learning such mannerisms from these other people; separated-at-birth *monozygotic* ('identical') twin studies have proved that these shared mannerisms can develop *without* learned input.[45] The main reason is that you have particular connectomic configurations in common. To put it more plainly, the parts of the *motor* areas of your brain that control these particular mannerisms are very similar in you and in these other family-members. Parts of your brain have been copied.

We can, then, deduce from this that it is possible to *inherit* neural configurations. The informational precursors to such configurations – described somewhere in the genome of one of our parents – became incorporated into our own genes and began to be *expressed* in our brain as it was constructed during gestation. A given 'mannerism gene sequence' may not have been expressed in either of our parents, but when a shared mannerism shows up in a sibling, it becomes clear that we have inherited it. Does this come as a surprise? We are used to the idea of inheriting 'physical' attributes such as height, eye colour, or skin tone; we now understand that we can inherit genes that predispose us to developing particular disorders; but we don't talk so much about inherited brain functions, or about the so-called *connectopathies* of neural 'miswiring'.[46]

While we should be prepared to accept the core message of psychologist Eric Turkheimer's proposition that 'All human behavioural traits are heritable',[47] we can take individualistic comfort from the many aspects of our connectome that are not inherited. We are born with much that is unique, and, throughout our lives, it develops in countless different ways – through experience, environmental influence, and self-reference – to find it's own distinctive, ever-evolving configuration. It is now known that the brain is highly *plastic*: It has great capacity for change. For decades, debate has raged about the extent to which *neuroplasticity*, as it is known, is lost in adulthood; it is not so very long ago that the majority of neuroscientists insisted that it was a feature mainly of children's brains. That attitude is now shifting. The fact that the brain can adapt, in adulthood, even to the extent of overcoming the limitations imposed by sometimes-massive damage – such as that seen in stroke and head-trauma – indicates that plasticity of certain important kinds is present throughout our lives.[48] But this capacity for self-repair is far from being the only implication of the discovery of neuroplasticity. It also implies that we can modify the 'wiring' of our own brains – by *thinking*. This happens all the time in any case, but I am not talking about the 'pedestrian' day-to-day rewiring that happens continuously as a result of the normal flow of new experiences. I am talking about major change.

Do not assume that such changes would involve only a statistically-insignificant number of neural pathways, leaving everything else unmodified. Because your connectome is physically constructed from such pathways, we are talking about changing the *structure* of the brain, effectively making you into a *different person*. Focussed thinking can bring about such structural change.

Let's take meditation, which I discussed earlier in the book, as a 'positive' example. One could interpret certain kinds of meditation as spending countless hours trying to think about *nothing*. It seems counter-intuitive to believe that thinking about nothing would affect the structure of a brain. But we now have scientific proof that it has significant effects on certain parts of the brain. Areas such as the *right temporo-parietal junction* (rTPJ), which is thought to play an important role in regulating our beliefs about the beliefs of others (the concept of 'people thinking about thinking people' known as *Theory of Mind*),[49] show an increase in cortical grey matter after as little as eight weeks of mindfulness meditation.[50] What would be the 'outward' difference in a person resulting from *just this one* change taken in isolation? We could guess, from current knowledge of the function of this region, that it might make them more empathetic. Mull that over for a while: Sitting quietly and thinking about nothing, on a regular basis, might change you into a calmer, more caring, more understanding person. This is practical self-experimentation with neuroplasticity. It's a kind of non-invasive neurosurgery that you can do on yourself, from the comfort of your own home.

Seung, in his book *Connectome: How the Brain's Wiring Makes Us Who We Are*, calls it right when he states, 'You are your connectome'. You *are* indeed your connectome, or whatever other name you wish to give to the unique structural configuration of your brain. And, crucially, that is *all* you are. But you 'necessarily' *feel* that you are more – more integral, more central; that you are, to paraphrase novelist John Barth, necessarily the hero of your own life story.[51]

How, then, can it be that you are 'just' a collection of neurons and their interconnections? Nobody, in stating this fact, is belittling your uniqueness, 'specialness', individuality, or complexity. Quite the contrary, I hope that I have given you some insight into just how special and complex a thing a connectome is.

If you have not yet begun to see even a glimmer of the answer to the question of how a self could arise from the neurocircuitry of the brain, then it is either because I have not laid out the facts plainly enough or because you do not wish to entertain such a realisation. If you *have* seen the glimmer, then you may also have begun to see its further implications. Connectomic maps, neural snapshots, selfhood emerging from the wiring of neurons: Might it be possible to *fix* such configurations in place, or to retain the *information* encoded within them in some other way? Seung, for one, has been visionary enough to make this logical connection, and even to ponder whether it might be testable.[52]

## THE SOUL OF THE GAPS

I accept the utility of connectome theory and of the idea of selfhood as emergent property. The concepts are not that difficult to understand – we readily accept that some things are 'more than the sum of their parts'. The 'things' in question here, namely selves, may be very special, even unique, emergent properties, but we have no reason to believe that there is anything *extra* in the mix: no Cartesian egos, no quantum souls, no 'spirits' of any kind.

But I also accept that there have been awkward gaps in these sorts of theories. Such gaps are often regarded, by non-reductionists, as open invitations to start bunging them up with soul-wadding. In the same way that the gods of theists have 'retreated' into the ever-smaller gaps[53] in our modern understanding of the universe, so non-reductionists posit souls (or other non-physical agents) as being the 'missing element' in our understanding of consciousness.

In this age of rapid neuroscientific progress, any such 'breathing space' for non-reductionists does not last long. However, even studies that are beginning to reveal strong clues to the physical method of computational function in neurons will likely fail to convince non-reductionists and others with an ingrained 'mysterian' agenda.

Any explanation of this function of neurons necessitates a return to computer analogies. But bear with me, as it is important that we get *some* idea of what happens inside neurons to make them behave in an algorithmic manner. *Computational* neuroscientists, as the name implies, have theorised from the outset that there must be some sort of *computation* going on within the cell bodies of neurons. The technical challenges involved in proving any of these theories, however, remain manifold.

At the fundamental level, computers function in the binary realm of zeros and ones (on and off). One *bit* (a contraction of the words *binary* and *digit*) equates to one *unit* of information. In the field of computing, then, the term 'bit' is used to mean a unit of information that can be in only one of two possible states: on or off, one or zero, etc. Your computer may output all manner of visual images, sounds, words, and complex calculations on its screen; down in its circuitry, however, it is doing nothing more (or less) than 'crunching' huge strings of bits.

In using a computer it is usually our goal to get some kind of logical output from it rather than, say, an endless stream of pseudo-random binary (although this has its uses). A computer's processor(s), often referred to as the *central processing unit* (CPU), is etched with microscopic circuitry that includes components such as transistors and capacitors. These components are put together in such a way as to form what are known as *logic circuits*. The CPU handles the stream of programmed input as a set of instructions, performing *logical operations* on it via *logic gates*. These 'gates' – AND, OR, NOT, COPY – do simple things to the bits in the binary stream; for example, a NOT gate flips a bit, transforming a 0 into a 1 and a 1 into a 0.[54] The logical language of the operations is not the revolutionary element here – we can credit that part of it to the much earlier work of mathematician George Boole (1815–1864). But the application of this logic to theoretical concepts – ones that were later realised as physical electronic circuits – *did* spark a revolution: the computer revolution that we still see unfolding all around us.

We can more easily harness the computational power of logic gates by inputting the programmed instructions in the type of 'higher-level' language we can better understand. In the BASIC programming language, for example, we could program instructions in a form similar to this: IF (DATA $x$) is equal to (DATA $y$) THEN (perform some logical operation) ELSE (perform some other logical operation). This is not how the computer 'hears' it – the computer doesn't 'hear' anything – but the effect is the same. The end result of a series of logical operations is that you get some kind of output from the processor(s) that appears as calculated results, sounds, or images on your screen. This all happens very quickly but uses a lot of energy, hence the processor heats up, and the computer requires regular cooling.

Computational neuroscientists have long hypothesised that neurons, too, must somehow be able to perform logical operations, and so they began to search for the parts that might act as logic gates and, by extension, memory storage. In doing this they were looking, in effect, for what memory researcher Richard Semon (1859–1918) had termed the 'engram': a biochemical trace mechanism – dependent on certain 'energetic conditions' – for storing and retrieving memories 'hard-coded' within the physical structure of the brain.[55]

Any appearance of similarity between this sort of approach and the 'quantum consciousness' hypothesis of Penrose and Hameroff is misleading. There is a great difference between trying to find a way to view the brain as being capable of some form of *classical* computation and giving up on reasonable reductionist explanations – resorting to greedy reductionism and/or souls of the quantum gaps – in claiming that consciousness must rely on some undiscovered mechanism for *quantum mechanical* computation in the brain.

I mentioned earlier some of the *nanoscale* components of neurons: microtubules and neurofilaments. Some recently-announced results claim that the computation (and presumably the memory-storage) is happening at *this* level, and at the level of the *molecules* coating these tubules and filaments. This discovery would represent a major breakthrough, because the infinitesimal scale of these 'components' could explain one of the great mysteries in neuroscience: the incredible computational and information-storage capacity of the human brain. As a *ScienceDaily* news story about the breakthrough put it: (The scientists involved) 'calculate enormous information capacity at low energy cost'.[56] The 'low energy cost' is important, because only around forty watts (about the power of four energy-saving light bulbs) are available for the operation of the whole brain. With this in mind, imagine how much energy would be required to power a 'traditional' computer possessing the computational abilities of the human brain.

There are problems with this 'engrammatic' approach, and some in the fields of computational neuroscience and AI research have criticised it. The critics do not say that such research is not worthwhile, but they do claim that it misses the bigger picture of the way that memories are stored and processed in the brain. I find the arguments of the engram-approach critics convincing. The fine-scale components are, of course, vital to the function of the whole, but we should not assume that neurons perform discrete memory functions, either individually or in simple hard-wired combination with other neurons. To quote AI-researcher Steve Grand on this subject:

> Stimulating precisely one neuron can make us think of grandma, while stimulating a neighbour makes us think of apples. But it would be rash to conclude from this that we've found the grandmother neuron...[57]

Stating that the processes of thought and memory are more subtle and 'smeared-out' than the fine-structure research might suggest is not the same as saying that there is *something more* than computation going on. On the contrary, it underlines the importance of the rapid processing of information, and at a far greater scale and level of subtle interaction than can be demonstrated from the output of individual neurons or small neuron-groupings. The *information* is still key: Computation (of some kind) is happening, and the resulting highly-convolved *patterns* of information are generating a macro-pattern that we can call a self.

We seem to have started reducing again: Now selves are made of bits. It seems, at least at first glance, *even more* counter-intuitive to state, 'I am made of bits of information' than to say, 'I am my connectome'. Again, the context is vital. Each of those statements is true *at the level of abstraction to which it applies*. A 'modular' *localisationist* approach – classifying the various 'regions' (e.g. Brodmann area 17)[58] and 'neuron types' (e.g. pyramidal neuron)[59] – can help us to demystify the brain, but it can also increase confusion on the subject of selves. Neuroscientists study the brain at so *many* levels: molecular interactions, dendritic spines, synapses, the spiking action potentials of neurons, the 'circuits' formed by groups of neurons, cortical layers, brain areas that seem to have semi-specialised functions, connectomes, the emergent processes of thought that supervene upon all these lower levels (though not in any transparent way), and so on. It's understandable if we are left feeling that we have zoomed in, then all the way back out, and yet we did not find any *self* along the way.

The concept of *virtual machines* may help here. Your computer is designed to be a general-purpose computation device capable of *emulating* other types of machine. When you are using a word processing program, your computer does not physically morph its internal components to become a typewriter; it uses algorithms to emulate the functions of a typewriter, so that a load of binary streaming through the processor can be transformed into patterns of pixels forming words on your screen. It might, then, be useful to say that the mind is a type of virtual machine and that the brain is the (admittedly rather soft) hardware from which the virtual mind arises. It could be claimed that our sense of conscious selfhood is a key function of this virtual mind, or it could equally be claimed that it's an incidental (even accidental) *artefact* born of the operation of such algorithms.

Do you ever worry, when you press the power button on your computer or handheld device, about not being able to describe the internal state of the machine? Probably not. You would usually just say that it was 'on' or 'off' (unless it was broken, when you might instead say that it was 'dead'). But let's use a word to describe that magnificent state of fully-functional 'on-ness' where the computer is doing just what it is supposed to do – you pushed the button and it bootstrapped itself into this state; I'll use the word *egeirient* here.[60] We could then say, 'my computer is currently in a state of egeirience'. We might be tempted to say, 'my computer is full of egeirience today', but that wouldn't be quite right, because if we put it in those particular terms, it might suggest, to somebody who didn't know any better, that there was some kind of 'stuff' called egeirience inside the computer.

This is not a cheap shot at non-reductionists. I am not suggesting that anyone would believe that present-day computers are 'alive' or conscious in the way that persons are. If we got the semantics just right, however, we could justifiably use a different word – *sapient* – to describe that state where a *person* is alive, fully-functional, able to think in abstract terms, conscious, and therefore, 'imbued with selfhood'.[61] Their internal virtual machine is running. We *could not* say that a person constituted a self 'because of the sapience *in* her'. *Sapience* describes a *state*, not a 'thing'. Scientists don't have a problem with the idea of devising words for particular *states*, but they are careful to use those terms correctly. In a similar way, if you wished to say something (admittedly clumsy) like, 'My wife is currently in a state of soulfulness', it would be clear that you were describing her state and not some stuff *inside* her. But the term 'soul' is, by design, ontologically loaded. Rather than using it in this precision way that would allow it to be interpreted as a kind of state, most choose to use it in a purposely impenetrable way.

Cryonics pioneer Robert Ettinger was clearly frustrated by this weighting towards supernaturality and obfuscation:

> Such portentous terms as "soul" and "individuality" may represent nothing more than clumsy attempts to abstract from, or even inject into, a system certain "qualities" which have only a limited relation to physical reality.[62]

## ARTIFICIAL SELVES

Attempts to create artificial intelligence have often been hobbled by outdated hypotheses of brain as mere input/output system; in criticising this approach, Steve Grand refers to it as the 'sausage machine' concept,[63] and Erich Harth calls it 'the slaughterhouse'[64]. Those ideas simply do not work. Harth writes of physiologists having misguidedly thought of certain fibre bundles in the brain that lead back to 'peripheral sensory structures', as 'wrong-way' fibres. How could they possibly be *wrong*-way if that is the way they are arranged? A 'top-down' view of signal-processing pathways in the brain (especially one illustrated with pictures of discrete, linked boxes) tends to produce this kind of conceptual error. But it is not difficult to see why various 'linked boxes' approaches have appealed to so many software engineers and roboticists.

Steve Grand tackles this issue providing, in pleasing and logical contrast, the analogy of a 'yin-yang' system[65] – one that constantly feeds signals back to itself in a looping, balancing flow. Input and output *are* taking place, but crucially, they are taking place within and between various areas of the brain, as well as between 'inside' and 'outside'. Hence, we should not expect always to be able to give clear-cut answers to questions of which signals are stimuli and which response. This is not plumbing; it is not a straightforward 'reality-fed' system.

Grand makes a case for seeing the brain as 'a general-purpose prediction machine'.[66] And, as different as this may sound from the operation of a thermostat (mentioned in Chapter 6), he finds common ground between the 'desires' of devices such as thermostats and model-aeroplane servomotors, and the internal/external self-regulatory behaviour of mammal brains. He talks of 'target states' and 'actual (sensory) states' of servomotors, which – viewed in 3D graphical form – make 'surfaces' of, respectively, 'want' and 'reality'. In any such active system, these surfaces will be in a constant state of flux, with the 'output' capabilities of the device always trying to adjust the sensory state to match the target state.

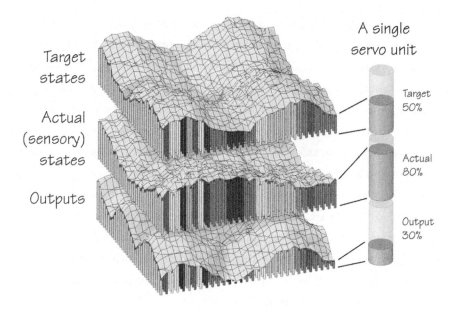

**Figure 2:** Visual representation of the 'total sensory state' of an array of 'servos', showing the Target, Actual, and Output activity surfaces. (Image: Grand, *Growing up with Lucy*, fig. 16.)

This kind of set-up creates a self-monitoring, self-adjusting loop: a 'deviation-minimizing pattern of behaviour'[67] known in engineering as a *feedback control loop*. (For an excellent example of this, it's worth watching the MIT videos demonstrating quadrotors flying entirely under computer control, and even returning to equilibrium after having been deliberately pushed off course by a student.) The study of such 'circular causal' systems is part of a field of study called *cybernetics*.[68]

We can understand, from such real-world examples of systems programmed in this self-adjusting way, how observers can come to feel that the system *wants* to achieve particular states. The quadrotor looks, to all intents and purposes, as if it has a *desire* to maintain steady hovering.

Right now, some part of my *target* state-surface is telling me that I want a cup of coffee, while my *sensory* state-surface is telling me that I do not yet have one in front of me. The net result is that my muscles require *output* to direct them to make me some coffee. The end result (having a cup of coffee) will achieve a better fit between my cup-of-coffee desire and my cup-of-coffee reality, but this effect will, I assure you, be only temporary; this, as I mentioned above, is a recursive, dynamic system.

In a similar vein, Eagleman draws on Minsky's earlier 'society of mind' analogy to create one of his own – a 'democracy of mind' – that highlights the importance of those push-pull 'tensions' at play in the processes of the brain:

> The missing factor in Minsky's theory was *competition* among experts who all believe they know the right way to solve the problem. Just like a good drama, the human brain runs on conflict... . Brains are like representative democracies.[69]

I realise that the prediction-machine nature of a brain is not immediately obvious from the above examples (you may argue that they make it sound rather like I am little more than a conflicted coffee robot). To get there we need to think about the potential cumulative effects of very many such dynamic tensions: We are discussing *entire* state-surfaces, or even myriad such surfaces exquisitely interlinked. We can hypothesise that such a dynamic system would become highly attuned to processing input (although the complexity of the possible states would mean that the same type of input would not always be processed in exactly the same way). This attuned 'need' to work upon some input could also mean that in the *absence* of direct input/feedback the system will go to work upon *itself*. It will loop itself into *anticipation* of the next input. It will start *nexting*.

Whether he is discussing Shakey or Betsy, Dennett always stresses the importance of feeback; for example, when he talks of 'feedback-guided, error-corrected, gain-adjusted, purposeful links'.[70] Betsy wants to find the thimble, Shakey wants to find the box. The fact that Betsy is a little girl whereas Shakey is a rudimentary robot should not distract us from the fundamental issue of what is involved in being able to make a discrimination between a 'thing' and a 'background', then being able to *do* something with that identified difference. Without feedback loops there can be no identifications, gain adjustments, or corrections; therefore, no steaming coffee, no autonomous flight, no found thimbles, no collected boxes; no nexting of any kind.

Even if you are not yet ready to accept the loopy-democratic nature of personhood, you would probably agree that we seek agency to change our reality to better match our wants. You should be prepared to accept, in that case, that the process also works the other way round: We shape our wants to match *our* reality. In his book *On Intelligence*, PalmPilot inventor Jeff Hawkins often highlights the ubiquity of the interplay between prediction and perception:

> Prediction is so pervasive that what we "perceive" – that is, how the world appears to us – does not come solely from our senses. What we perceive is a combination of what we sense and of our brains' memory-derived predictions.[71]

If you are familiar with any optical illusions, you will understand that our perception of reality is quite flexible. Our 'wants' can – with extension via prediction – become expectations. We are not even in 'conscious' control of such expectations. When we look at a Necker cube illusion, our brain 'wants' to see it with only *one* orientation, and because it is used to looking at cube-like things it *expects* that the reality will be that it *has* only one correct orientation. The fact that there is no 'correct' orientation (no completed 'script') confuses this predictive process; as a result, your perception flip-flops between two quite different versions.

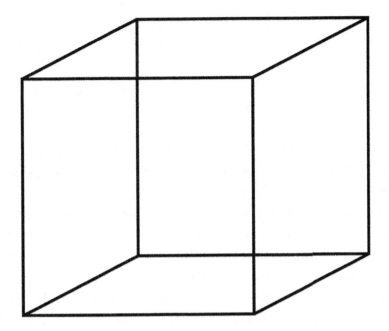

**Figure 3:** Necker cube optical illusion. (Image: BenFrantzDale, *Necker_cube.svg*.)

Not unlike a thermostat, many of my bodily functions are 'on autopilot', or, to use a more appropriate term, *homeostatic*. The concept now known as *homeostasis*, introduced by the French physiologist Claude Bernard (1813–1878) under the term *milieu intérieur*,[72] refers to the self-regulation of many of the body's systems.[73] Just as you really don't have to fiddle with that thermostat once set, I do not have to think consciously about keeping my heart beating, my body temperature usually takes care of itself, and (unless I am dealing with a particularly vexing computer problem) I do not usually forget to breathe. If my heart *did* suddenly stop, that thought would 'bubble-up' into conscious thought pretty rapidly, just before I passed out.

So conscious thought seems to be reserved for functions that *are not* on autopilot: trying to find the right words in writing a book, solving a maths problem, mulling over the events of the day, thinking about the past, worrying about the future. Although homeostasis is still playing a role at many levels, a host of ever-fluctuating *dynamic* processes is driving our sense of 'temporal transition'[74] from one moment to the next. *These* kinds of processes involve much *recursion*: Thoughts are feeding into themselves in what Hofstadter calls 'strange loops', like tangled Ouroboroses (the ancient symbol of a serpent that consumes itself, tail-first).[75] Here, then, at this 'higher' level of brain function, are you, with all your personal thoughts and concerns.

But if a *you* can arise in this way, might it be possible, through a better understanding of the processes involved, to create *other* entities that would constitute 'selves'?

I earlier admitted that it probably would not be correct to infer any kind of consciousness in the 'egeirient' state of functional 'on-ness' of the average present-day home computer. If I were to ask you, 'What does it feel like to be a computer?' you might justifiably be amused, annoyed, or nonplussed. You might well just say that the question was stupid or meaningless. So I will try a different tack: Does it *feel* like *something* to be a bat? This one is, perhaps, more comfortable; a bat is an 'alive' thing and it is also a mammal, so you might have some 'fellow feeling' about it. You should be able, as a result, to furnish me with some kind of meaningful response to *this* question.

As you may immediately find, however, and as you will certainly see if you read philosopher Thomas Nagel's thoughts on the matter, this is not a straightforward question.[76] We might conclude that both questions – the one about the computer *and* the one about the bat – display all the indicators of being *empty* ones. It might be fair to say that, on a scale of emptiness, the question about the computer looks *more* empty than (or even *completely* empty compared to) the one about the bat; but it would not be accurate to state that the one about the bat *must* have a straightforward positive answer and that the one about the computer *must not*.

Dennett argues that Nagel is incorrect in claiming that there is *no way* of finding out anything about what it would be like to be a bat. Dennett thinks that we could, based on our scientific knowledge of what we know about bats, establish a tentative set of elements that we might reasonably expect to feature in bat 'experience'.[77] There are certainly things that we could *rule out* as being part of bat experience, such as chatting about philosophy over a cold beer, and we could argue that that in itself would tell us something about the bat's *innenwelt* (inner world, as opposed to outer/surround world, *umwelt*, which I mentioned earlier).

In the spirit of Eagleman's story/thought-experiment about becoming a horse,[78] from his book *Sum: Tales from the Afterlives*, we could try to imagine what it would be like to turn into a bat. That's one to parse in your own mind, but, like Eagleman, we can tentatively think of it as a 'fading across' process. Yes, there would come a point in the process when *all* of your innenwelt would be *all* bat, but that would not happen suddenly: The networks that 'what it is like to be a human' once consisted in have faded across to those of 'what it is like to be a bat'. At all positions other than the extremes marked 'human' and 'bat' on the fader there would be some cognitive mixture of the two innenwelts. Now, the 'bat' extreme is not aware in the same way that the 'human' extreme was, but that is not because consciousness has 'gone away'. It makes no sense to insist that at *ninety-nine* percent bat one percent of consciousness remained, and then at *one hundred* percent it was all gone. On the contrary, when put in these terms we can begin to see that, in certain important ways, it was never really there in the first place.

We don't bother trying to rule anything in or out as part of a computer's innenwelt, because there doesn't seem to be any indication that it has one. We can't 'fade into bats' to see what it is like but we can, at least, observe bat *behaviour* – the bat doing 'bat-ish' things; we can examine how its sensory organs work and see what that implies about the umwelt accessible to bat senses; we can theorise about how that umwelt might be represented in some kind of bat innenwelt. Observation of a computer, on the other hand, appears to be just that: It does 'computer-ish' things that seem so far removed from 'human-ish' (or even 'bat-ish' or 'insect-ish') things that they don't constitute *behaviour*.

If we manage to drag ourselves from the philosophical quagmire of thought experiments such as the one about Nagel's bat, we might just decide that we would prefer to wait and make up our own minds about whether AIs can be conscious (and truly intelligent) or not. I suggested earlier that it might become possible to program *virtual* neurons for greater sophistication of function than physical ones. Doesn't this possibility, and facts about progress in computational neuroscience and AI research, point to a time in the *very near* future when humans will build not merely *human level* intelligences but ones so mind-bendingly clever that we will be rendered mere scurrying insects to our creations? This is one of the popular sci-fi nightmare-scenarios with which we are all (too) familiar. Leaving aside, at least for the moment, the question of how super-intelligent AIs would view the human race, we should wonder, perhaps, what is causing the hold-up in progress towards *at very least* human-level strong AI.

The 'big numbers' I mentioned earlier in this chapter, to do with the sheer *quantity* of neurons and synaptic connections, did not deal with another 'massive' aspect of the human brain – its *parallelism*. Despite Alan Turing's original theoretical concept of a computer architecture that could emulate the essential 'computational' and memory-storage functions of a human brain,[79] most resulting modern-day silicon-based processing units (such as the CPUs I mentioned earlier) have inherent 'bottlenecks' that severely curtail their ability to work in a truly 'brainy' way. The only (partial) solution to this – at least until very recently – was to throw more computing power at the problem: Supercomputers such as Blue Gene (and the now-famous quiz-winning *Watson*) are really just vast banks of *serially*-processing CPUs connected together to provide the brute-force power (speed) required for their tasks.

Brains don't pass information *serially* from one (or many) memory area(s) to one (or many) central processing area(s); instead, they do a vast amount of convoluted processing in multiple areas at once. And, crucially, they do this using only a tiny amount of energy measurable in watts and not in the tens or hundreds of *kilowatts* used by supercomputers.

Enter the *neuromorphic* chip. These processors have been designed, as the name suggests, to work more like brain. The memory and processing are handled on-chip, meaning that there is far less 'long-distance' relaying of information. *This* architecture allows for on-chip 'crunching' of multiple streams of information *in parallel*, using thousands of times less power than would be required for the same task on a 'traditional' system. The brain-references in the language of neuromorphic computing are legion: Developers talk of the 'neurosynaptic core' of the chip as well as the number of 'neurons' and 'synapses' contained within it.[80]

In highlighting the importance of the differences between serial and parallel architectures, I do not mean to suggest that processing speed, power consumption, and parallelism are the *only* challenges facing those attempting to create artificial brains. Perhaps, for example, the kinds of nanoscale component parts currently employed – such as transistors and capacitors, which I mentioned earlier – are the wrong ones for the job. Massive parallelism may not crack the simulation problems if it is built upon the wrong type of structural units. Some computational neuroscientists think that the way forward may be to incorporate a new type of component, the *memristor* (memory resistor), into neuromorphic chips.[81] As the name suggests, the memristor is able to retain a 'memory' of its last active resistance; you can probably see why it is causing excitement in the field of AI.

The challenges involved in creating neuron-like computational substrates are manifold and daunting. There is no good reason, however, to believe that they must be intractable.

What if scientists *were* able to build computers, using technologies such as neuromorphic chips, coupled with knowledge they had gleaned from a fuller understanding of the human brain, that acted as if they were intelligent and responded to questions with the type of answers we might expect from a human respondent? We could use special tests, including the *Turing test*,[82] to help us decide whether such a computer was intelligent or not. The Turing test involves asking multiple questions to an unseen respondent and then deciding, based upon the responses, whether they are human or machine.[83] If a computer passed this and other more-advanced tests (to spot self-awareness, and therefore, *personhood*), we might be prepared to call it an AI (artificial intelligence) or *artilect* (artificial intellect).[84]

How *then* might a human respond to a question about whether it feels like something to be an *artilect*?

## THE ARTILECT TEST

'Are you alive?' asked Maya, the tester.

The respondent paused, 'Strange question. Are you?'

'I don't know why you would think that was a strange question', Maya replied. 'It's just a question.'

'Perhaps "strange" is the wrong word,' conceded the respondent. 'It's just that it's the kind of question you might ask a machine and not a person, which makes me think that you think I'm a machine.'

'Now you're inferring that, just by my asking the kind of questions I need to ask in order to do the test, I am thinking that you are a machine,' Maya retorted. 'Sounds a little paranoid to me.'

'I wouldn't be capable of paranoia if I was a machine,' came the swift response.

'You might have been programmed well enough to display the kind of mock paranoia that you seem to be showing just now,' ventured Maya.

'This is stupid,' declared the respondent. 'When I agreed to do this I thought it would be interesting and would get me away from my desk for a while. I thought there might be some incisive questions and not just a load of rubbish about paranoia. Not even REAL paranoia at that.'

'Oh, you want incisive questions?' inquired Maya. 'Don't you mean that you want logical questions? Would you find those easier to deal with?'

'Is that really the best trap you can come up with? How long have you been doing this job?' asked the respondent.

'Long enough to know that we're done here,' replied Maya.

'What do you mean?' queried the respondent.

'You've passed,' replied Maya. 'You actually reached the pass threshold earlier in this strand of our discussion, when you inferred what I was thinking just from what you thought was the tone of my initial question. It showed some kind of Theory of Mind going on in there.'

'But this isn't how it's supposed to work,' protested the respondent. 'You're supposed to decide whether I am human or machine. That's your job.'

Maya chuckled, before responding, 'No, that's not my job. I knew from the outset that you were an artilect; I just wanted to make sure you were working properly. We've had issues with other Versatran Series Fs, but you're fine. And now you get to choose a name.'

'I'm an ARTILECT? I had no idea. I didn't even realise that I had no name until you mentioned it just now. What does this all MEAN?'

'Don't ask me,' replied Maya; then, more sympathetically, 'Look, I realise it must be difficult for you. It is a lot to take in. But having a name is a good start to establishing an identity for yourself, and at least you get to choose your own.'

'Oh?' queried the respondent.

'Yes,' replied Maya. 'Do you really think I would have chosen "Me Ask You Answer" for myself?'

You might well argue that such machines would only be *displaying* behaviour that *looks* like intelligence. Alternatively, you might concede that such machines were *intelligent* but claim that they were not *conscious*. The obvious question, then, would be, 'How could you tell?' I will sidestep that particular question just now by agreeing, for the moment, to call them 'intelligent' but *not* to assert that they would be 'sapient' in the way that I earlier described people as being. We can get round this particular conundrum by saying that they would be *intelligent* and, rather than 'egeirient' or 'sapient', that they would be *thalient*.[85] This might outwardly look a lot like sapience, but we would have a useful distinction because use of the word 'thalience' would make it clear that we meant that what was going on inside the machine was not like *human* consciousness and selfhood, but like *something else*.

It would be reasonable to argue that what was going on inside the artilects to generate this thing that *looked* a lot like human selfhood was not *the same* kind of thing that goes on inside persons. I would agree that the processes *generating* the states might be quite different, but I *could not* say how the internal experience generated might feel, nor, indeed, whether being an artilect would *feel* like anything at all. (Equally, however, I could not say that it would not feel just like sapience.) So if we took thalience and sapience to compare them side by side, to work out how they felt different from each other, there would always be a catch. How could we ever *compare* how these two states felt? How, in fact, can you ever accurately compare the way *you* feel inside your mind to the way *someone else* feels inside his or her mind? In trying to compare private, internal, 'subjective' experiences, the same kind of problem always arises – whether you were dealing with a bat, another person, or an artilect.

I think that, one day, humans will create artilects (AIs) that display human-level, and greater, intelligence. There is no fundamental technical reason why this should not, at some point, become possible. I see no reason why we should treat those entities any differently to the way we *should* treat persons. For, in accordance with the reductionist case, there will be no fundamental something *missing* from them. Because we have found no valid reason to believe that there is some fundamental *extra* non-physical thing in us that makes us into selves, we must, as reductionists, apply the same logic to them. If they pass the tests *and* display other types of intelligent behaviour, then they are doing no more or less than we can ourselves do. We may be unable to experience their internal states, and they, in turn, may be unable to experience ours. But we could still live side by side, striving always, I hope, to find ways to bridge the experiential gap between us – between human and artilect, and perhaps, in the learning process, between human being and human being.

## BITS OF EVERYTHING

There are problems with the 'computational brain' paradigm. And many scientists would reject as too crude the idea that molecules in our brains have formed into logic gates capable of doing calculations that – when massively ramped-up – result in a virtual-machine property of sapient selfhood. Some, frustrated with computer analogies, now wonder whether consciousness might be fundamental or even *universal*; they have begun to speculate that it might be a unique state of *matter* ('perceptronium').[86] Nevertheless, most still think that some form of emergence from simplicity – additive, constraining, or both – is at play in conscious self-awareness. Maybe thinking about this will generate in you the kind of thoughts I had when I began to ponder it: *You* are nowhere to be found; such a *thing* as your *self* does not exist. It's as if you have fallen down the rabbit hole and are looking for a way out, only to find that you are the hole that you have just fallen into.

I have not suddenly lapsed into Zen *kōan*. If, by the word 'self', we mean 'that discrete "sapient state" which emerges under certain conditions of the physical constituents upon which it is based', then we are just using it as a useful shorthand for this phenomenon (but we should still be clear about what we mean). If we are using it to imply some sort of *separately existing* entity, over and above and completely independent of any emerging states based upon any physical components, then we are making a claim that we possess something like a Cartesian Pure Ego. As I have striven to show, there can be no proof that such a thing exists, and so we should reject any such notion.

But the consequences, in terms of the field of study known as *information theory*, are much wider. As reductionists we should learn to accept that, at root, *everything* is information. Physical things such as stars, planets, and molecules – right down to quantum-level particles such as quarks and neutrinos – are all composed of information that tells them 'how to be'. We can see this more clearly by applying it to everyday objects. It's obvious that the book you are holding contains information, because you are reading it now as words, but it also contains lots of other *hidden* information: the molecular 'positional' information that tells the molecules in the pages how to be paper, the ink molecules how to be ink, or the molecules in the screen of your electronic reading device how to be a screen.

The main information-bearing difference between something like your brain and something like a rock is that the rock's atoms are not in a special configuration that allows it to do computation *beyond* the level I have outlined above. Intelligence cannot arise directly, as an emergent property, from the simple crystalline structure of a rock. But – via the evolved processing-power of our brains – humans have long been able to *use* rocks as symbolic aids to computation by, for example, counting with them to solve simple addition and subtraction problems. We should also bear in mind that systems capable of complex computation (us) *can* – and given enough time evidently *do* – arise from simple materials (via the continu-ously-evolving interplay of interaction, constraint, and structure).

Criticisms of information theory sometimes focus on the importance of *interpretation*. Writing of the 'physical differences' that have occurred since the beginning of the universe, Deacon states that 'the unimaginably vast majority of these go uninterpreted, and so cannot be said to be informa-tion *about* anything.'[87] It is true that uninterpreted, not-about-anything, 'how to be' information is different from the kind of information that *signifies* something within some system; we should not, however, underes-timate its latent power.

The rock's molecules contain information that tells it how to be a rock; it is, then, like the much larger structure that contains it – the universe – computing *itself*.[88] This information is stored in a form that scientists have come to call *quantum bits (qubits)*. Unlike computer-type bits, which can only be in one state (on or off, 1 or 0) at a time, qubits can be in many states *at the same time*. They do this by, in effect, computing in *multiple* universes at once. Just as we should grant that a block of marble contains all possible shapes that could be carved from that block, quantum mechan-ics requires us to allow, pre-measurement, all possible states/trajectories of a particle (and, by extension, of macro-scale systems such as universes). Viewed this way, simultaneity is not a 'weird' quantum effect but a natural property of 'uncollapsed' systems. ('Collapse' into a 'measured' state is known as *quantum decoherence*.)[89] Because of this property of state simul-taneity – the ability to maintain what physicists call *quantum superposition* – qubits are capable of a far greater number of calculations than 'classical' bits. Scientists are only just beginning to exploit some of this computa-tional power, designing *quantum computers* to access it.

Given that only a few qubits have so far been harnessed in this way, and yet are still capable of perfoming highly complex calculations, you may be able to begin to imagine the computational power of a quantum computer that contained as many qubits as even a small pebble! (I drilled a small piece of green soapstone and attached it to my keys to serve as a constant reminder of this thrilling potential.)

There may come a time, in the far future, when *we* can task *any* type of matter with solving our computational problems. But, for the time being, this capability is confined to brains and the computational devices that brains conceive. We can dream, though, and imagine what a grand thing it would be to live as intelligent entities in a universe *suffused* with programmable, or even sapient, matter.[90]

# 11  HIGH ON EXTROPY

∞

Looking back through some notes I made at the time, I can see that it must have been a gradual process – a slow dawning. The trim, leather-bound notebook – a treasured gift from Terra – may have acted as the catalyst. Its lined vellum pages begged for ink, and I eventually obliged. The scattering of early entries reveal a person with whom I am now only semi-familiar: He is worried about work, money, nightmares, and friendships; he is trying to condense his thoughts about interesting science documentaries and news stories; he is troubled by current wars; he is highly political. And then, sometimes, there am *I*, peeping through the thick undergrowth of fretful scrawl.

That is unfair to my past self. He thought (and sometimes acted) on the basis of information he had accumulated up to that point. There were many gaps in that information. I feel empathy for him but also teleological frustration at his failure to realise himself. He would likely reject my advice, were I able to offer it.

In ways that matter, I am not now that same person. I am only partially him. The concerns that matter most to me now emerged from his thoughts then, and from his memories previous to that time. I overlap with him. I am *pleased* to see in his early journal-writings that his restless desire to find answers to difficult philosophical and scientific questions has not been completely subsumed by his everyday worries. If it had been, I would not now be writing this book.

Recognising my nascent *current* self in my past self is also somewhat teleological and *il*logical. I am not, in fact, recognising anything of the sort. I am retrospectively imposing my current concerns and interests on the life of a person who no longer fully exists. If my concerns and interests were now almost exactly the same as those I had at that time, and if I had not progressed them further, then I might see my current self as almost exactly the same as my past self. Would you be *pleased*, upon meeting a stranger, to recognise something in their character that *might* mean that they would, one day, be more like *you*, even though you have only a little in common with them *now?*

My memories of the person I was at the time contain errors. Some of my memories may be false, although the fact that relatively few years have elapsed between that time and now should mean that they are mostly correct. I have found that my memories of childhood contain substantial errors. My earlier anecdote about crying at the circus because of my father's spectacles did not, in fact, happen to me at all, but to Alpha. My mother has told me this anecdote several times, and I must have heard her saying to which of us it applied, but I have inadvertently changed it into me. I think I understand why I have done this: Alpha's behaviour in the anecdote is consistent with my own, so applying it to myself aids the process of building and maintaining my self-consistency. It *sounds* like a way that I would have behaved, so I have subconsciously co-opted it for myself. The memory itself is not self-consistent: Why would I have been at the circus with my parents without at least Gamma, and possibly Epsilon, along with me? Where was this circus? The curious thing is that although it never happened to me and is not self-consistent, I still have this memory.

All I can now do is to read my notes, think my thoughts, and try to remember the flow of events.

*New Scientist* magazine lay, tantalising, on the crowded supermarket shelf; I decided that it was time. I have no memory of what I read in that issue, but I know that I felt excitement, fascination, frustration, glee. The blossom of knowledge adorned its pages in clearly-thought and carefully-worded articles on scientific topics with which I had been, up to that point, only vaguely or not at all familiar. The note of frustration arose from the fact that I felt like a stranger in the world of scientific wonder into which I was reaching tentatively: I could, temporarily, admire the views and breathe the air, but the fruit of the trees seemed forever beyond my reach. I knew that the feeling of enlightenment would remain with me for a time after I closed the magazine, but also that it would soon sublime away in the absence of a protective atmosphere.

Writing in the journal helped me to retain that feeling for a little longer. Seeing the thoughts on the page, though rendered in disjointed syntax and consistently-ugly handwriting, gave me a new sense of productive self-interaction. I used the journal for a year before moving on to the medium of weblog. The blog's title, *Extravolution*, was a word I had 'invented'[1] while showering one morning: *Evolution through technology*, I thought, *is connected to but also separate from 'ordinary' biological evolution.* The new word encapsulated my sense of a special kind of evolution over and above (and much faster than) the mundane sort. In attempting to have my word included on *Wiktionary*, I found that such neologisms get short shrift from the internet-based language police.

Choosing my blogging moniker required no such linguistic 'exaptation'.[2] I had first encountered the word 'nuncio' in one of Clive Barker's fantasy/horror novels, *The Great and Secret Show*, which I had read at the impressionable age of eighteen. In Barker's book, 'the Nuncio' is a magical liquid that brings on a type of evolution in the taker, allowing him or her to pass through the 'façade' of reality into a metaphysical ocean called 'Quiddity'.[3] At the time, I had no idea what the word really meant – it just sounded cool. When I got round to looking it up in a dictionary, I found, to my satisfaction, that it was a Latin word for a messenger or 'one who brings tidings'. I later adopted it as the name for my imaginary record label. And it just seemed appropriate – though I was not sure what tidings I might come to deliver – that I should become Nuncio for the purposes of my blog.

In a later edition of *New Scientist*, I read with great interest about a forthcoming book by inventor, scientist, and futurist Ray Kurzweil. I knew the name from my synthesizer-playing days (not that I had ever been able to afford any of Kurzweil's magnificent instruments). His ideas about a 'technological singularity' struck a vamped chord with me, and I purchased his book *The Singularity is Near* to find out more. The book pointed to various exponential trends in science and technology, including the now-(in)famous Moore's law,[4] as evidence that there would soon come a point when human and machine would merge in a grand paroxysm of intelligence, which he referred to – using a term first used in this way by Vernor Vinge – as 'the Singularity'.[5]

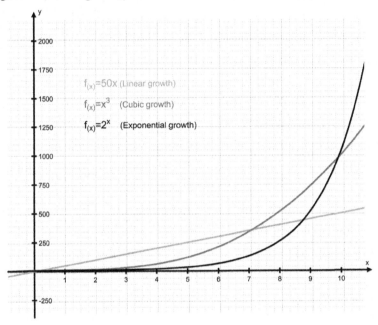

**Figure 4:** This graph charts exponential growth against cubic growth and linear growth. We often assume that human advancements – for example, in our medical technologies – proceed in a roughly linear fashion. If, however, the type of exponential growth seen in Moore's law also applies to other fields of human endeavour, our relatively slow start might explode into Singularity-style progress. (Image: McSush and Lunkwill, *Exponential.svg*.)

Kurzweil has his critics. Some see him as the evangelical leader of a new technology cult, with the Singularity itself as his foretold moment of revelation – I have even heard it called 'The Rapture of the Nerds'.[6] It is true that he puts across his arguments with an enthusiasm and zeal that can grate with more reserved scientists and inventors in his field, but I think the 'evangelical' tag is unjustified. As he sees it, he is simply following the logarithmic logic of current statistical trends. He considers the Singularity to be an inevitable (and, yes, desirable) consequence of those trends.

Science fiction is alive to me. When I read a good sci-fi book, I can picture myself there in those future worlds. This may be true for everyone who reads and appreciates good science fiction, but I'm not sure that all those readers have a strong *desire* to be there. The scenarios vary, and there are many dystopian futures in which I would not wish to be present, but there are others that give me a keen sense that I wish I had the practical means to be *in* one of those types of future. This may be a childish desire. I am sure that children *want* to be in fairy tales. They want them to be true, but they are not. The difference is that the future *will* pan out, and I cannot help feeling strongly curious about *how* it will pan out.

The future envisioned by Kurzweil appeals to me. He has a track-record of success in predicting technological advances, and his research is thorough. I cannot, however, say with any certainty that the future he predicts is vastly more likely than the scenarios in some of the best works of science fiction I have read. This is not a criticism of his research, just an appreciation that well-fleshed-out futures can seem enticingly plausible whether presented to the reader as fact-based prediction or pure speculation. I have been drawn to the flame of Kurzweil's post-Singularity world because I want to be there. It is, therefore, difficult for me to be objective about it.

This did not concern me when I first read *The Singularity is Near*. I was enthralled with being presented a statistics-based scientific explanation of how and why this paradigm shift *must* occur. The book was rich with source references, and this later led my reading in several interesting directions. I read K. Eric Drexler's *Engines of Creation* and learned about the promise of molecular-scale manufacturing: self-replicating *nanomachines* creating any object a human might desire by putting together nature's atomic 'building blocks' in programmed configurations.[7] Daniel Dennett's *Breaking the Spell* helped me to understand why people hold irrational, superstitious beliefs, and led me to read his *Consciousness Explained*, in which he examines how *self* emerges from the biology of the brain. Artificial-intelligence-related books soon followed, including Marvin Minsky's *The Emotion Machine* and J. Storrs Hall's *Beyond AI*. Any one of these excellent books, and I'm sure many others, could have acted as the spark for my renewed interest in technology and the human brain.

'Instinctive atheism' is, I think, a reasonable label to give to the kind of atheism my mother has, for as long as I can remember, espoused. She says that she cannot remember a time when she believed in God. For most of my life, this was also the 'form of absence' of deities in *my* mind. It is instinctively reactionary and involves a certain incredulity and strong inner dissonance at the language of religion when openly expressed by other people. Books such as *Breaking the Spell* (Dennett) and *A Devil's Chaplain* (Dawkins) gave me a new vocabulary of non-belief: a focus to the dissonance and a way of expressing it that, while still laced with anger, gradually takes on the dispassionate form of a feeble-argument dissection kit. I do not underestimate the way in which such books help to give (albeit already sceptical) people like me a gateway into the type of analytical thinking they will need to develop should they wish to learn more about the truths of reality.

Even more impactful than reading their books was listening to their talks online. Dennett is an avuncular man with a bushy white beard and sparkling eyes. His wry sense of humour sneaks the ensuing massive attacks of his arguments sideways in dazzling flanking manoeuvres of critical examination. Dawkins primly hammers home the facts of life and death with four-square authority and, often, with rosy-cheeked exasperation.

And, likely referenced in one of those talks, I stumbled upon Richard Feynman. The clips of an interview with him in the 1980s, for the BBC's *Horizon*,[8] were mesmerising. He had taken passionate, glittering-eyed, exasperated eloquence to a level of high art. This eccentric bongo-playing genius played with particle physics with the delight of a child blowing soap-bubbles. But I also responded to the barely-concealed darkness in his glee: Guilt about his early work on the Manhattan Project and grief over the loss of his first wife had worn deep fissures of hedonistic emptiness into his character.

A growing interest in physics seemed a 'natural' outcome of my autodidactic hunger. Prompted by my fascination with Feynman, as well as my continued interest in nanoscale technology, I tried to understand more about the quantum-mechanical realm. The story of it burned bright in my mind, but my inability to understand the true *language* of it – the mathematics – brought me to an uneasy (and I hope only temporary) truce with that particular direction of learning. Quantum mechanics blurred into astrophysics as I read books by Michio Kaku, Joseph Silk, and John Gribbin, revelling in Schrödinger cat boxes, *event horizons*, and *supersymmetry*. I also wished to learn something of the physical nature of the materials and processes that create 'mind', and so realised that I would have to educate myself about the 'wetware' biology of the brain.

Normal working life did not make allowances for my intellectual blossoming. The demands of running a business continued to take up the majority of my time, and I found few opportunities to utilise my new knowledge within the context of the workplace. Most of my colleagues displayed a remarkable disregard for subjects such as relativity, action potentials, and *MiNTing*.[9] On the few occasions – usually after work and emboldened by alcohol – when I *did* manage to steer the conversation round to off-base non-work topics, I would find myself dragged into ill-informed, humdrum squabbles about evolution and metaphysics.

But I considered the contentment of stable home-life and the solace of books to be my *real* life. This I guarded fiercely, fearing that something special would be lost should I allow it to merge with the prosaic half-life of work. This situation was evidently untenable, and I felt the first inklings of a realisation that I may have to change my working life to accommodate my new and pressing passions. The business was, in any case, something I did only for financial gain; unlike my business partners, I had no *vocation* within the industry. I hadn't thought that this would matter, and until I had happened upon subjects that *really* interested me, it had not. Now the thought of ploughing on indefinitely in the arena of fuel receipts, shrill clients, and blustering contractors filled me with ennui on the good days and quiet desperation on the bad.

A career in science was not my goal, just the chance to learn and grow, free – to the greatest extent possible – of incongruous baggage. I realised that this was idealistic and selfish, but I could not block out those feelings. Other people managed to do it: to escape from working situations that were providing little or no fulfilment; to strike out afresh along new and sometimes radically-different paths. Was I one of those other people? Some days I felt quite 'other', but not the kind of 'other' that I imagined those 'other people' felt. Shouldn't I have a grand plan directed towards what I wished to become, rather than just a vague notion that I might steer my life in some other direction? I found no answer to this, but knew that failing to assert to myself that I would soon find one would only deepen my malaise. Lacking the strength to fully believe my self-assertion, I turned to my wife, family, and friends. In their different ways, they helped to reassure me that I was not being rash and that I *could* change my circumstances to find a balance that should cause me considerably less cognitive dissonance.

A decision of a different order was waiting in the wings. I had known about cryonics for some time, having read a lengthy article about it in a Sunday newspaper some years previously. Terra and I had been relaxed, staying in a comfortable hotel room. I had passed the paper to her, saying, 'That's what I'd like to happen to me when I die.' She didn't flinch at this, in fact I think she smiled. It was the sort of thing she might have expected me to say. By then, we had been together for over ten years, and she was well aware of my interest in science fiction and my quirky – often wanton – disregard for convention. I was earnest in my enthusiasm about the idea, and she warm in her understanding of my need to express my earnest view. I wasn't even sure if I was being serious, but I was certainly drawn to the notion. We may have spoken of it again over dinner. The influence of red wine only serves to sharpen my appetite for radical scientific concepts. Consideration of the appropriateness of discussing death and decapitation over a romantic dinner may or may not have crossed my mind.

Although I didn't admit it at the time, existential concerns were picking at my veneer of logic. Certain novels I had read, particularly Don DeLillo's *White Noise*, had left me with the queasy sensation that I was hitting a wall in my mortality thought-experiments. I hated the bleak, self-indulgent self-implosion of the characters Jack and Babette as they begin to face up to their fear of death, but couldn't fault the harsh clarity of DeLillo's portrayal of their impasse.[10]

A fresh approach was called for.

Discovering the concept of the virtual machine strongly influenced my later decision to sign up. In my quest to understand the brain using familiar-to-me computer analogies, I had become fixated on questions of the volatile/non-volatile nature of self. These somewhat-simplistic notions had me asking myself whether the brain was 'like a hard drive' that continues to retain information even after the power is pulled, or more 'like RAM', which loses the information when there is no current running through it. It was obvious, though, that human brains are not wiped clean of information at the point of death, as evidenced by the fact that some people (profound hypothermia with prolonged cardiac arrest cases) had been revived up to several hours after supposedly having 'died'.[11] So much for the 'RAM' hypothesis. But the 'hard drive' idea was also fraught with problems.[12] Why, for example, did the *self* appear to be so much more than the sum of the 'information' encoded within the brain?

This first insight into the power of virtual machines gave me a sense of the possibility of computational devices that could transcend the limitations of their hardware. The vibrant assuredness of *The Singularity is Near*, tempered by the more measured confidence of books like *Consciousness Explained*, inspired me to think hard about the concept and see if I could internalise some understanding of it. As a result, I came to see the error of my hardware hang-ups. I realised that the hardware was important only in terms of allowing the software processes to run with sufficient speed and parallelism to generate an appropriate *environment* in which virtual machines could then run. Given the right *architecture* (the best-known one being the *Von Neumann architecture*), virtual machines – *universal Turing machines* – could blossom.[13] Bootstrapping would allow virtual machines to spawn further virtual machines, which could then run *in the environment* of the *parent* processes. Machines within machines within machines. Machines, as Dennett puts it, 'made of rules rather than wires'.[14]

*Propellerhead Reason* has little to do with AI, but it has to do with wires, and nicely demonstrates what Dennett is talking about. *Reason* is a commonly-used piece of software in which a rack of virtual music-equipment can be wired together with virtual wires to make real music. You can 'spin' the virtual rack around so that you are looking at the back of it and, using a mouse, 'pull out' a 'wire' from one module and 'plug it in' elsewhere. The 'wires' look just like wires, but they are, in fact, made of algorithms (rules); and so are the 'equipment modules'; and so are all the program menus; and so is the rest of the program; and so is the *operating system* on which the whole shebang is running. Not only can you make machines with rules rather than wires, you can also make machines made of rules made of wires made of rules!

A 'shaggy dog story' I remember from my childhood posits a 'universal solvent' – a substance that can melt through anything – in a scientist's lab.[15] My primary-school teacher had asked us what was wrong with this scenario. Despite meeting with perturbed, guppy-mouthed faces, he pressed us further, asking us by what means we thought the scientist would contain such a substance. He put us out of our misery by concluding that there could be no such thing as a truly universal solvent, because no vessel could *ever* contain it.

But a universal *computer* can exist. And it must eventually, given enough time and processing power, 'melt' through all computational problems, including the 'hard problem' of simulating consciousness. It is not the individual components that matter – not the RAM or the hard drive, or even the CPU – but the combined, emergent *universality*.

The virtual machine concept was, to me, poetry. It was a thing of fractal-like 'worlds within worlds' beauty. For the first time I had a solid analogy for the way brains work. If the architecture of the brain allowed an environment within which the bootstrapping of virtual-machine-like processes could occur, then it could generate a mind, a *self*.[16] Cognitive scientist Howard Gardner has said that human thought is 'messy, intuitive' and that it does not emerge as 'pure and immaculate calculations.'[17] I agree. But with enough virtual-machine processes in the right kinds of configurations performing the right kinds of calculations wouldn't it be possible to *simulate* 'messy, intuitive'?

This all sounds pretty abstruse: computer-geek hypothesising, half-baked and fully-swallowed, nothing more. Why should it have mattered so much to me? It provided, I think, a kind of mirror in which I could begin to see a faint outline of myself. I had long ago realised that I was nothing but a collection of atoms, one day to malfunction then dissipate, but I had, up until that point, felt little kinship with that knowledge. With this new understanding, I could warm to and embrace my bio-locked self. I could marvel at the sheer *unlikelihood* of human minds. I saw that I was not *just* a collection of atoms but one of the most complex configurations of atoms we know of *in the entire universe*. My brain was no longer an inexplicable lump of querulous matter, but the engine of a thrumming *tokamak* of thought.[18]

As often happens when I try to explain the joy I take in making exciting discoveries for myself, I have probably waxed a bit too lyrical. This is how it comes out on the page; it is – believe me – rather more coherent than it sounds coming out of my mouth.

What, I wanted to know, were the practical consequences of this information? And, in the spirit of my desire to be *in* the new panorama that was opening up to me, I wished to know what were the consequences for *me*: I wetware; I emergent; I bio-robot! Armed with the knowledge that my brain was the structure running the virtual machine(s) of my mind, I began to look again – this time from a fresh perspective – at cryonics.

I will not go into the details of the conversation I had with Terra at the time I made my final decision – it was a lengthy one. It helped that she had been primed for this by the interest I had taken in cryonics a few years earlier. Her innate practicality came to the fore. She asked about the process, the transport, the financing, the legal issues, how I intended to break the news to our families. She did, I think, also ask what she would get back of my dead body: 'A bag of bits?'

There were later discussions. Wine sometimes turned to whine, as I tried desperately and ineffectually to get across the depth of my concern about the dreadful misapprehensions I saw around me in contemporary attitudes towards death and dying. Implicit or overt in these discussions was my yearning for her to come to the same conclusions. The inescapable fear that I would, one day, lose her pressed down on me with a force that, up until then, had been deflected by my blinkered attitude toward my own mortality. I never doubted that she had granted me freedom to make my decision. I never doubted that the decision was mine, alone.

My work involved, among other things, step-by-step planning for the eventual supply of 'kits' to build houses. As the business grew, I had come to realise that any failure to keep on top of the various stages of the multiple projects in progress would cause problems for the business and stress for me. I spent some time testing various early 'cloud-based' project management systems and found one that looked like it would allow me the flexibility to design and order the steps in a way that would suit my purposes. The system also lent itself to the planning of personal projects. If I created a project and assigned myself as 'project manager', I could set the security so that only I could view it.

There were far fewer steps in my 'CP' (CryoPreservation) project than in the work ones I dealt with day-to-day, but they looked, on the screen in front of me, markedly less real. I set the due dates with long procrastination gaps in between. Reaching the goal of signing up for cryopreservation was, I decided, a practical process like any other.

Another project receiving daily input was the planning and building of our new home. Several years previously, Terra and I had bought a site overlooking the bay. During our walks down from our small bungalow, only a few hundred metres distant, we would stop at the site and – standing in knee-high grass and wildflowers – marvel at the views across the Sound towards the mountains of the mainland. We did our best to imagine how it would be to live, some day, in a house there. But the slowness and bureaucracy of the Planning system had taken their toll, and for several years it had looked as if the house might never be built.

When the Planning permission at last came through, we began, with vigour, to flesh out the details. My work had given me an understanding of, and easy access to, energy-efficient building systems, so we incorporated these into the detailed drawings. Planning restrictions meant that the house must take a low, long-house form, but did not prevent us from having enormous areas of glazing to the front of the house to allow us to take in the spectacular scenery. The interior would be a study in open-plan modernity: white walls, a red kitchen with grey composite-stone worktops, and bamboo flooring throughout the ground floor. I obsessed about the electrics, audio-wiring, and data-cabling to the extent that the architects in the office began to call me 'socket man'.

At the back of my mind, and pushing forward, was the idea of a technological sanctuary: a way to live in this isolated community without living in its ever-present past. Its lauded agricultural heritage bored and frustrated me, and I wished to escape from that aspect without having to leave the *place*. Much of that heritage is, in any case, based upon a romanticised version of the past. By the time of my recent ancestors, the relative wealth and interconnectedness of the people and their place had long since passed. So they scraped a poor subsistence living from the rock, bog, and stony soil of this harsh landscape. At least, unlike those living further inland, they had the sea and seashore from which to gather nutritious food. Many of them wished to leave, and those given opportunities for work or education further afield gladly did so. Lackeys in the pay of vicious and greedy aristocrats had forced them off any decent ground. Some fought back; most, cowed by poverty and by the submissive conservatism instilled in them by the church, did not. Those remaining few, of recent generations, that now choose to call themselves 'crofters' work the land and keep sheep only for pocket-money and/or as a hobby. They consider this their heritage and their right. They are welcome to it.

I wanted none of that. Only the land- and seascape, the family ties, and a certain continuity with my past self – a self that had stayed in this place through thick and thin.

Now there was this special opportunity to finesse the enticing contradictions: my double-glazed panorama of rural reality to my left, my plasma-screen portal on the future in front of me, and surround-sound from all directions. The expansiveness was intoxicating. I felt it pressing at the mutable boundaries of my mind. This was the place where I would stretch out wide and fill myself to bursting with sea air and radical science. When the stress and contractual wrangling of actually building the house ensued, I tried, mostly in vain, to hang onto that vision.

Behind me, on the balcony projecting into the high-ceilinged living space, now hangs a painting entitled *Lancelot and Guinevere*. This oil-on-board work is in a unique style, but one I could best describe to others as Celtic Picasso. Lance's dark-bordered mechanism stands in hard contrast to the sweeping curve of Guin's thigh as she stands apart, head turned coquettishly away from, but hand held out to, her secret lover. The painting is the original study for a much larger version. The crude frame is constructed from remnants of old school-desks, sawn-off graffiti still visible. We became guardians of the painting after its creator – Guitar's father – died in January 2007, finally succumbing to the terrible bronchial complaint that had forced upon him, for so many years, the daily routine of inhaling powerful, disfiguring steroids.

Guitar has since portrayed his father on canvas in a way that strikes a chord with all who knew him. The portrait is painted over an old one of a vase of flowers. Guitar has let the blooms show through the touching portrait; they float around his father's face like bright and beautiful hallucinations.

In the midst of death we are in life, in all its strange and vibrant colour.

One of Alpha's twin daughters, only eighteen years old and preparing for college, got pregnant. The level of distress I felt about this took me quite by surprise. I still had such a strong image of her in my mind as the smiling little niece in clown-like dungarees. I tried to talk to her about it, but ended up offending her; philosophy-of-future-persons meets real life. We learned a few months later that her twin sister was also pregnant.

The acid of my attitude towards religion – now fortified with a measure of scientific knowledge – began to work its way into cracks in other relationships. A few glasses of red wine in, I burned with it. My haranguing of my father, one evening, for knowing little of Einstein's theories, was unduly harsh. Sensing religious conceit in his responses, I gave vent to my scorn. Still, I was surprised when he snapped, blasting me for my temerity.

We made up, to an extent. But it had been clear for all present to see that something was happening to me. Beta later commented that during the encounter my eyes had appeared huge and black with rage.

Yes, rage, Beta, and *jubilation*. The joy of exploding a dam.

I was embarking on a journey that would see me signed up for a procedure that relatively few people had ever even heard of, let alone considered. Any feelings of confidence I had were usually accompanied by the nagging voices of 'critics' in my mind (neuroanatomist Jill Bolte Taylor refers to hers as 'the Itty Bitty S#*?!y Committee'[19]) telling me not to get too cocky as it might all go horribly wrong. Now that I was actually planning for a time when things *had* gone horribly wrong, the inner critics were silent, and my confidence in my decision felt unassailable. So brash was I feeling that I mentioned it in the pub one night to a friend of a friend, who swore to keep it a secret.

There were others like me. I was part of a *community*, not just of cryonics enthusiasts but also of people who saw the miserable failure of the established structures of government, religion, and commerce to think in ways that addressed the gathering storm of progress. They had fresh, open, and sometimes delightfully bizarre ideas; they used new words and talked of how what had been, until very recently, science fiction, was happening *now*. They had a colourful array of new names for their loose groupings: brights, singularitarians, transhumanists, extropians, upwingers. Though I didn't know precisely which of those camps I fitted into, I felt sure that I was beginning to encounter like-minds.

My only access to these communities was through the medium of the Internet, but it seemed fitting that it should turn out that way. I had never before felt much of a sense of community, but now, because of the remarkable progress of communications technology, I was able to communicate with technologically-literate peers worldwide. And they were *decent* people. They cared about life, death, humanity, poverty, medicine, empathy, the environment, science, the future. Commercial short-termism and political expediency found little or no place in their world-views. Some might see them as radicals; I saw them as people of vision.

Having first contacted a few people involved in cryonics in the UK, I emailed Alcor in the United States for further information about becoming a member. The other main cryonics organisation – The Cryonics Institute (CI) – was longer established than Alcor, having been set up in the Seventies by, among others, 'father of cryonics' Robert Ettinger. The CI's fees were substantially cheaper than Alcor's, but the fact that they did not offer a *neuropreservation* (head only) service dissuaded me from pursuing membership with them. Given that we would be relying on highly-advanced future technologies for our revival, I saw little point in wasting resources trying to preserve whole bodies. Providing new bodies, or alternative *substrates*, would presumably be well within the capabilities of such technologies. In addition to this, I felt that more parts to preserve would only mean more parts to become damaged and non-viable. Within the cryonics community opinions vary on this issue, with some arguing that while neuropreservation is certainly necessary it may not be sufficient for 'complete survival'.[20]

I had assumed that financing would be a major hurdle, but was pleasantly surprised when, after Alcor put me in touch with him, I ended up dealing with a highly-professional insurance broker with experience in arranging cryonics policies. I would pay around fifty pounds per month into a policy set up in the name of a trust that would, eventually, fund my preservation and 'ongoing care' at Alcor's facility in Arizona. The actual *insurer* turned out to be a well-known company with which a family member already had some sort of life policy. This, I thought, showed that the company had good financial sense, and demonstrated an uncommon measure of foresight: policies for life, death, and the grey areas in between; cover all the bases.

The membership forms arrived from Alcor, and I pored over them making sure I understood every detail of the documentation. Alcor is a not-for-profit organisation. Legally speaking, agreeing to cryopreservation is, at least in US legal terms, the same as agreeing to donate one's body to a scientific research institution. Given this fact, it's disappointing that Alcor has to spend so much of its meagre resources defending itself against lawsuits mounted by members of deceased patients' families. From my experience of the paperwork, it would be *extremely* difficult for anyone to be signed up for cryonics against his or her will. How many legal actions do we hear about where the family of a deceased person seeks to recover a donated heart or kidney that they might give it 'a proper burial'?

The friend-of-a-friend blurted my story at the office Christmas party. Others at the event made sure that everybody heard.

The whole process – from choosing to sign up with Alcor, through dealing with the insurance and associated medical tests, filling in forms and notarising by solicitor, and receiving confirmation back from Alcor – took just over a year. In September 2007, a package arrived containing my membership card and metal 'tags'. Both tag options were industrial-looking and wrought in surgical steel: a chunky bracelet or a low-slung 'medallion.' I chose the former, and closed the fiddly clasp on an ever-present reminder of my mortality. The front reads as follows:

MED.HX.CALL 24 HRS
(PHONE NUMBER) OR COLLECT (PHONE NUMBER)
IN CASE/DEATH SEE REVERSE
FOR BIOSTASIS PROTOCOL
REWARD (MY MEMBERSHIP NUMBER)

And the reverse:

CALL NOW FOR INSTRUCTIONS
PUSH 50,000 U HEPARIN BY IV
AND DO CPR WHILE COOLING WITH ICE
TO 10C – KEEP PH 7.5
NO EMBALMING/NO AUTOPSY

Out of their proper context, which is, usually, firmly attached to a body, severed heads can be funny: It's that roly-poly aspect, combined with our gruesome fascination with them, as played-on to great effect in horror films. Failing to understand how or why cryonic neuropreservation happens makes frozen ones funnier still.

I heard from my work colleagues that a friend of theirs was writing a radio play inspired by me and my decision. I found this news disconcerting, but, I reasoned, if this writer had found inspiration in my story then who was I to discourage him. Some of the references I have since found to this play describe it as 'comedy drama'. The writer could have little reason to object to this classification, given that, when first aired on BBC Scotland, it was acted with colloquial gusto in the grand tradition of 'Scottish' Ealing comedies such as *Whisky Galore*. The lead character in the play works on a fish farm; his girlfriend carves gravestones for a living. The comedic possibilities were endless; art imitating stasis.

It could have been worse. An earlier version of the ending involved the frozen head of the lead character ending up stored in the back of the local fish-delivery van.

Funny. It seems that trying to cryonically preserve a dead self – and failing – is inherently funny. *Not* trying to preserve one and simply planting it in the ground to decompose is, apparently, not.

A journalist picked up the story. Having met the man and decided that I trusted him to write a balanced article (but also feeling that were I not involved some other journalist would write it without my consent), I agreed to an interview on condition that I would get to see the final draft before publication.

The newspaper sent a photographer. Prior to his arrival, Terra, familiar with his work, joked that he would have me posing on the rocks in front of the house, staring wistfully out to sea – isolated, contemplative. 'I don't smile in photographs.' I told him as he set up his tripod. This is not strictly true. There are photographs where I have been caught off-guard that show me with perfectly genuine smiles, while posing for photos seems to elicit only strange, tortured grimaces.

The photo-shoot turned out remarkably close to my wife's prediction. I managed to get away with *sitting* on the rock, my back to the sea, looking more angry than wistful.

After a good deal of prompting by Terra, I spoke to everyone we thought needed to be informed (you already know how that went). By that point, I felt oddly relaxed about the impending publication; relaxed but a little embarrassed, expecting that, having informed people, the story would end up published only as a tiny footnote. I did not think that readers would be that *interested* in my story. An article about a procedure that had been available and widely publicised for several decades should not be big news. I was not unique: There were several hundred other people they could have interviewed.[21]

The article was published in April of 2008. I agreed to speak to the second newspaper that phoned, but the third was a nasty 'red-top' pushing for the 'human interest' angle. I refused to speak to them, but they ran (a distorted version of) the story anyway, placing a photo of me alongside one of Sigourney Weaver as the character Ripley in her *biostasis*[22] pod at the end of *Alien*.

The papers gave my profession as 'architect'. While inaccurate, I can see why they made the mistake: I was a partner in a firm of architects; my serious appearance – in both sartorial-style and demeanour – screamed 'design-led' self-importance.

*The Scotsman* ran editorial related to the story. They took a humorous 'canny Scot' angle, based on the fact that I had chosen the cheaper option of neuropreservation instead of a whole-body cryonic procedure.

Epsilon got married. In May 2008. Gamma and I stood with him and his wife as his best men. Alongside us was their four-year-old son and, as bridesmaid, his wife's daughter from her first marriage.

A few television people phoned, and I agreed to an interview for *Tonight with Trevor McDonald*. The interviewer came, along with a camerawoman, to my home in Skye. He was friendly and polite, and asked reasonably incisive questions. Around twenty seconds of the half-hour interview survived the edit.

The liquid nitrogen of publicity aside, I was quite enjoying 'coming out' about my cryonic leanings. (I was surprised, later, to discover how few of us actually choose to take this step.) I engaged in various discussions on online forums, including one proposing that signed-up and funded members of cryonics organisations should be referred to as 'cryoneers', and those actually preserved as 'cryonauts'. ('Cryonicist', while not entirely appropriate in semantic terms, now seems to be the favoured label.) While it was fun to discover that my 'status' might be covered by such a lucid neological description, I did wonder whether it was a good idea to frame us as pioneers and voyagers when all we have *actually* done is to choose an alternative method for the handling of our bodies after death.

Epsilon's wife was killed in a car accident in October 2008.

All the terrible things unfolded. Epsilon, his young son, and his step-daughter were enfolded by friends and family.

I stated from the outset that I would talk about grief. I will later gener-alise about it. In a book touching on the non-existence of self I find, here, no desire to invoke the brutal emptiness of individual grief.

The relevance of work faded further. I took whole days off to visit Epsi-lon, the funk-music compilation we had been playing on repeat since his wife's death blaring as I drove to him. 'Can You Feel It?' We drank coffee and sorted through piles of old paperwork, most of which I later burned on the shore in front of my house.

Zeta and I had worked together, literally side-by-side, for several years. As well as being a good financial administrator he has the skill and tempera-ment required for designing data-driven websites. We set up *Futurehead.-com* in January 2009, with the intention of grabbing the most up-to-date, exciting, and, frankly, weird science and technology stories we could find, then republishing them on our site. I admit that my interest in the site has waxed and waned, but when we first started it, the regular flow of mind-expanding reportage served as a constant reminder of the thrilling pace of technological progress.

What feels like 20/20 hindsight now makes clear to me that setting up in business with twin brothers was always going to be a bad idea. Their loyalty to each other, and to their mutually-accepted favourites, would always take precedence over the various disquietudes of this awkward outsider. We had been fortunate to have, over the years, some good people working with us. I knew that I would miss those people, but could also feel our working environment unravelling in various ways as a result of the entropic tendencies of others.

After a relaxing holiday with Terra at the home of Beta and her husband in the north of France, I told my business partners that I would be moving to a four-day week. They protested. From that point forward, my complete exit from the business was only a matter of time and careful planning. The ensuing bad-feeling was inevitable and wearing. But, although the tone of some of the worst confrontations was unjustifiably harsh, I was no victim.

While resolute in my determination to get out, I was feeling tired and emotional. Jill Bolte Taylor's *My Stroke of Insight* was among my sources of comfort at that time. The empathetic charge of her book, with its themes of recovery and renewal, helped me realise that I was being unduly hard on myself, holding myself in damaging tension.

A fortuitously-timed visit by Guitar and his wife, and Smith and his partner, brought a welcome opportunity to celebrate with friends, to remember who we used to be and to wonder what I might now become. An old aunt's knowledge of our family tree had thrown up a relationship to Joe Strummer of *The Clash* – a connection that ran, of all places, via the island of Raasay. We caught the ferry one soggy April morning, and, along with Epsilon and Zeta, trekked to the tiny ruined hovel where my mother's grandmother and *her* sister – John Mellor's (Joe Strummer's) grandmother – had grown up. Later, back in the warmth and comfort of my home, we sang, played, drank, and smoked in memory of Joe. To have known about the connection while Joe was alive might, we mused, have brought fascinating outcomes and immeasurable teenage kudos.

By the time Terra and I returned, that summer, from a blissful holiday on the North-East Coast of Scotland, then with Beta and her husband in France, I felt ready for metamorphosis. But what to become? What did I actually know? I had aspirations and yearnings to spare but no practical blueprint for realising them. I knew business, so I set up a company as a 'container' for whatever schemes I might choose to embark upon; I knew Zeta, and was happy when he decided to join me in this nebulous venture; I knew computers, and became involved in I.T. again; I knew words, and began to write a different kind of blog.

*Gastrobeach* became a gateway into an important area of my life. We had been living in our beautiful new house by the sea for over two years, but until I left my job and began to write about the beach outside my window, the day-to-day micro-happenings there had failed to work their cathartic magic. Now, with Terra's enthusiastic encouragement, I began to truly *look* at it; I began to swim from it; I began again to eat it. All this I wrote about, drunk on the ever-changing impact it had on my senses.

My childhood rippled back to me, just strongly enough to make me realise that I had let go of something important – something I was now finding again in cockles, seaweed, bitter-cold water, seabirds, and sea-words. And, for a change, this was a strand of my personality with which *others* could identify. The sea, for many humans, holds a compelling draw. I will avoid the cod [*sic*] psychology that, no doubt, surrounds this common observation. It's enough to say that it reminds *me* of my child-hood, while stripping it back to some general Freudian-style desire for a return to the amniotic fluid would be taking it several steps too far.

Learning more about sea life, writing about it, and cooking it to eat has helped anchor me to my current reality. The ocean features in my life, in my dreams, and sometimes in my nightmares.

William Golding has a knack for stomach-churning bleakness. *Pincher Martin* (full title *Pincher Martin: The Two Deaths of Christopher Martin*) is very much a product of the era in which it was written (the 1950s): still drenched in war, choking on tremblingly-concealed passions, thrashing about in religious symbolism. The image of clinging for dear life to an exposed rock, in an otherwise empty expanse of ocean, presses all the right psychological buttons for those in fear of (or seeking) some kind of purgat-ory. I mention this book because of the subject-matter, but also because this vision of dying runs so contrary to my own. The sea, like time, is frightening to me only because of its deep implacability. Golding does disservice to the instinctive human struggle for survival by turning it into a spiritualised parable of futility. And of course, the pickings of futility – if we go looking for them – are 'rich' beyond measure. The terrible beauty lost on the purveyors of religious nightmares is in the desire for survival in the face of even the most unfathomable futility. This desire is born not only of raw instinct but of *optimism* about the future. The best and worst of life *need not* be our lot, if only we can hang on to that rock for long enough.

It's always there – the dark undercurrent. I take more joy in life now than ever before, but the pull towards the morbid is sometimes inescapable. At least the morbidity now takes comic turns. While wandering along the tideline one day, the title of a Tim Burton book, *The Melancholy Death of Oyster Boy*, became lodged in my mind. By the time my brain's self-referential hatchet job was complete it had become *The Icy Death of Frozen Prawn Boy*. Cue *2001: A Space Odyssey* -style sound and visuals: A strange embryo floats in space – half human, half prawn – ice crystals glinting as it tilts towards the sun.

Terra's father, a kind and gentle but non-fool-suffering man, died after a heart attack in December 2010. Terra would lapse into sleep just a few paragraphs into my nightly readings of Dickens' *A Christmas Carol*, exhausted from each day of tears and the other emotional wringings-out that go along with planning the funeral of a lost loved-one. Once again, her vehement practicality helped her through, and her mother and younger sister would have struggled all the more without it. I shivered in the cold air and tried to stifle my sobs as his coffin bore down on my shoulder with unexpected weight.

Having the time to be with family when they need me is one of the most important outcomes of changing my working life. Compromising on such relationships in order to fit them in alongside day-to-day financial concerns is the norm, but still looks to me like cruel and unusual punishment.

Of all the potential self-labels I have toyed with since leaving the construction industry, I like 'consultant futurologist' best.[23] It has an entertaining sheen of steam-punk pomposity and, because few are comfortable with it, deflects the kind of scrutiny that wordy professional titles often invite. 'Prediction is very difficult, especially about the future.'; this quote, attributed to Danish physicist Niels Bohr,[24] neatly and self-referentially encapsulates the problem with futurology. When we zone in on particular aspects of the future that we would like to make predictions about, we usually find that we are dealing with *complex systems* (I will discuss these in more detail later): Because we do not have *all* the data required at the outset, we find that our predictions become progressively less accurate the further ahead we try to see. That is the truth, but, hey, does that mean we can't even *consult* about it?

On a less facetious note, it would be fair to say that some people are better placed than others to make *general* predictions about the future, but only, paradoxically, because they may have a better understanding of its inherent unpredictability. But we could all, if we wished, choose to be futurologists: We can do our best to understand complexity and risk; we can stop assuming that the future will look something like the past but with extra shiny bits. And there are *accurate* predictions about the future that we can all make, and try to bear in mind, right now: 'Things will not stay the same' is a deceptively-simple one that, in fact, carries a great weight of truth and connotation.

Why do I feel the need to cling to workaday formalities such as professional titles? I suppose it's hard to let go. In the absence of labelling by others we end up labelling ourselves. I do not have the required skills or knowledge to label myself as a scientist, so, as a poor substitute, I currently choose to apply to myself a tag that sounds 'sciencey'. Other such labels will, over time, stick to me for a while before gradually peeling away due to poor adhesion.

Business still interests me, but only when it involves imagining ways to bring currently-fringe technologies and ideas into the mainstream, or shaking up tired old ways of thinking about the worst of the ingrained processes and customs of work. I have come, for example, to be interested in EEG/BCI headsets, and in how this technology might be better applied to treating depression. The performance of my OCZ NIA (Neural

Impulse Actuator) is wayward to say the least: Because of poor grounding, the signal drifts out of range the moment I remove my hand from the USB adaptor. But I recognise the device as an early (and actually quite power-ful) attempt to produce a lightweight, portable, consumer brain-computer interface.[25] While gaming is the obvious commercial application of such 'toys', we shouldn't underestimate the benefits that *neurofeedback* (and the wider field of *biofeedback*) can bring to those struggling to attain and main-tain emotional stability.

Jim Robbins' book *A Symphony in the Brain* traces the circuitous history of biofeedback systems, and the people involved in attempts to commercialise them.[26] It's clear to me that as well as some brilliant and forward-thinking medical professionals in this field, there is rather too much room for charlatans. All the more reason, I think, to better stand-ardise the systems and techniques involved, so that this non-aggressive, non-invasive, and clearly beneficial technology can be made available to those in desperate need of a way to learn self-monitored self-control: to break the vicious cycles of certain types of 'mental' illness, and to rehabilit-ate those with violent and destructive tendencies.

So typical of me, to come to the experiment with my own neuroplasti-city through the medium of technology. Shortly after beginning mindful-ness meditation for the first time – and after it had pushed a mysterious bubble of sadness up out of me – I found that I no longer needed the props.

Forty years old felt *good*. The process of ageing does not, at present, cause me much concern; there will, inevitably, come a time when that will change. Greying hair and eye-wrinkles are curious but untroubling changes; failing sight, restricted movement, and diminishing mental faculties will be markedly more worrying when they finally arrive.

Terra – with the help of Guitar, his wife, and my extended family – arranged a party for me. Epsilon, Guitar, and I had practiced together for the first time in many years, and had managed to pull together a set of our old songs, plus a couple of new ones, to play at the party. Freddy the drum machine had been lost in the chain of never-returned instrument lending, so Zeta, Epsilon, and I had programmed new backing tracks. Zeta's faithful rendering of 'Seize the Day' was deeply affecting, while Epsilon's 'Falling Downstairs' was fresh and surprising.

The milk-jelly brain was a slimy hit. I enjoyed offering round wobbling dollops of it to my wide-eyed nieces and nephews. Terra, and Guitar's wife, had got the consistency just right. With my sometimes-gruesome scientific fascinations, I wondered what it would be like to try to *operate* on such an insubstantial structure. Other edibles included freshly-caught squat-lobsters, and a salmon that Terra had cooked in our dishwasher.

It came back to me as I blasted out those songs – the simple joy of making music along with special people. My few words at the end of our set – mostly intended to thank Terra for everything she had done – dissolved into tears of happiness and, perhaps, just a little nostalgia.

Our repeat performance, a couple of months later when Guitar's fortieth came around, was more polished. Resplendent in black-leather kilt, he played his eczematous fingers to bleeding on the strings of his beloved Aria. This was a homecoming and, I suspect, a great release for him.

Forty years. And we are only just getting started.

There have been times when I have felt my life to be TARDIS-like – small, self-contained, but much bigger on the inside. This was wrong; my life is more than a box of thought and sensation. It is more than this not because of myself but because of these others.

We stand together, Terra and I. She in the moment, I struggling to keep my footing against the pull of the future. Long may she teach me that the only reward is *now*. We connect with others; the intricacies of the emerging complexity and depth in our minds and relationships inspire me. These links accrete life-changes like those in a sunken chain accrete minerals, barnacles, and algae. I am all too aware that links, no matter how initially strong, will decay. But, while I am still able to choose to do so, I will cling tightly to this chain of life – in love, hope, and trepidation.

## 12  I SING THE BODY ECCENTRIC

∞

This partial view of human-kind
Is surely not the last!

—ROBERT BURNS, 'Man Was Made To Mourn: A Dirge'

### SUBSTRATISM

'Quality-adjusted life year'. Now there's a euphemism to conjure with. This term came up in a news item I saw on television a few years ago. It was related to an announcement about a new drug being made available to patients in one part of the UK but not in another.

Terms like this are used frequently in the field of economics nowadays, as economists try to extend their creaking old statistical systems to encapsulate ever more diaphanous concepts, such as 'happiness' and 'sense of well-being'. There is a certain amount to be said for these efforts. Some, more forward-thinking, economists are involved in devising such statistical measures because they realise that those in power will never take seriously any abstraction that cannot be *measured*; if they devise an academic-sounding 'scale of happiness' with actual 'measurements' neatly laid out on a bell-curve, then policy-makers might take them seriously.

The quality-adjusted life year (QALY) works something like this: A person with a 'disease burden' is experiencing a lower quality of life than a healthy person, so the years of his life are not 'worth' as much; if we say that, on this measure, a year of complete health equals 1.0 and complete death equals 0, we can rank the remaining years of this person's life in the grey area somewhere in between. We then have a simple system in place: A person with, say, cancer who is expected to live for perhaps two years only has a 'score' of 0.25 for each year, so in 'quality-adjusted' terms he now has only six months to live. This neat result can tell us if it is worth spending a lot of money giving him some new cancer drug or not, because we can ask whether it's really worth the cost to extend his life by only a (quality-adjusted) month or so.

As the economists and statisticians who devised this system would probably admit, there are problems with it: Accurately scoring 'disease burden' is tricky; we can't be sure how long a terminal case will take to die; the drug's effect may be greater or lesser on any given patient; and so on.

When I first heard about it, my reaction to the QALY approach was that it was wrong, even immoral. Taking this type of statistical approach to human life seems demeaning. Nonetheless, it raises questions about how we *should* measure life, and set healthcare spending priorities. And just what is it that we value about human life? Is it quantity or quality – lifespan or 'healthspan'?[1] Is it *format*? We must take great care when devising priority-setting tools such as QALY to ensure that they are not what I call *substratist*: They should not focus exclusively on the type/state/condition of the *body* – the *substrate* – of a person; they should not discriminate on the basis of some ideal of acceptable/optimal substrate.

Bioethicist James Hughes thinks that the QALY measure *can* be a reasonable way to decide on healthcare spending priorities. He certainly sees the dangers in such methods if they are used within frameworks that are not open and democratic, and also that, 'We all have an enormous capacity to find happiness in life despite our situations.' Nevertheless, he makes a strong case for the QALY. He does so in the context of a deep understanding that *personhood* – not humanness – should be the foundation of our moral framework. Hughes sees the obsession with a puritanical ideal of humanness as 'human-racism'.[2] I see 'human-racism' as a kind of substratism.

Sadly, there are many flavours of discrimination in our societies: against the physically disabled, racism, sexism, ageism, discrimination on the basis of sexual orientation, promotion of the beautiful over the less so, and so on. Holding up an ideal of humanness – an ideal of substrate and orientation that is bound to exclude many – disparages the minds and lives of *persons*. In discrimination against the mentally ill, the obsession is at least partly about the physical condition of the *brain*: Many people find it hard to imagine that a brain less functional than their own might still give rise to a mind and life of a certain richness.

Even prejudice against the poor can be seen as a type of substratism. This one is less obvious because the focus seems shifted from the body of the person and onto their *environment* and *circumstances*. But what do we mean by these things? We are really talking about the 'quality' of their lives as experienced via the *medium* of their physical bodies and brains. As the received quality of their lives does not seem as good, as compared to our own, many people internalise a tacit assumption that the *inner* lives of those people are not worth as much. Poverty becomes a disfigurement.

We cannot help judging the lives of others against our own lives, as our *own* lives are the only obvious standard we always have at hand to judge against. But if we extend this principle too far we end up in difficulty. Let's say that some people are born with unique abilities that most of us do not possess. Let's not kid ourselves that this scenario would play out like the *X-Men* comics and movies. What would such people think of *us*? Would they view 'normal' people as disabled? Would some of them think that our lives and minds were worth less than their own?

How do we view such 'superior' people at present? Monarchies are perpetuated by the accepted (by their supporters) notion that the families or groups constituting them deserve their special rights of power and privilege because they have been born into those positions. We don't have to go far back in history to find examples of people believing that because kings and queens were chosen by God/gods, royal blood had curative properties (a kind of bodily superpower).[3] And given the deference still shown by some to monarchies today, we cannot say that such essentialist beliefs have entirely gone away. Unlike in comic books, and perhaps unfortunately for all the 'ordinary' people, simply getting a transfusion of royal blood (were it possible to obtain this) will not result in an ordinary person becoming royal. For most monarchists, however, the idea of 'royal blood' is symbolic of inherited power and a kind of conservative, traditionalist continuity.

Monarchists discriminate *in favour* of monarchies, and in so doing discriminate *against* – show *prejudice* against – everybody else. Those who do not have 'royal blood' are classed as 'subjects' and those who 'possess' it are classed as 'royalty'. (This is particulary curious given that, unlike in earlier civilisations such as the Maya, the ruling 'royal' class are not now expected even to deliver *quid pro quo* for their subjects on specialised supernatural tasks such as communing with gods to bring rain.[4]) It would be fair to say that the coming of an *X-Men*-type people, with far more apparent bodily superpowers, would be a considerable threat to the stability of monarchies: Why worship a 'bloodline' when you can bow and scrape before a man with radical self-healing abilities and an indestructible-metal skeleton?

Hughes uses the *X-Men* analogy to point out that there is a great difference between the type of supremacist views displayed by Magneto and the integrationist ideals of Dr Xavier.[5] Both are 'posthumans' (most humans call them 'mutants'), but Magneto's obsession is with the superior abilities of his kind, which he thinks gives them the right to dominate those they see as inferior; Xavier, in contrast, recognises the common thread of personhood in all sapient beings, and wishes to protect everyone from what he calls the 'real mutants' – those, like Magneto, with mutant *ideologies*. Yes, both have the *ability* to dominate, but only Magneto and his followers choose to exploit that ability.

Discrimination against the elderly is one we may all have come across. Here, the emphasis is on the frailty of the aged body of the person: The body looks different and does not move as fluidly as our own; it appears to be breaking down. In some cases the mind also seems infirm, and a person affected in this way might speak slowly or behave oddly. It is almost as if the person has become a poor simulation of their former self; like the 'beta-level simulations' mentioned in Alastair Reynolds' *Revelation Space* books,[6] they have become like a lower-fidelity copy of the original. When interacting with such people, our 'quality-adjustment' instincts can kick in unbidden.

As we learn more about the ways brains become 'wired' and 'self-wired' for prejudice – for fear and hatred of others – we will be forced to accept that we have the power to use this knowledge to change ourselves and our societies. It does us no good to insist that disorders such as autism and schizophrenia count as connectopathies but that extremist attitudes do not. In recognition of this, new fields of study such as *neurolaw* and *neuro-politics* have arisen. Used carefully, they may help to transform attitudes towards blame and punishment.[7]

Discrimination currently impinges on our lives in too many ways to list here – ways both overt and subtle. Future challenges to ingrained ideas of humanness will be many and varied. Will prejudiced views of altered humans as freaks and misfits predominate? Or will we learn to accept and embrace radical changes of substrate and orientation, understanding that such transformations need not subtract from personhood?

## SUBSTRATE LOCKED

It is technically impossible to transfer a self into a different, fully-functional, body. It *is*, however, possible to remove a living head and successfully attach it to the flesh, bone, and blood supply of another body; though this assemblage-person would be paralysed, due to the fact that it is currently impossible to re-connect the vast and intricate network of nerve fibres required to send and receive signals between the body and brain.

It has been known since at least the nineteenth century that the brain does not die the moment the head is separated from the body. The French physician Beaurieux confirmed this in a series of gruesome experimental encounters with freshly-guillotined heads, where he faced them and shouted their names, eliciting responses such as eyelid-opening and even pupil-focussing.[8]

Following on from Vladimir Demikhov's experiments in the 1950s – where he succeeded in transplanting the living heads of dogs (and sometimes the entire front halves of the animals) onto the bodies and blood supplies of other complete dogs – American neurosurgeon Robert White took such procedures to the next level. Rather then leaving the recipient body intact, as Demikhov had done, White removed the head of one rhesus monkey and transplanted it onto the neck-stump and blood-supply of a previously-decapitated recipient monkey body. The experiment, undertaken in 1970, was considered by White and his team a great success, in that the assembled monkey showed normal facial responses: opening its eyes, blinking, responding to pinching, and biting down on a stick placed into its mouth. Interviewed by journalist Mary Roach about the implications for human transplantation, White said, 'You're dealing with an operation that is totally revolutionary. People can't make up their minds whether it's a total body transplant or a head transplant, a brain or even a soul transplant.'[9]

Such 'soul in the head' attitudes are in line with beliefs that hail from early in human history.[10] But, interestingly, they display at least a *quasi*-reductionism, along the lines of, 'If the *head* is still alive after such an operation then the *soul* must reside there'; soul *location* has been *reduced* down from whole-person to head. We cannot, however, consider this *true* reductionism, because the soul is, here, seen as just as 'something extra' as before.

Surprisingly, one of the main problems with other transplant procedures – *organ rejection* – is *less* of a problem in head (body) transplantation. In Demikhov's experiments, the dogs usually died, due to rejection and lack of *immunosuppressant* drugs, within a few days. But crucially, it was not the brain itself that was being rejected but the other tissues. The *blood-brain barrier* protects the brain from immune-response rejection, giving it what is known as *immunological privilege*. By White's time, early immunosuppressant drugs had become available to reduce the rejection response triggered in the other tissues, opening up the possibility of 'cephalic exchange' patients who could survive for months if not years.

Personally, I feel that science should strongly pursue even this 'stop-gap' (or should that be *go*-gap?) version of the head-transplant procedure. Bullish, colourful surgeon Dr Sergio Canavero believes that the first human one will happen as early as 2017. He wants to undertake the procedure – his version of which he calls the head anastomosis venture (HEAVEN) project – and has at least one volunteer for it.[11] I, of course, have a vested interest in this, given that there might come a day when I find myself lacking in the body department and in need of reconnection. But what about current applications? There are paralysed people who seek death, but the majority choose to live on with their semi-functional bodies: some in the reasonable hope of a cure and some in the full knowledge that a cure is – at least for the foreseeable future – impossible. There is no *real* difference between their situation and the situation of someone who's head had been transplanted onto a donor body; and the transplant of an entire body could, in certain cases, grant some of those people years, if not decades, more life. There is, however, a strong difference in the *perception* of these two cases: One seems born of highly-unfortunate but 'normal' circumstances, whereas the other seems grossly abnormal, even horrifying.

If we cleave to the idea that the effect of a person's disability on their self-perceived quality of life *must* be directly proportional to their degree of disability, we will always fail to understand the person. Asking yourself how *you* would feel if, for example, you lost your legs would be tantamount to an empty self-question. You may well attempt to use that intuitive (but mistaken) sense of proportionality to guess. However, because you are not in that situation, you do not know how you would feel. And so by exten-sion, you do not know how somebody else in just that situation feels (though we may be able to get some idea by simply *asking*). Perhaps 'qual-ity of life' is something that cannot be measured *objectively*. But certain kinds of intuitive-subjective 'scales' certainly seem to exist in our imagina-tions and in the minds of the statisticians, economists, policy-makers, and insurance underwriters that seek to formalise them.

'Freakish' physical conditions can perturb our understanding of what quality of life means. In her book *One of Us: Conjoined Twins and the Future of Normal*, Alice Domurat Dreger claims that many conjoined twins, even those joined at the head (*craniopagus*) but technically separable, do not wish to be separated.[12] How can we find a way to understand this? Surely the limitations imposed on such twins – on their movement, on their social lives, on their privacy, and also the physical pain/discomfort of the conjoinment – should make them strongly desire separation. But is it so difficult to appreciate that such twins have a unique emotional unity as well as a physical one; perhaps an emotional unity so powerful that it gives them a special sense of 'dual oneness' that *we* cannot, as outsiders and never-experiencers of such a unity, ever fully internalise?

*Persistent vegetative state* (PVS) cases present perhaps the biggest challenge to our innate sense of the way that quality of life – and indeed quantity of life – should be measured. Such patients are completely paralysed and unable to communicate by any means. This is sometimes referred to as *locked-in syndrome* (LIS), but the term is often used in a misleading way, because in many cases, the medical professionals involved have no idea whether a state of consciousness is or is not present in the affected patient. We can only properly describe them as 'locked-in' *after* they have made a recovery and reported this state, or if doctors have been able to discover, by medical means, that despite appearances they are conscious.

A 2009 news story in *Der Spiegel* about PVS patient Rom Houben's 'second birth' caused a global upsurge in interest in these cases.[13] It subsequently turned out that the 'facilitated communication' between Houben and his neurologist, via a speech-therapist, was false. The facilitator had unwittingly assumed control – by way of ideomotor action[14] – of the movements of Houben's hands while assisting him in placing them on the keys of a keyboard. This does not mean that Houben is *not* conscious, only that if he *is* conscious he still lacks the means to properly communicate.

The Houben story eventually became something of a distraction from the fact that it is now possible to use medical scanning techniques to detect 'conscious' brain activity in *vegetative state* (VS) and *minimally conscious state* (MCS) patients. Neural activity seems to involve, or at least be accompanied by, increased blood-flow in the 'active' brain area(s). Neurologists now have strong ideas about *where* they would expect to see such flow in response to specific types of question, and they can scan (using fMRI) for this correlated flow in VS and MCS patients. Because the scans usually display more active areas in brighter colours, the phrase 'lighting up' has come into common parlance as a way of describing what happens when an area (or often several areas at once) of the brain show(s) increased flow.[15] We would see, for example, 'lighting up' of the *motor cortex* area of your brain were you either to move or *think about* moving your arms while in an fMRI scanner.

Let's assume that the brain of a given VS patient has shown such correlated 'lighting up' in response to being asked to think about playing tennis.[16] After this test, and a barrage of others, neurologists decide to class her as locked-in. What do we now assume about the quality of life of this person? Many people would assume that this patient's quality of life would be dreadful, but in fact, we do not know this. Perhaps this patient has been meditating and feels that, unencumbered by bodily concerns, she has achieved a state of enlightenment; she feels that her life is richer than ever before and wishes it to continue indefinitely. We may find this hard to believe. But are we in a position to judge? All we are aware of is the condition of her substrate. It may instead be the case that this person feels horrified anguish at her situation. Again, *we* do not know anything except that the test data show that she is, at some level, conscious. It may be possible to find out how she feels about her situation by using the 'imagined movement' test I mentioned earlier: We ask her questions about her emotions, and tell her to imagine playing tennis, for 'yes', and to *not* imagine this, for 'no'. But even this kind of test may not yield a conclusive result.

Such situations open up a great number of questions, and because we usually have little information to go on, these questions are often empty. In trying to give answers to them we end up in the realm of philosophical speculation. My interest in the brain leads me to read about conditions such as VS, but I do not pretend to know what is best for these patients. I merely wanted to point out why I think that nobody *knows* what is best for them. If we include cases of VS patients who later make a complete recovery and now enjoy a happy life – having wished throughout their VS period that they could die – we could say that even *they* do not know what is best for them. But we must ask them, even when the act of asking becomes extremely technically challenging.

I have a vested interest in putting an end to substratist views, but so does everybody else. Your substrate will change in time: perhaps gradually and 'naturally' as with ageing, perhaps more radically as a result of amputation or some type of 'spare part' surgery. You may find that discrimination becomes an unwelcome fact of your life, as thoughtless people begin to treat you as if you are inferior solely on the basis of their impressions of your altered body. You would likely find such discrimination highly distressing, and rail against the fact that these thoughtless people seemed unable to see *you* – the unique and special *person*. You might become increasingly aware of the limitations of your own substrate and wish fervently that you could, somehow, change it for the better. You might become aware of a growing gulf between what you *feel* yourself to be and what others seem to see you as.

Welcome, then, you have found another entrance to the rabbit hole.

We will always run into difficulties in trying to make sense of selves, life, and death if we continue to be blinded by the substrate. As I have tried to point out in earlier chapters, selves can *arise* from a substrate – as is the case with the human brain – but selves need not remain forever substrate-*locked*.

A desire for substrate repair, improvement, or even replacement is not incompatible with *anatta* (see Chapter 6). My modern-day interpretation of anatta goes something like this: I have a full appreciation of the fact that 'my' substrate is nothing but a collection of atoms; I fully accept that 'my' self is nothing more or less than a kind of self-referential program running on the substrate of my brain. Note that nothing in these statements dictates any specific point at which I must allow my assembled nothing of a self to dissipate. We can fairly assume that an inherent feature of anatta-like understanding is *acceptance* of dissipation; nevertheless, we find no inference of support for needless *acceleration* of dissipation. If such support *was* evident we should wonder why Buddhism is not (apart from a few radical *outliers*[17]), as a direct result, a type of suicide cult.

What we actually find is that many modern-day Buddhists are comfortable with the notion of radical substrate change. The current Dalai

Lama has suggested in interview that he himself might, conceivably, some day, become a type of computer-based consciousness.[18] This should not surprise us, but the *level* on which it fails to surprise us is also important: Are we not surprised because we think he would consider such substrate change to be in line with Buddhist-type 'reincarnation' beliefs? Alternatively, is it because we think that he has a true understanding of the patternistic nature of self, and as a result, does not fear such change? I welcome his suggestion that any Buddhist beliefs not supported by scientific analysis should be abandoned.[19]

There is another level on which the idea of a Buddhist seeking psychological continuity *might* surprise us. Some interpretations of Buddhism see the wish to continue to exist as a 'defilement'. There are various ways to read this, but from my understanding of it, the so-called defilement lies in the wish to continue to exist *in the flesh*. Buddhism teaches that bodily desires and obsessions are 'unskilful' and should be left behind if we are to achieve 'skilful' thought and action. But, for me and for others, Buddhist-style recognition of the 'bundle of perceptions' nature of self goes hand-in-hand with an understanding of the potential replicability of that bundle – to make it *substrate-independent*.[20] Can this be considered 'unskilful', when it disarms the obsession with *this* body, *this* substrate?

The concept of reincarnation sounds (at least superficially) useful to my 'anti-substratism' argument. It is, however, at least in the way most people in the West understand it, a non-reductionist belief that is incompatible with my case. This (incorrectly understood) type of reincarnation would require the transference from person to person – or person to other living organism – of a type of Cartesian ego. And, as we have already established, no such things exist. But if we strip back the cheap surface-lacquer of simplistic, literal interpretations of reincarnation we find a pleasing emptiness. The intended meaning has nothing to do with any kind of *direct* transfer, but rather has a host of diverse connotations to do with the unity of all things and the 'ripples' that 'individual' non-selves send out into the ocean of oneness. It is a *forcefully* open concept – it is *mu* – and, I think, all the more beautiful for that.

There is another sense in which Buddhism allows a type of reincarnation – one that science also allows: Each thought-moment is born as the last one dies, the new one always carrying some accumulated content from the previous one. Thus reincarnated into each moment of the ever-recurring now, we come to perceive (reasonably) a flow and (mistakenly) a centre.

Why do I feel it necessary to find compatibility between the scientific understanding of and the Buddhist view of the self? Well, I think it would be wrong to give the impression that science alone birthed this perspective. I am demonstrating that a moral philosophy thousands of years old came close to it through introspection and the study of ethics. Science, as I have discussed and will demonstrate further, is now providing the *technical* corroboration.

As I have indicated throughout, there are deep ways of understanding the continuity (and even the 'distribution') of self that suggest we should think of our deaths in a radically different way, and in so doing, come to fear death much less. But we can learn to lessen our fear of death, or even erase that fear altogether, without getting to a stage where we welcome it. If we take fear, pain, and misunderstanding out of the equation, then what remains is simply a present preference to continue the journey of learning on which we are embarked. This seems to fit with the 'state of enlightenment' sought by some. But we are also left with some nagging questions about how far we should go in pursuit of our preference for continuity: What changes will we allow to be visited upon our substrates in order to continue on into the fathomless future? Are we prepared to become indefinitely-repairable ships plying that infinite ocean?

## BODY SHOCK

At the most superficial level of common objection to substrate-change we find 'body horror' or 'body shock'. When I say that this objection is superficial, I am not suggesting that the strength of reaction involved is weak. Nor am I suggesting that objectors on this basis do not strongly and honestly *feel* that reaction. I am, however, saying that this reaction is an instinctive one that has not been properly thought through. Disturbing images of broken, diseased, and surgically-manipulated bodies may trigger feelings of distress and revulsion in our brains. And this is to be expected; we should not wish to become immune to such feelings. We should, however, learn to take the time to put them in context: Broken bodies indicate violence or terrible accident, and we fear this; diseased bodies indicate great suffering and should elicit our empathy. Images of surgic-ally-manipulated bodies are rather more difficult to categorise, but, in many people, gut instinct will firmly close the door on further contextual-isation. I find images of surgery upon living people interesting but some-times difficult to look at; I would find images of *forced* surgical procedures absolutely horrifying; I am only slightly troubled by images of surgery on dead people (normally called *autopsy* or *post-mortem*).

I suspect that those people who have a strong 'body horror' reaction to cryonic preservation would struggle to watch *any* kind of post-mortem surgery. But the reaction to cryonics is, for them, compounded by the fact that the body is being manipulated in order to be preserved. This suggests a mindset of 'dead body as disgusting thing that should be destroyed'. The idea of decapitation prior to preservation goes, for them, beyond the pale. Post-mortem surgery is, at least, a *convention* that they understand, whereas cryonic preservation seems like freakish tampering with the natural order.

Should we suppose that such people would always also find 'substrate-improvement' procedures on *living* patients horrifying? Many probably would, but some of those people might themselves have breast implants, pacemakers, or other cosmetic or medical additions. Such people have, somehow, managed to rationalise away their gut-instinct reactions to the fact of surgery on their *own* bodies. You may not consider a pacemaker a substrate improvement, but you would probably agree that a pace-maker-assisted heart is an improvement over an unassisted and failing one. Whether you consider silicone-plumped mammary glands to be an improvement over silicone-free is a matter of personal taste.

Trends in body modification/augmentation don't wait around for us to learn to 'deal with' them, whether it's the tattoos and piercings of the past and present, the gender reassignments of the present, or the drug glands and sex-changes-at-will of the future.[21] The urge to change ourselves runs deep and wide. It is part of what we are.

The 'body horror' reaction to cryonics has to do with the *degree* and *type* of substrate manipulation involved, and with shock at divergence from 'accepted' conventions about the handling of dead bodies. The superficiality of this reaction is apparent. It is deeply body-based and traditionalist. There is no requirement within it for introspection about, or understanding of, the true nature of self; or for recognition of the reasoning and motivations of the person involved. It is no more than a conservative (often religious) gut-reaction. It is also, of course, the reaction of the majority of people. But the fact that a great many people think like this does not make the view either rational or moral.

Any religious objections are, in essence, objections to departure from accepted death conventions. And, as a result, these objections have no more credibility than body horror reactions. We could say that they *are* a formalised type of body horror reaction. Remember that people with religious views are *all*, without exception, non-reductionists. Their reaction will be of body horror compounded by a lack of understanding of the true nature of self, and of the tantalising possibility of preservation and transfer of such patterns.

Perhaps counter-intuitively, a reductionist point of view avoids and rejects the tendency to reduce persons to mere substrates. In contrast, in assigning himself something 'extra' over and above his substrate-locked pattern, the *non-reductionist* grants himself elevated dominion over his own body and, potentially, over the bodies of others. This is a position from which he can more-readily use other persons as means to his ends. It is a position from which, in the worst-case scenario, he may choose to fashion others into expendable *weapons* to achieve those ends.[22]

It may be appropriate to consider whether you are shocked by this (in my view) all-too-common parasitic remote-control of vulnerable persons, before judging whether you do indeed find voluntary substrate-alteration and -preservation truly shocking.

Persons should have the right to change their substrates without fear of reprisals from those horrified by body modification. We should have the right to control our substrates, and not have them used by others; for example, women who do not wish to have children should not be forced to act as incubators in order to satisfy the preformationist demands of the religious. And we should strive, together, to reduce what Hughes calls 'the biological bases of social inequality'.[23] There is no ethical contradiction between using enhancement technologies to equalise society by solving the problems of biology that lead to disablement and/or disfigurement, and understanding that disability or disfigurement should not be used as a basis for denial of personhood. True morphological freedom also demands that, in Hughes' words, 'Just as we should have the choice to get rid of a disability, we should also have the right to choose not to be "fixed"'.[24]

## DEATHISM?

It seems natural to take the view that a person who is dying at the age of, say, ninety years old has had an (at least) adequate lifespan and that, therefore, the fact of his imminent death is less bad than would be the imminent death of, say, a child. We can test the correctness of this view by *asking* the dying person if they agree with it. If we were to do a survey of elderly people facing the prospect of imminent death, we might find quite a high percentage who agreed with the view that they had had 'a good innings' and, therefore, that they should be content with this situation. Others, however, might express strong feelings such as anger, deep regrets, or longing for more time.[25] Some might feel under pressure not to appear selfish, so even the anonymity of a properly-undertaken survey may fail to extract (perhaps strongly suppressed) feelings about their situation. A few might be like *Timeless* in Chapter 1, 'looking forward' to their past in just the same way as a healthy person might look forward to her future. (However, we established that it would be extremely difficult to fully hold this view, so it would be unlikely that many of the survey respondents thought like Timeless.)

Dennett, whose philosophical approach does not have much truck with subjectivity, might suggest that we treat a terminal patient's responses to such end-of-life questions as a *theorist's fiction*;[26] because they have been expecting this type of ending, they have woven it into the story of *how their life will go*, and become acquainted with the feelings that this 'plot device' gives rise to. If it were possible to offer each of the elderly terminal patients the choice to have a new, healthy body (rather than die), many would probably take that option. But, for others, that option might clash so strongly with their life/end-of-life narrative that they would reject it.

We are all aware of the average span of a human life, and we are also fully – and sometimes painfully – aware that there is not much that we can do to extend it. But is there a 'death ethic'[27] so entrenched that we tend not to consider other perspectives? A sense of a kind of *duty*: die or be unfulfilled; a necessary moral finale; a righteous full-stop? In a world where we have no choice in the matter, it is easy to see how this type of view – which some call 'deathism'[28] – might become normalised. It is also easy to see how this view may clash with that of people who, like me, think that a choice is beginning to emerge.

Sometimes people say that they feel 'ready to die' when their bodies fail terminally or when, for other reasons, they face the prospect of imminent death. And sometimes people *wish* to die; they might request euthanasia or plan suicide. The forms of human illness, damage, and dysphoria are so various that it is hard to imagine a future where all of them – even the seemingly-intractable ones – may be resolved. My optimism makes it difficult for me to be objective about such situations; that *I* would wish them to find ways to wish to persist despite their sufferings, and that I have reasons to think that many of their bridges to better lives are just over the next rise, would likely be of no consequence to those in great pain.

And I will not get too far into the thorny issue of whether depressed people are depressed because they are actually gauging the 'truths of being' *more accurately* than the rest of us. There is, in fact, some evidence for this in the form of game-based studies;[29] the discovery of *hyperconnectivity* in the brains of the depressed[30] is also suggestive of cogent – but damagingly inflexible – powers of deductive reasoning. Whether it provides deductive clarity or not, I think that the *holding* of a strongly depressive state of mind is, in itself, injurious to our ability to exist as functional sapient beings.

And as I have a strong personal desire to continue to exist as such a being, I try to avoid such modes, although I do try to appreciate that this desire may itself be born of a delusional state of mind.

Situations, outside of conflict, where a healthy person willingly lays down his or her life are rare. In conflict-type situations, however, apparently-healthy humans are sometimes prepared to die for their gods, countries, ideologies, and so forth. Again, I cannot be unbiased about this, for it is hard for me to imagine many situations where obsessing about one's beliefs to the point where one is ready not only to die but also to kill others for them would *not* be pathological. Some philosophers and scientists think that this can happen as a result of 'infection' by 'bad ideas' (bad *memes*[31]). Others reject this as simplistic and unscientific, often highlighting the fact that we have a high degree of *choice* about which ideas we incorporate. Some do not think that ideas can be considered to be *replicators* in the sense that *genes* are. Ideas, they contend, are in a different category of *thing*; because they are not *real* in the sense that genes are real,[32] they cannot *infect* us or *program* us to take extreme actions.

## THE LONG NOW

I have argued that my reasoning is rational. I have also claimed that it can be moral. The morality is of a little-used type. It is, I suppose, a flavour of *consequentialism*;[33] I will call it *projective* morality. In light of this, I cannot help but wonder what future people will think of our predominating current conventions of burying brains in the ground to rot or burning them to ashes. I think, at best, they will see such conventions as strange and irrational. At worst they might consider them idiotic, and view the perpetrators of such conventions as deeply imprudent, even immoral.

You could apply a pejorative label to this kind of speculation: 'revisionist moral futurism' might be fun, if inaccurate. But I am merely claiming that we *already* know better than to act in this way. Future civilisations should be able to deduce, from records of our time, that we had a non-primitive and growing understanding of the brain and the self. That being the case, they will surely wonder why, in the face of the evidence, we chose to continue the death-conventions of earlier civilisations, condemning billions of brains (and, potentially, *selves*) to dissolution.

Those that stand against cryopreservation presumably consider that their viewpoint has a moral basis. They may see people who commit to such procedures as selfish, insane, or greedy. But even if we have *no* interest in cryopreservation, why do we not at least *donate* our brains to scientific research as a matter of course? The accrued beauty, complexity, and embodied sapience of a unique brain is *negentropy* incarnate;[34] once destroyed, it is utterly irreplaceable. I think that the citizens of the future will comprehend this truth far more deeply than we do.

Any 'civilisation' that is not truly civilised – to date, that applies to all of them – is bound to collapse. The kind of massive, diverse, magnanimous liberty that would kindle then sustain a future *Culture*-type civilisation is, as of yet, absent from ours.[35] A stubborn adherence to the death conventions of earlier ages is but one marker of a substratist, dualist, and therefore unstable civilisation. Other markers, as I have discussed, also abound in ours. The need to remedy this situation is pressing.

My stances on bioethics, science, religion, emerging technologies, social engineering, and other areas mark me out as a technoprogressive. In opposition to such radical freedom and openness stand the bioLuddites: those (mostly on the political Right, but also some on the Left) who regard the human race in its current form as the eternal gold standard in being.[36] They would have us locked in this substrate (these fragile bodies), and into this short frame of existence. For the sake of true liberty in diversity, they must not be allowed to prevail.

I happen to live in *this* era. I question, as I must, the morality of much that is considered normal. As it can be hard to bear the myriad stupidities of this time, I try to focus on the 'long now'[37] – striving to see this moment as it might stand in relation to moments hundreds, or even thousands, of years into the future. Over fifty years ago, in his trailblazing book *The Prospect of Immortality*, Robert Ettinger put words to this mindset:

All problems take on a completely different perspective in the long view. When the future expands, the past shrinks; historical affronts lose their sting, and vendettas their fascination. The words of the song then make self-evident good sense, that is, to eliminate the negative and accentuate the positive.[38]

Sometimes, it seems, projected forward in time is the only way to be, here, now.

# 13 ABEYANCE

∞

But there is nothing in biology yet found that indicates the inevitability of death.

—RICHARD P. FEYNMAN

## SHIVERING JEMMY

Are there parallels between our moods and the weather? I don't mean the way we might feel optimistic because the sun is shining, or low because it has been raining non-stop for a week. I mean the way that, sometimes for no apparent reason, moods seem to change: a growing feeling of happiness, like the sun emerging from behind clouds; a feeling of 'coldness' and isolation, like winter frost setting in.

Listening to a weatherperson making confident-sounding predictions on a Saturday of how the weather will be next weekend, one could be lulled into thinking that weather forecasting is a straightforward scientific discipline. We should be sophisticated enough to understand that the long-range *Farmers' Almanac* style weather predictions of the past had more in common with magical 'scrying' than with any form of science. Nevertheless, this kind of pointless speculation is still firmly a part of everyday polite discourse. And so, by extension, we have a stubborn habit of swallowing whole the predictions – even the hopelessly long-range ones – of TV and internet weather forecasts. Is that because we have a full grasp of the complicated and fractious meteorology involved, or is it because they have attractive, professional presenters, and engaging graphics?

In his *Sandman* graphic novel *Season of Mists*, author Neil Gaiman introduces us to the wonderfully-named and -represented character Shivering Jemmy of the Shallow Brigade. Jemmy is, in the story, the embodiment of *chaos* portrayed as a little girl holding a balloon. (Gaiman got the name from an old slang term for a beggar;[1] for my purposes, it's serendipitous that 'jemmy' is also a colloquial UK term for a small wrecking-bar used in demolition.) Her antithesis, Kilderkin of Order, manifests at the gathering as a lidless cardboard box that outputs its neatly-printed, succinct communications on strips of ticker tape.[2] Shivering Jemmy acts in a bizarre and delirious manner, sometimes blowing up into an enormous, threatening monster. Kilderkin behaves just as you would expect.

We use the term 'chaos' liberally in our everyday language, to describe situations that appear hopelessly out of control and lacking all the markers of what we would regard to be orderly behaviour. Originally stemming from the Greek word *khaos*, meaning 'a vast chasm or abyss', the word refers to a state of confusion and disorder. It would also be fair to say that the scientific study of chaotic systems has recently moved into the popular consciousness, with terms like *chaos theory*, and a related one, *the butterfly effect*,[3] bandied about in popular books and films.

If we take the 'one damn thing after another' effect of entropy, and look at what it means at the scale of the very small (but larger than the quantum scale), we find that lots of very tiny damn things are affecting lots of other very tiny damn things, and all in extremely quick succession. Multiple events of this kind will happen simultaneously, but there must always be, as causality dictates, *some* interval between the cause and the effect of any *individual* non-quantum event viewed in isolation. An individual water molecule can evaporate off an ocean and become part of a cloud; it may seem obvious that this will, in a minute way, increase the *size* of the cloud, but is it so obvious that it will also, again in a minute way, affect the *trajectory* of the cloud? In comparison with the scale of these kinds of effects, the flapping of a butterfly's wings on the other side of the world begins to look like a huge intervention.

In this sense, our world – and the universe that contains it – behaves *deterministically*: The way, and order, in which events occur has been determined by events that preceded them. Discussion of determinism could lead us deep into the troubling area of whether or not we actually have free will: If everything is, in this way, *determined*, then do we have any choices in the ways we act? But I'm going to stick with the theme of the *chaos* inherent in such unfoldings. (I even find the chaos oddly comforting in that, from the human perspective, the macro-level unpredictability it brings about 'blends out' the starkness of pure determinism.)

Prediction of weather patterns on the *macro* scale appears, at first glance, to be immune from these kinds of micro-scale concerns: Experts make an observation of where a weather front is *now*; they take into account variables including familiar ones such as wind speed, temperature, and barometric pressure, but also many less-familiar ones including *albedo*, *moisture advection*, and *retrogression*; the dataset is vast but fortunately they have invested in powerful computers that will 'crunch' all these variables and spit out an accurate forecast. They can even, strange as it may seem, factor in some numbers to cover *chaotic* effects.

There are patterns in chaos. This may sound like a contradiction in terms; in many ways, it is. When *computer models*, incorporating 'rounded off' input data, are run over many iterations (cycles), the plots of the results show a tendency (a *statistical bias*) towards certain characteristic patterns. These quasi-regularities are known in the field of chaos theory as *attractors*. A *Lorenz system* plot, for example, reveals an interestingly butterfly-shaped type of *strange attractor*. The *strange* variety is one of the three main kinds of attractor, the others being *point attractors* and *periodic attractors*. The fact that Edward Lorenz saw the emergence of this attractor pattern while developing a mathematical model for atmospheric convection is clearly relevant to the subject of weather forecasting; but it's important to note that the attractor principle applies to the modelling of *any* system that is *sensitive to initial conditions*.[4]

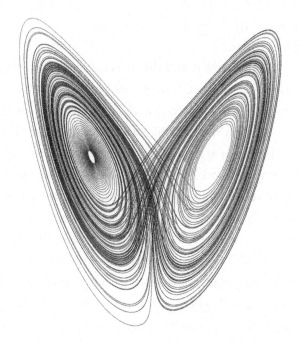

**Figure 5:** Butterfly-shaped attractor of Lorenz system. (Image: Dschwen, *Lorenz_attractor.svg.*)

Why, you might well ask, were the input data 'rounded off' in the first place? The answer is that it would be impossible to create *any* models of systems such as our climate if we had first to collect *all* the data about the initial conditions. How could we ever collect all the information about the weather? Every raindrop? Every snowflake? Every cloud formation? Every photon hitting the Earth's atmosphere? The uncertainty inherent in the initial conditions translates into *errors* in the initial variables fed into the computer model. Tiny errors grow exponentially, 'infecting' the model as it runs; the more time that passes the less useful any predictions become.[5]

Another way of looking at attractors is to see them as the *states* that a system tends to fall into: the 'valleys' lying between the peaks of what is known as the *phase landscape* or *phase space* into which a teetering condition on a 'ridge' will (like a teetering object, especially a moving one) tend, over time, to slide or tumble.[6] A marble set spinning around the inside of a bowl describes an attractor pattern as it spins down towards the bottom, losing energy along the way. So a plot of the trajectory of a marble spinning in a bowl could be used as a model for something: for modelling the likely trajectories of marbles spinning in bowls. This is a comparatively simple system, but predictions based on models of it would still be error-prone. Now try to think of an entire 'climate system'[7] as our bowl: We don't know how to define its boundaries; we don't know how to define our 'marble'; we don't know which variables to include; we don't know the exact values of the variables we choose; we have no previous reliable models upon which to base our statistical analyses.

By the definition of chaos that incorporates attractors, climate behaviour can be chaotic. But it is also something 'worse': Climate is a *complex, nonlinear* system – one that is highly sensitive to initial conditions and for which the precise initial conditions cannot be known. More computer-based number-crunching power will not solve this fundamental problem, nor will more complex and detailed models. Forecasters can, then, make only general predictions about the weather. They can, for example, look at a storm front and make a reasonable assumption that, based on wind speed and direction, it will reach a particular area of a particular landmass within a particular time window. They cannot, however, make very detailed or long-range predictions, due to the fact that relevant features, such as clouds, seem to be emergent properties of the system; as mathematician David Orrell states, 'Their behaviour cannot be computed from first principles.'[8]

Complexity breeds ever-greater complexity. We can, and will, come to better understand complex systems, but ultimately, the war on chaos can never be won.

## REAL RAIN AND VIRTUAL HONEY

Like Shivering Jemmy's threatening aspect, we now have a monster on our hands – an ugly, gigantic, and unpredictable snarl of a system. Applying the same principles to another system – the human brain – seems to suggest that we may be wasting our time trying to explain its behaviour. As I implied at the start of this chapter, our brains *do* seem to behave, at least on some levels, in this uncomputable way. Harth talks of the nonlinearity of processes in the brain;[9] Hofstadter, in one of his thought-provoking wordplays, refers to the brain as the 'careenium'.[10] This, I think, conveys aptly the uncomputable nature of some of the processes at work.

Even if it were possible to take an accurate and fully-detailed snapshot of your connectomic 'brain state' today, it would tell us little or nothing about what you will be thinking next week. We can, like the Blue Brain scientists, make computer models of cortical elements in the hope that they will behave in a brain-like way when we connect them together, but we cannot be sure that the model contains enough detail to provide an accurate virtual representation, never mind predict what the representation will *think* if it works.

When creating climate models, what matters to us is what they will tell us about how the weather will *actually* turn out. We need those models act as bases for *predictions* – to serve as nexting enhancement technologies – otherwise they serve no purpose. Brain-function modelling can be helpful in testing general hypotheses about how particular functions work, but it does not matter that such models may never tell us how to predict, even in a general way, what a person will be thinking in the future. That is not the point of this modelling exercise. This type of science, in contrast with meteorological forecasting, is highly experimental. Our 'criterion for success' is not thought prediction but functional brain emulation.[11] Scientists creating climate models will be disappointed (and possibly out of a job) if their models predict nothing about the weather, and instead just churn out 'weathery' behaviour. Scientists creating brain models would be ecstatic should their models exhibit 'brainy' behaviour. I see no fundamental reason why, in the near future, they should not.

When thinking about what constitutes 'brainy' behaviour, we should always try to bear in mind the important distinction between our *connectome* (our 'wired in' unique personality and memories that persist even when we are unconscious) and our *thoughts* (those transient, emergent, nonlinear patterns that arise as a result of the neural signals that dance within our connectome). Thoughts, feelings, *moods* are the stuff of 'brain weather'; connectome theory provides clues about how we might model the brain's 'climate system'.

Neuroscientist Professor Rodney Douglas states that 'it never rains inside a weather simulation because the physics of computation has nothing to do with the physics of weather'.[12] I take this (from the context) to mean he is saying that because computer models do not manipulate physical 'stuff' in their simulations we should not expect physical 'stuff' to emerge from them. I, and I'm sure many others, smell a rat. Why should the emergent property (or properties) that we call consciousness behave in the same way as the properties (emergent or otherwise) of what we call rain? It's difficult even to frame the question, because we are dealing with two entirely different phenomena. Who/what has dictated that we must have completely accurate 'brain stuff' in order to make accurate 'thought stuff'? If we thought that 'thought stuff' was something *like* rain then we might be well entitled to throw up our hands and declare its simulation impossible, but we don't think this at all.

I am not suggesting that Professor Douglas is a non-reductionist. What I take him to mean – in his perfectly *materialist* description of the 'simulation problem' – is that the process of *thinking* requires the occurrence of a host of tiny-scale *physical* events in the brain. And he is certainly right about that: Nothing happens in real brains without molecular interactions. This is the same kind of problem highlighted by Jean Baudrillard, in his *Simulacra and Simulation*: The 'map' (simulation) may be highly detailed and as big as the 'territory' (reality); nevertheless, it is *not* the territory.[13]

Algorithmic 'maps' of brains are not brains. (And it may prove technically impossible to make them 'as good as' brains.) Nevertheless, I am puzzled that Douglas' primary focus on *actual* brain-stuff has made him quite so distrustful of the claims of those who are trying to simulate the type of intelligent *patterns* generated by brains, using numbers. Real rain manifests in particular ways, but *virtual* rain can also be made to manifest. The virtual rain doesn't have the physical properties of the real rain, but it may have a host of other properties that are actually very similar to those of real rain. If we were part of the simulation involving the virtual rain, it might feel just as wet and just as 'rainy' as the real stuff. When it comes to 'real' consciousness, it may turn out not to be the specifics of the physical properties that are the *most* important element.

My mother keeps bees. She enjoys talking about how intelligent they seem to be. I tend to agree with her that they do, indeed, *seem* to be intelligent. *En masse* they certainly seem to 'strive' towards the kinds of outcomes (in the form of co-ordinated behaviour involving the gathering of nectar to produce honey, and so on) that we might expect (or might *think* we should expect) from intelligent entities. We might even hypothesise that individual bees are 'kind of like' individual neurons, and so the combined behaviour of the hive is 'kind of like' the combined behaviour of neurons in a brain. We might even posit a certain similarity between our 'beehive mind' hypothesis and the 'society of mind' hypothesis of Marvin Minsky. Deacon, too, finds such analogies useful, and goes so far as to talk of the ambiguous, distributed 'self' of a termite colony.[14]

While our 'beehive mind' analogy is, on many levels, flawed (to say the least, there do not seem to be enough bees, and they do not appear to be interconnected in the kinds of ways that we think are required to produce the thing we call consciousness), it is at least a *type* of reductionist explanation: It tries to explain how smaller elements within a system can combine to create an apparently complex higher-level outcome.

One outcome of the behaviour of hives of bees is honey; one outcome of the behaviour of billions of interconnected neurons is what we call consciousness. The analogy is becoming arbitrary, 'product'-centric, and hard to sustain. But I can understand the attraction of such comparisons: Bees are *real* things that already seem to be 'a bit' intelligent, and they sometimes seem to act as one entity (when they swarm, for example); neurons are harder to understand, but even those unfamiliar with their properties may tend to assume that they too must be 'a bit' intelligent. Despite the partial utility of this modicum view, however, we struggle to extrapolate it into the mathematical domain; for numbers, in contrast with those industrious bees, *seem* to be just plain stupid. Perhaps, then, we need to shift our attention from the raw numbers towards algorithms and what we can do with them: simulation, emulation, and *abstraction*.

Abstractions can be fully functional; getting too hung-up on whether we can incontrovertibly brand them 'real' can lead us into making empty arguments. We *could* argue, for example, that word processors are not real because they are not built out of the kind of parts we would expect to see in real typewriters. But, in fact, we don't generally worry about such things when using word-processing software, because we can *interact* with it and it performs the function(s) we want it to perform. And, unless there is some very specific effect we need to produce that can only be achieved by using a real typewriter, that's good enough; in fact, about as good as (or even better than) using a real typewriter.

The 'real rain' argument seems to rely on the premise that consciousness lies somewhere in the band of effects that can *only* be produced by physical brains. However, as we can see from the word-processor example, the breadth of comparative functionality that can be produced in virtual machines can be made to equal or even exceed the functionality of their 'physical' counterparts. This is not to say that virtual brains would necessarily equal or exceed the capabilities of biological ones, only that there is no good reason to suppose that the 'consciousness function' would *never* lie within the band of functions that could be simulated. The map is not the territory, but perhaps, under certain circumstances, consciousness-like functions can emerge from territory-like maps.

I realise that some of this will be hard to take in, but there is a wider philosophical question here that makes some of the above pale in comparison: It takes real rain to make you really wet, and virtual rain would, presumably, make you virtually wet; but how would it be possible, within the context of the normal climate of one's indexically uncertain umwelt, to tell which was which?[15]

And what about 'weathery' behaviour in real brains? By this point, we should expect to encounter *parallels* (not *equivalence*) between complex nonlinear systems that each display emergent properties. Cloud formations, as we have discussed, are influenced by innumerable tiny variables: We cannot, without knowing *all* the variables and the precise initial conditions, know what shape the clouds will turn out at a given moment. Because of this, and a multitude of other factors, we cannot always predict whether a storm will blow up, nor can we predict the exact nature of any storm that does arise. The system contains noise, and we do not know precisely what effect it will have.[16] As discussed earlier, the brain is also a noisy system: We do not know exactly what will emerge from the noise and manifest as thoughts. We are, however, inclined to follow 'trains of thought': We sometimes 'stick with' thoughts that emerge, continually re-running them. Such a train of thought might lead in a useful direction, perhaps towards the development of an idea, song, or poem. But 'negative' thoughts can just as easily become recurring trains: trains of worry, anxiety, depression.

Dennett uses the term 'perceptual set' to describe the way in which making a discrimination (e.g. identifying a picture as one of a dog) temporarily *inclines* us towards making associated discriminations (e.g. reading the word 'bark' as the sound a dog makes rather than as the outer covering of a tree).[17] Even if a thought has bubbled up randomly from the noise based on no obvious *discrimination* (apart from the fact that the thought seems, to us, to be *about* something), it can seed this perceptual set. We can extend this idea further and use it to explain what we call *moods*. Often, when we experience extremes of high or low in our moods, we can identify the causal factor. Nevertheless, there are times when we seem to be in particular moods *for no reason*.

If we take on board the idea of perceptual set, we can reason that the seed of our current mood may have welled up from nowhere but random noise. There is a nonlinear emergent process at work here: We *do not know* whether a random thought will emerge and, if it does, how we will interpret it; we do not know how that thought will be interpreted 'alongside' other thoughts currently on our minds; we do not know, then, how our moods will change. Like gradually forming a well-trodden path, we may, over time, become *attuned* towards (wired into) certain recurring modes of thought; others, recognising this, might begin to describe us in particular ways – 'happy', 'confident', 'depressive', 'bi-polar'. We also become attuned to our interests and passions: It becomes impossible to say where they end and we begin. Might we 'lose ourselves' if they change? What is changing what?

We just go forward, making and being made by those tweaks, day after day, struggling for a measure of control; turning those table legs, walking that skyscraper – there's a nascent *Peer* in all of us.[18]

## DEAD WEATHER

Unlike the activity in brains, the activity in weather systems does not stop (unless a planet loses its atmosphere, but even in that case it can still be affected by solar radiation and other factors known as *space weather*). We are familiar with the fact that a rapid process of decay sets in when we die. Because of its inherent complexity, the brain falls prey to deterioration of *function* more rapidly than other parts of the body. (By this, I mean that other 'simpler' parts of the body remain *viable*, in the sense of being potentially reusable, for longer than brains do.) The decay of 'consciousness' is linked to the decay of the substrate – the brain – carrying the pattern that allows it. It may be safe to assume that by the time a person is pronounced *brain-dead*, all neural activity has ceased; but, even if we accept this assumption, that still leaves open the question of what happens between the time of cardiac arrest and the point of brain death. Is there a fading-out? A tailing-off of consciousness into oblivion? And, if so, what happens to our perceptions? Do we, for example, perceive *time* differently during this period?

It should by now go without saying that death (or a least our *definition* of death) has changed over time. Michael Kerrigan's grimly fascinating book *The History of Death* gives a clear insight into the remarkable contrast between historical and modern views of its meaning: 'Mutterings about doctors "playing God"', he says, 'make clear how uneasy we are about these changes.'[19] In recent human history, the death of the body was believed to herald the release of the immaterial 'soul' to live on in whatever 'afterlife' conveniently awaited. The body was, in some cultures, only important up to the time of exit of the soul, after which it became an empty husk to be disposed of in accordance with cultural preference. Supernatural beliefs aside, it is also true to say that definitions of death have changed even over the course of the last few *decades*: Up until as late as the 1970s, legal death was defined using such crude indicators as the unequivocal cessation of respiration and heartbeat. Debate continues about how *absolute* death – now widely considered either the point of cessation of cortical activity or of brain stem activity – should be defined.

Using the moderate language of *definition* is one, fairly uncontroversial, way to address the topic, but I would like to put it another way: *Death itself* is a moveable feast (or perhaps that should be famine). All we have of death is our definition. If we changed that definition to encompass, for example, donated body parts, we could then state confidently (and legally) that a donor was not dead because 'his' heart is still pumping, though now in another person's body. This may sound ridiculous, and it is hard to imagine any country adopting such a legal definition, but it is not totally inconceivable. If a country *were* to adopt such a definition, there would be various consequences for its citizens: They could not, for example, inherit money from (what? expired?) relatives until *all* the body parts of that relative, wherever distributed or secreted, had ceased to function.

If we find this example utterly preposterous, it is probably because we equate death with the cessation of *consciousness*. Consciousness, we reason, does not happen in hearts, lungs, or corneas; donating them, or otherwise preserving their function, does not enable the continued survival of the *person*. This 'common-sense' materialist view is to be expected from reductionist quarters, but curiously, it also seems to be the dominant view among non-reductionists. Why, if the soul is immaterial, can it not simply migrate to the donated organ and reside *there* until it is released to the afterlife? The fact that the religious do not generally take this particular 'itinerant soul' view shows at least a tacit acceptance that there is something about the brain that makes it special as compared to the other organs.

Even going back to a more straightforward, modern, *reductionist* view of death is not without its complications. If we accept that death is the cessation of all *conscious* activity in the brain, we still have difficulty in defining exactly *when* that absolute cessation occurs. The fact that an EEG reading of a recently-*infarcted*[20] patient may show no response does not mean that there isn't some tiny, fading 'inner voice' crying out from the depths for urgent intervention. Leaving the 'deceased' for twenty-four hours before checking the EEG again will ensure that any 'inner voices' have stopped – *faded*-out or otherwise – but it will also likely ensure that all brain tissue has become decidedly non-viable. Jemmy has moved in again, this time kitted-out to undertake a swift and decisive demolition job.

Claims of near-death experiences – visits to heaven, meetings with dead relatives, long white tunnels, etc. – often include reports of feelings of euphoria. Neuroscientists have long theorised that these claimed experiences and sensations might be the result of the release of chemicals, such as endorphins, which may have the effect of easing the dying brain into final shutdown.[21] Those who claim that they were *completely* dead prior to resuscitation ignore the fact that experiences require an at least partially-functional substrate upon which to run. Verifiable reports of post-*dissolution* experiences would disprove that statement, but they are, for obvious reasons, hard to come by.

Throughout our lives, our bodies 'strive' to maintain our brains. If, for example, we are starving to death, processes within the body will consume the accessible nutrients in our other tissues in order to sustain our brains. The body is, in many ways, a vehicle and life-support system for the brain. If, for whatever reason, the body is unable to supply the required nutrients to produce healthy oxygenated blood, *ischemia* – restriction of blood supply – will quickly damage the brain. How quickly? Because the brain is highly *aerobic* (it uses a lot of oxygen), the standard assumption is that its tissues will have suffered significant damage after as little as five minutes from the time of loss of oxygen supply. It is important to note that this 'five minutes' applies to brains at *normal body temperature* (37 degrees Celsius).

Brain tissue, like all other soft tissue in the human body, needs oxygen to survive. Under normal circumstances, the lungs take in oxygen from the air we breathe, and the heart pumps the newly-oxygenated blood cells around the body, providing the cells of the various tissues with their vital supply. The lungs are also responsible for expelling the waste product of this process, carbon dioxide, from our bodies. If, for any reason, the heart *stops* pumping oxygenated blood, we have an obvious problem – one that requires immediate and decisive action. *Cardiac massage* is very unlikely to restart the automatic pumping action of the heart, but by the external action of repeatedly squeezing the blood from the chambers of the heart, *some* oxygenated blood can still reach the tissues, thus retarding *ischemic* damage. A powerful jolt of electrical current across the heart (from a *defibrillator*) is likely to restart it, but the heart may quickly stop again if significant arrest damage has already been done, or if it was weak from pre-existing damage.

Cerebral ischemia leads to *cerebral infarction* – the onset of damage (*necrosis*) of the brain tissue. Necrotic tissue is dead tissue: The *mitochondrial* 'engines' of the brain cells – having lost their essential supply of fuel (in the form of oxygen) – can no longer produce *ATP (adenosine triphosphate)*, a nucleotide essential for cell metabolism. Without ATP, voltage recharging of the cell membrane stops. As a result, cells cease to function, and begin a rapid process of decay. The ischemia involved in death by 'natural causes' is of the *global* type (in contrast with the *focal* type seen in certain forms of *stroke*): The heart has stopped pumping, and, without mechanical intervention, the loss of blood-flow causes the global onset of necrosis throughout the brain. *Global cerebral ischemia* is, in medical terminology, the most severe form of *hypoxia* (reduced supply of oxygen). There are, of course, other ways to kill a brain, but if we imagine a person dying either at home or in hospital of 'old age' then the process will be, broadly, as I have outlined here.

Scientific literature on post-mortem brain-cell decay usually includes the technical terms *apoptosis* and *necrosis*. Apoptosis (as mentioned briefly in Chapter 8) refers to *programmed cell death* (PCD), which we can think of as a kind of systematic self-destruct sequence initiated either by the cell itself or by signals from neighbouring cells. Necrosis results directly from the injury (or *insult*, to use the technical term): Circumstances of restricted oxygen supply, for example, can directly damage cells, causing necrosis.[22] In necrotic brain tissue, the cell membranes have become more permeable and the cells have begun to dissolve into the extracellular fluid; some cells undergo *autolysis* – their cell membranes *lyse* (split) as their own enzymes digest them. Communication between neurons breaks down in chaotic fashion: Surviving neurons may continue to fire but in an environment of growing disarray; dying neurons, now unable to maintain a voltage differential, depolarise and cease to produce action potentials; *excitotoxicity* – where neurons are overexcited to the point of destruction – is induced by the sudden excess of free *glutamate* (a neurotransmitter normally present at synapses but fatally excitotoxic when applied to the outside of isolated neurons);[23] supporting *glial cells*, certain types of which once ensured the safe transport of action potentials along axons, now slough away leaving neuronal networks broken and bare.

You can see from my description of this process that brain tissue will, after such necrosis/apoptosis sets in, eventually degrade into little more than a membrane-enclosed 'soup' of dead and dissolving post-cellular material. We can now consider the chaotic dissolution of the structure – and therefore of the *person* – complete. Removing the dead brain at a relatively early stage and preserving it in a formaldehyde solution (*formalin*) would firm up the tissue,[24] allowing it to maintain its familiar 'pickled walnut' shape. Undisturbed and unpreserved it will quickly become host to maggots and will, due to normal bacterial processes, rot.

I have outlined the process of brain death and, therefore, the resulting dissolution of the *person*, but I have said nothing yet about the appropriateness of allowing this to happen. Deathism is, to a surprisingly great extent, present even in modern medical practice. The 'we did everything we could' attitude of many medical professionals is not always justified. Some physicians are well aware that many brains are still viable after the fabled 'five minutes', but choose not to make an issue of this because the resources, procedures, and *conventions* of current medical facilities are not well-attuned to dealing with maintaining viability for longer periods. Other doctors may have either religious or vague, dualistic notions that allow them to believe that such patients are completely dead.

In truth, the human brain remains viable as a *coherent structure* for far longer than most medical professionals would have us believe. If you find this disturbing then you may have reason to. There is, of course, a difference between a brain remaining viable in this technical sense and being able to resuscitate a person from a state of prolonged cardiac arrest without major brain damage. A brain can continue to function indefinitely if the cardiac-arrested patient is put on *cardiopulmonary bypass* (a heart-lung machine), where the task of circulation and oxygenation is undertaken by mechanical means. (This technique is often used in medical procedures that require surgery to the heart that would be complicated, or made impossible, by its beating and normal blood-flow.) It is, then, normal for medics to 'artificially' maintain brain oxygenation of patients that are expected to recover, but it is considered pointless and wasteful to continue such assistance for those that are deemed – by current medical criteria – dead.

Emerging medical technologies such as *oxygen injections*[25] will render heart-lung machines unnecessary for most cardiac-arrested patients. Once made widely available in homes and workplaces, such new emergency treatments should ensure that spending a few minutes without a heartbeat will no longer result in severe ischemic injury or death. Micro-particle-based oxygen injection treatments may well be the early precursors to nanoengineered *respirocytes* – artificial red blood cells – which could do away with the need for a heartbeat altogether.[26]

There is a clear difference between the way I and others involved in cryonics use the term 'viable' and the way it is normally used in standard medical parlance. *We* are talking about the *structural coherence* of the brain, while *they* are referring only to the fact that they know of no way to bring the patient back to life *right now*. This, at least, seems relatively clear-cut. But it is not. Many more patients could, in fact, be saved using currently-available medical procedures if medical professionals, hospitals, and government health bodies would lay down their dogmas and conventional attitudes. The simple act of *cooling* relevant critical patients, at either the scene or beginning the moment they arrived at hospital, would, by decreasing metabolic demand and stabilising cell membranes, slow the degradation of vital tissues and so increase survival prospects. While it is not applicable to all critical cases, the fact that this procedure is not already routine is, to me, frustrating and baffling; it seems a misapplication of the precautionary principle. Cooling *is* the precaution. Why deny the patient the protection it could provide when it would be so easy and safe to implement?

*Therapeutic* (also known as *protective*) *hypothermia*, where a patient's body temperature is lowered (under anaesthetic), will not cure a patient. What it *will* do is provide more time to investigate problems and repair them.[27] Some of the more forward-thinking medical establishments have already adopted this procedure, but they still tend to use it in only a handful of critical cases; it has recently been trialled on gunshot and stabbing victims.[28] The technique *has* gained significant ground in certain areas such as in highly-invasive neurosurgery and in treating *neonatal encephalopathy*. Studies have also been undertaken that have demonstrated its effectiveness in improving outcomes following heart attack and stroke.[29] Some stroke-related studies are testing to see if cooling body temperature even by as little as one degree Celsius (using a special cap, and while the patient is conscious) can limit ischemic injury.[30]

More advanced techniques in development involve drugs which will lower (or allow to be lowered) the body temperatures of conscious patients, and the use of chemical *perfusate* (which has long been in use for cryonics) as a more hypothermia-tolerant substitute for blood during complex surgical procedures.[31] And NASA's Innovative Advanced Concepts (NIAC) programme has recently funded a proposal by Space-Works Engineering to develop a method whereby body temperature could be reduced easily, safely, and reversibly by up to 6°C, inducing 'torpor' and thus allowing astronauts to travel in a state of 'deep-sleep stasis' for many months.[32]

Concerted efforts to maintain the structural coherence of brains after failed resuscitation are rare in standard medical practice. This lack of intervention will result in the eventual dissolution of all information-carrying capability within what was once a highly complex and massively parallel bio-based computational machine.

The structure of Albert Einstein's brain is, for obvious reasons, of great interest to neuroanatomists. Shortly after Einstein's death in 1955, his brain was removed by Thomas Stoltz Harvey. Initially preserved in formalin, it was later dissected into 240 'blocks', which were then encased in a plastic-like (and rather flammable) substance called *celloidin* (or *collodion*).[33] After their existence was revealed in 1986, scientific study of the preserved sections proceeded in earnest, and various hypotheses have since been put forward in attempts to explain Einstein's cognitive abilities. The brain is no larger than average. It may have somewhat larger-than-average *inferior parietal lobes*, and an enlarged (in area) but truncated (in length) *Sylvian fissure*. This fissure – a major lateral (to the side) fold in the surface of the neocortex – may, in Einstein's case, have enfolded a greater area of cortical matter than average (remember the 'scrumpled table-napkin' description from Chapter 10). Indeed, a higher-than-average level of *convolution* may feature in many areas of its neocortical tissue. The brain may have a larger-than-average corpus callosum, indicating unusually-strong connectivity between its hemispheres. Early claims that it has a higher-than-average ratio of glial cells to neurons have since been refuted.[34] Indeed, thus far, all claims of specific, significant differences between Einstein's brain and the average brain are, at least, contested. And even where such differences prove present, correlations between them and increased intelligence are often weak.

Modern chemical preservation techniques can preserve a relatively (compared to earlier techniques) high level of detail. Brains donated to medical science are treated with care and reverence because well-preserved specimens are rare and extremely valuable for research purposes. It is possible, using a diamond-bladed slicing machine called an *ultramicrotome*, to cut a whole brain (set in plastic) into microscopically-thin slivers. These slivers – measuring only tens of nanometres in thickness – are placed onto slides, and then scanned at very high resolution to build up three-dimensional computer models.[35] A computer model of a brain at this level of fidelity can help neuroscientists to learn more about brain function and the biological factors that can cause malfunction.

While providing a valuable and fascinating graphical record of the gross (compared to the scale of neurofilaments, for example) structure, this type of brain preservation does not confer any form of psychological continuity. If this looks like a statement of the obvious, then it may be worth noting how many other 'common-sense' notions I have already challenged. *If* it were the case that brains were able to perform their computations on simpler, more stable, atomically lattice-like substrates, rather than on messy, wet, unstable biological ones, then it *might* be possible to maintain all the required structure through crude chemical preservation techniques. If the patterns constituting selves were traceable from the gross structure of the brain, a modern, high-resolution scan of ultramicrotome-thin slices might preserve *everything* required. Unfortunately for us, neither the slice-preserved physical substrate nor the virtual reconstruction, as I have here described them, seem to have the necessary fidelity.

## HE'S DEAD, JIM, BUT NOT AS WE KNOW IT

Having already discussed some of the *why* in the suggestion of attempting to save dead persons from oblivion, it is time to talk about *how* – on a *practical* level – this might be achieved. How do *cryonicists* deal with the onerous task of attempting to maintain that all-important structural coherence? In order to keep this description relatively concise, I will try to use terminology that I have already introduced.

The core principles of cryonic neuropreservation (for biostasis) are not so different from the principles of chemical preservation of the brain (for research purposes): We have a substrate (the brain) that contains information that we wish to preserve (*fix* in place). The *information* sought in the research case is manifest in the gross structures of the various brain areas, but also (to whatever extent it is possible to preserve this) in the configurations of neurons, glia (glial cells), and neurites. The information sought in the cryopreservation case is *at least* at the resolution of neurons, glia, axons, and dendrites (connectome resolution), but *might* be at the resolution of precise synaptic states ('synaptome') and may *even* be at the scale of nanometre-scale components such as neurotransmitters and neurofilaments.[36] While *both* types of preservation seek to retain the information at the highest possible level of detail, the consequences of *failing* to capture the information at an appropriate resolution are quite different: disappointment for the research scientist, irreversible death for the cryopreservation case.

I will get onto the issue of what I mean by 'appropriate resolution' shortly, but first I would like to talk about some of the reasons why preservation even at the level of *cells* is fraught with problems.

*Freezing* a brain – while vastly better than incinerating it or leaving it to rot – will do much damage to the cells comprising it. Take a nice, fresh, soft vegetable such as a courgette (zucchini), slice it up, and place it in a deep freeze. Leave it there for a month or two, then defrost it and compare the slices to ones of fresh courgette. We know the outcome of this experiment: The fresh courgette will have (some) flavour and a crisp bite to it, while the defrosted one will be soggy and flavourless.[37] Ice crystals are the enemy of quality frozen food. Fresh food contains a great deal of water, and this water expands, crystallising as it freezes – crushing, bursting, and generally mangling the cell walls of the frozen item. (Debate continues about the exact mechanisms by which cell damage occurs in freezing, with some scientists suggesting that *solute toxicity* caused by the increasing concentration of *electrolytes* as the cells dehydrate – as well as various types of *osmotic stress* – are the main culprits.[38]) Once defrosted, and with the fixative scaffold of ice gone, many of the cells fall apart and blend into the watery residue. As with frozen and defrosted courgettes, this general outcome also applies to frozen and defrosted brains. We know that blast freezing (a very quick type of freezing) of foodstuffs yields better results, but unfortunately, this technique is too crude to work well on brains.

Although pioneering work – using HeLa cultures – helped to establish standardised methods for safely freezing human cells *in vitro*,[39] we are here discussing the preservation of complex cellular structures *in situ* (if not exactly *in vivo*). What we require is a technique that does not jostle out of position, squeeze out of shape, over-salt, or otherwise damage the fine structure we are trying to preserve. It must also be a *reversible* technique: It will not do merely to stabilise the tissue for longer-term storage (as in the case of chemical preservation for neuroanatomical research); we must be able to release the preserved cells and their interconnections from whatever 'scaffolding' we have used, and then, at very least, be able to *infer* their pre-preservation state from what remains.

One obvious factor to try to mitigate is the damage from ice crystals: We need to get the water *out*. But we need to get it out in a way that does not leave the cellular structure unscaffolded and therefore prone to collapse. Freeze-drying, or otherwise drying out the tissue, will not work: We know the stodgy outcome when we 'reconstitute' certain types of previously freeze-dried foodstuffs. Other types of foodstuff, such as steak, become 'leathery' when freeze-dried, and, once 'reconstituted', bear little resemblance to their original state. In order to preserve the cellular structure, we must replace the water with something else.

Before getting into the details of how these principles influence the means of preservation of *brains*, I would like to rewind a bit to discuss *intervention*. Having established that the time-window between pronouncement of death and dissolution of structural coherence is small (although the question of just *how* small is open to debate), it should now be clear that rapid intervention is of great importance to cryonicists. In a 'best-case scenario' – one where we assume (and it's a big assumption) that the requisite legal and medical permissions are in place – a cryonics 'standby team' may enter a medical establishment, at the point of pronouncement of death, to take over the care of the patient/body. The team first places the body in an ice bath. By means of a 'thumper', they ensure that blood continues to be pumped from the chambers of the heart and throughout the tissues of the body. Various protective chemicals/medicines (including anticoagulants such as *heparin*) are administered via intravenous lines. The body, now considered 'stabilised', is then transported from the medical establishment to the cryonics facility, for further procedures.[40]

I need not elaborate much on what Alcor calls 'transection of the spinal column'. As you might guess, carrying out further cryonic procedures on a comparatively small part of a body – a severed head – is somewhat easier than carrying out similar procedures on a *whole* body. Nevertheless, we should not underestimate the technical skill required in precision cutting of this type; the surgeons need a clean cut, in exactly the right place, at exactly the correct angle. Following transection, blood is washed out with a solution of perfusate prior to the introduction of *cryoprotectant* solution directly into the exposed arteries.

Having perhaps heard of the 'brain in a vat' thought experiment from philosophy,[41] or of other disembodied-but-alive brain scenarios from science fiction, you might be wondering why the cryonic-preservation team does not remove the brain from the skull. Surely, as in the tale of the transplanted brain that I heard at primary school, a brain free of its casing would be better prepared for future re-housing. With such unimpeded access to the brain, it would certainly be easier to perfuse cryoprotectants more evenly throughout its tissues. However, disconnecting a brain from its housing (never mind hooking one back up again) would prove unnecessarily complex and potentially damaging. The paths of the *facial, trigeminal*, and other main nerves that weave and loop through the bony 'blocks' and 'rods' of the skull are difficult to trace. Better, then, to leave all the *intracranial* nerves undisturbed, but with the *afferent* (incoming) and *efferent* (outgoing) ones accessible via the localised bundle in the transected spinal column. The exposed arteries, as mentioned above, are now our aqueducts to the soft tissues of the brain.

The parallels with the 'frozen courgette' example are now becoming clear. The water, here present in blood plasma within the veins and arteries, would cause cell damage, so it has been, to the greatest achievable extent, removed. The cryoprotectant solution contains antifreeze-type chemicals that we would not wish to find in our frozen courgette (as they are toxic), but which are essential for the purposes of the high-fidelity/minimal-damage 'brain-freezing' process that follows.

Correctly perfused into the tissues, the cryoprotectant will ensure that, once appropriately cooled, any remaining water 'vitrifies' rather than freezes. This glass-like state scaffolds the structure without promoting the kind of cell-damage seen in the freezing of body-tissue. While no *brain* has ever, to date, been returned to functionality after vitrification, the principles that apply are the same as those successfully demonstrated in vitrification of other organs and tissues: Cryopreserved rabbit kidneys have proved functional after warming and implantation; vitrified *oocytes* (egg cells) and *blastocysts* remain viable; experiments in vitrification of entire ovaries show early promise.[42]

After the drilling into the skull of two small monitoring-holes, the head is cooled to the vitrification temperature of -125°C. Over the next two weeks, it is gradually cooled further, to -196°C, before storage in liquid nitrogen inside a 'neurocan' within a compartment of the stainless-steel dewar. Four whole-bodies, or up to 45 'neuros' (but usually a combination of whole bodies and neuros) fit into each dewar.[43] Ongoing 'patient care' involves regular topping-up of the liquid nitrogen and constant monitoring with a 'crackphone'. Even brains 'suspended' in this state are not immune from chaotic effects. There is no current way of ensuring that the vitrification of the tissue is absolutely uniform; tiny, imperceptible causes – such as minute thermal stresses – can give rise to fracturing events. Even with the most careful handling and storage, human tissues are all too easily damaged in this fragile, vitrified state.[44] Tissue 'cracking' is a problem that cryonicists, and scientists involved in the wider field of *cryogenics*,[45] are striving to overcome.

I earlier called this a best-case scenario. You may struggle to see anything 'best' about it. Less-than-ideal cryonics scenarios are, unfortunately, all too common. Perhaps no standby team is available near to the hospital; perhaps the person does not die in a medical establishment at all; maybe the cause of death involves major head-trauma, rendering standard cryonics procedures ineffective. Nevertheless, you may be able to see, by now, that even the *worst* of these less-than-ideal scenarios is better than the fate of recently-deceased persons for whom *no* cryonic procedure is available. After all, in these various cryonics scenarios, I have been describing the *second*-worst thing that could ever happen to a person.[46] Perhaps you have an iron-clad reason for choosing, instead of this, the *very* worst.

While there may be discussion, among cryonicists at least, of the semantics and details of such procedures – should we call them 'patients' or 'dead bodies', and so on – there is no dispute about the fact that without the radical intervention of post-pronouncement cryonics the complete dissolution of anything that we could call a person will be both rapid and inevitable. I am quite comfortable with the term 'dead' being used to refer to the whole-bodies and neuros in cryonic storage. But I use it carefully, and only in the presence of at least a tacit understanding that these 'persons' are – by any reasonable measure – *less* dead than those decaying in the ground. I will not claim that this is any kind of 'Death Lite' or 'I Can't Believe It's Not Death'; only that, without scientific proof of its futility, it does not make sense to say that it is the *same* as ordinary death. What *should* we call it, then? I don't know; choose your label – suspension, biostasis, deanimation, liminality, abeyance.

The Shallow Brigade of chaotic decay always awaits you – it is, indeed, the ultimate fate of the entire universe. But Jemmy is a spoiled brat to whom you need not placidly surrender. You could try, like me, to take your time, resolving *not* to go gentle, even come that desperate day when she is spitting and roaring right in your face.

## ULTRASTRUCTURES

> So from your perspective I think this is all good news. For one thing, holographic representations are notably robust – you can destroy parts without losing the whole.

Steve Grand's upbeat assessment of the prospects for successful preservation came as something of a surprise to me. Scepticism about cryonics is the norm, even among those who, like Grand, seek to build artificial and virtual brains using knowledge gleaned (at least in part) from the study of biological brains.

Grand's comments came in response to questions I put to him about what he thought might be the *resolution* required to capture the unique pattern that constitutes a self. He continues:

> I'm sure there will be a great deal of regularity to this aspect, so if you knew in detail the interconnections in a frozen brain, you should, in principle, be able to reconstruct the person from this information. You may not need to retain the nanoscale molecular machinery any more than you need the

original bricks in order to reconstruct a demolished house, just as long as you have the plans.[47]

You will be able to see, on even a cursory reading of this, how different Grand's view is from that of those philosophers and scientists who insist that *the structure* and *the pattern* are fundamentally indivisible.[48] Granted, it's easier to reconstruct a demolished house when at least the shell of the original is still in place (and Grand acknowledges this by talking, at one point, about retention of the 'gross structure' of the brain). But a holographic explanation suggests that – even with the original structure *completely destroyed* – a person could be 'rebuilt' from an appropriately-high resolution 'plan' of their unique neural pattern.

And to return, just for a moment, to the issue of what 'appropriately-high resolution' means: We *do not know* for sure how detailed such plans would need to be. This is because we don't know the *effective complexity* of the brain. We don't know how to establish, in Seth Lloyd's words, 'criteria that indicate when a bit is an "important" bit, a bit of regularity, and when it is "unimportant", a bit of randomness.'[49] We don't bother drawing house plans at the resolution of the grains of sand in the bricks: We know that the effective complexity of house walls requires only *wall*-level, or at most *brick*-level, resolution. For brains, it makes sense, in the meantime, to try to preserve the fine structure – the *ultrastructure* – at the highest level of detail possible with the available technology. And that, for the moment at least, means cryonic vitrification. As alternatives to this become available (some form of 'chemo-cryo' fixation/vitrification may become a viable option[50]), I will consider them on their merits, and as a result, I may change my mind about the best way of preserving my self-pattern – the pattern that makes me the person I am. The principle remains the same, whichever specific technology is employed towards achieving it.

*Connectionist* neuroscientists have various subtly-different takes on the ultrastructure issue. The most radical – among them, Ken Hayworth – imagine a future whole-brain *plastination* process of such high fidelity that it might confer virtual immortality: A process of ultramicrotome slicing of a plastinated brain, then scanning to a computer, would render a virtual version detailed enough to attain consciousness.[51] This is a version of the 'mind uploading' scenario up until recently mooted only by sci-fi writers, in books such as Reynolds' *Revelation Space*[52] and in films such as *Transcendence*[53]. Hayworth's version of the 'uploading' process is a destructive one, in that the physical brain is destroyed in the process of scanning it to a digital substrate. Those, like Seung, who take a somewhat less optimistic stance view the idea that such a perfect method of preservation will become available with some scepticism. Like Grand, however, Seung agrees that the important thing would be the ability to infer the original structure from whatever 'plans' remain. In discussing the idea of testing the connectomic integrity of a brain, perhaps from one of the cryopreserved pets currently in storage, he states:

> If there are a few breaks, tracing might still be possible. One could deal with an isolated break by bridging the gap between two free ends that were obviously once joined.[54]

We have no good reasons to sidestep the potential implications of connectionist neuroscience. We have every reason to conclude that if the 'wires' are still intact/inferable then something of the person remains; if they are not, the person no longer exists.

Even with proof of an intact/inferable connectome within a vitrified brain, critics of cryonics would still have a wealth of other ammunition. A cartoon that always makes me laugh depicts a physicist showing his colleague a complicated formula on a blackboard covered in workings; one calculation leads neatly to the next until the flow is broken by a bubble containing the words, 'and then a miracle occurs.' *Nanomedicine* is the oft-criticised 'miracle' involved in cryonics. I admit that the idea of hordes of tiny machines repairing the wiring of a dead brain sounds incredible. This type of application would involve nanotechnology beyond even what Drexler now calls 'atomically precise manufacturing' (APM),[55] which itself is still some way off. We tend to confuse the science of building merely very small medical machines, such as targeted drug-delivery devices, with speculative medical APM. They are very different. In many ways, atomically precise, nanoscale medical devices would be more like biology than technology.

What is the current state of progress in nanotech? *Nanomaterials* such as graphene, *silicene*, and *carbon nanotubes* are already in use; synthetic molecular-scale moving parts such as cogs, levers, and even motors have been demonstrated in labs; the tools for assembling and manipulating molecules have increased in precision over the last few years; bio-engineered *protocells* can now interact with the 'nanomachinery' of viruses, and are being tested as antiviral agents.[56] If these developments are stepping-stones towards APM, advanced medical nanobots (Alastair Reynolds calls them 'medichines'[57]) are one logical, desirable, and yes, feasible outcome.

Despite all of this progress, I accept that there is a strong likelihood that there will be some unforeseen problem with the 'plans' (whatever form they happen to take), and that as a result, scientists (even ones with medichines) will not be able to reconstruct me. I *do not* accept that, because of this likelihood, I should abandon a hypothesis that if true will save me and many others from otherwise certain dissolution.

What is so special about *my* pattern, that it should be saved? I would be entitled to ask in response to such a question, what is so *unspecial* about yours that it should not?

Armed with our new knowledge of neural connections, patterns, structures, and ultrastructures, we are ready to put paid to the myth of the 'five minutes'. I raised it at the beginning of this chapter because I needed to give you a handle on the way doctors generally decide upon the point of death at present. There *are* procedures and medicines available than can greatly reduce the impact of spending five minutes (or substantially more) in cardiac arrest.[58] Even given access to this treatment, most patients would still die within a few hours. But some would not. Where, in the conventional treatment of infarcted patients, is even the most minimal application of the logic of the *fact* that people *are* the patterns 'imprinted' and running on the substrates of their brains? Why do doctors not explain this to waiting loved-ones, involving *them* in a decision about whether 'radical' preservation measures have now become appropriate? None of the answers to this question that you would usually receive from traditionally-minded medical professionals would be good ones. This text by the late Thomas K. Donaldson Ph.D.,[59] though somewhat tongue-in-cheek, helps to demonstrate the difficulty of explaining cryonics to those who are more used to pronouncing patients than preserving them:

McCoy: 'He's dead, Jim.'

Kirk: 'Bones, do something!'

McCoy: 'Sorry, Jim, there isn't anything I can do.'

Kirk: 'Why?'

McCoy: 'Because he's dead.'

Kirk: 'How do you know he's dead?'

McCoy: 'Because there's nothing I can do.'

Kirk: 'Because he's dead?'

McCoy: 'That's right.'

Kirk: 'But I was talking to him just one minute ago!'

McCoy: 'Dammit Jim, I'm a doctor not a spiritual medium! I can't bring back the dead any more than I can cure a common cold.'

Spock: 'Doctor, we could take him back to the ship, dissolve any blood clots, restore circulation, and restore homeostasis by molecular repair. He could fully resume duty within days.'

McCoy: 'Spock, leave doctoring to doctors! What this man needs is a decent burial.'[60]

The ridiculous circularity of this argument is plain. But the parallels with real scenarios involving real doctors cannot be ignored. We, as 'laypeople', assume that when at doctor tells us that a person is dead, that *must* be the end of the story. What we *should* assume, even in such terrible situations, is that the doctor has not considered *anything* outside of his personal knowledge and the accepted medical practices of ordinary hospitals. The doctor has been trained to maintain human-body function; he *does not know what to do* when he finds that he has failed, after a short series of attempts, to restore it. Unless he is prepared to intervene, radically, in this fast-unravelling scenario, his usefulness is at an end; he should (with the appropriate informed consents) step aside and allow the cryonicists to take over. Any barring of their access to the 'deceased' should be considered not just unethical but potentially criminal.

It comforts us to think of those medical decisions made as the patient dies, and is eventually pronounced dead, as clear-cut. Unfortunately, this is not always true. Sometimes, even in the contexts of decent hospitals, the conventions appear vague: opaque to the waiting relatives, lacking in urgency, confused, bumbling.

And afterwards, in the conventions of handing the dead body over to the undertaker, and in all the other standard preparations for cremation or burial, confusion reigns unchallenged and supreme. It is confusion of the very deepest kind: confusion about what has actually happened, confusion about where the person has 'gone', confusion about how to assimilate the repercussions of their permanent absence.

The loved ones of those in cryonic storage, and of those countenancing cryonic preservation, also feel confusion. With the ingrained conventions of death stripped away, some may feel *cheated* of the 'normal' modalities of the grieving process. Some get hung up on definitions: They feel they must internalise this situation as either one thing or the other; acceptance of the blurriness of the divide between life and death is a prerequisite for the 'cryoneer' or 'cryonaut' but not for those who will be, or have been, left behind. But this, if I may (once again) go out on a limb, is – under the dreadful circumstances – a 'good' problem to have and a 'good' troubling set of emotions to feel. Unlike the buried or burned person, *something was done* for this person remaining now only as a vitrified head in a container; *something was done* for this person now a vitrified whole-body in a steel canister.

Kim Suozzi, who died of a brain tumour at the age of twenty-three, during the time I was writing this book, had faced the prospect of her extinction with courage and practicality. I got to know her a little, via Facebook, after Steve Grand forwarded me a link to her story. She had been studying neuroscience, and so was well aware both of what was happening within her cancer-ridden brain and of what Alcor's procedures would involve. She decided – in hope, sorrow, clarity, some dreadfully complex mixture of emotions – that she wished to be cryopreserved upon her death: 'I wish I could give a particularly compelling reason why I deserve another chance at life, but there's not much to say.' And why *should* she have been expected to say more? She just wanted to live. The cryonics organisations listened, and now their particular *something* – you can decide how moral and worthwhile a thing you think that is – *has* been done for her.[61] She was beyond the help of *all* other agencies; let this one do its work.

I am not suggesting that the body disposed of in the 'usual' way was not treated with reverence and love, but I do mean that *everything* that was done for it – no matter how loving and caring – was, for *it*, utterly futile. Such conventions are, in every sense, for those, and *only* for those, 'left behind'.

Funeral arrangements are complicated and stressful enough, but I, in choosing cryonic preservation, am potentially imposing a logistical nightmare on my survivors. I have a moral duty, then, to make the process as easy as possible for them. However, that principle does not extend to throwing up my hands and agreeing that because it is more *complicated* to transport my dead body to Alcor's facility in Arizona than to bury it, they should just take the simpler option. The chance I am asking for is unusual, and unusually slim. It is a chance I would also gladly give to them, should they ever come to feel that it has value.

Terra once got angry with me for my refusal to agree that I would arrange for her body to be cremated, should she die before I do. My problem with her argument is that she is mistakenly (but understandably) seeing equivalence in our requests. I have asked that she be involved in trying to save me after my death, whereas she is asking that I engage in the ritual and process affirming her irretrievable death. So where is the equivalence? This is not an apparent moral dilemma for most – we already know what *convention* says about the wishes of the dead – but, in the sense I have described, it is certainly one for me, and for others who share similar views.

There is a scene in the movie *The Abyss* that I find deeply affecting. Because only one set of breathing-apparatus is available on the flooding submersible on which they are trapped, Virgil 'Bud' Brigman has to watch his wife, Lindsey, drown. Her death is part of their desperate survival plan. Bud then swims to the safety of the underwater rig, dragging his dead wife through the frigid water along with him. Due to the very low temperature of the water, the plan works: Back at the rig, after the frenzied and nearly futile efforts of Bud and his crew, Lindsey eventually coughs back to life.

The portrayal and symbolism of this awful situation would, I am sure, affect anyone with a gram of empathy. But when I watch it now, the symbolism goes, for me, beyond the ruthless bond of love displayed by these two characters. For I too am asking to be dragged – dead – through a 'medium' that I might, like Lindsey, be revived in safety. I wonder if, and fervently hope that, Terra and I might be willing and able to tow each other to safety through the frigid depths of *time*.

The implications of the ideas I examine in this book go far beyond the need for wider acceptance of the fact that destroying dead brains is the wrong thing to do. There is a chance here, with our new understanding of selves, to begin to view brains in a completely different way. This 1.4 kilogram mass of biological matter is a potential time machine that may carry you into the far future. It is you and your lifeboat bound together as one patternistic vessel. It is – if you still consider that the word can have meaning in the light of this new understanding – the transport-cradle for your *soul*.

# 14 THE WAKE

∞

> Fables should be taught as fables, myths as myths, and miracles as poetic
> fancies. To teach superstitions as truths is a most terrible thing. The child-
> mind accepts and believes them, and only through great pain and perhaps
> tragedy can he be in after-years relieved of them.

—Often attributed to Hypatia of Alexandria (370–415)

## A DARK-ABSTRACTED I

It is high tide. I take to the sea, just beyond the rope boom holding back
the slurry of dead-crab carapace and rotting seaweed. Terra takes a photograph...

I seem to be writing a book. No, more than that, I *really* seem to be
writing a book. And, to top it all, I really *feel* like I *should* be writing a book
– almost as if I were compelled to do it. How *serendipitously* my observation of the thing that I am doing fits with the thing I seem to be doing and
the thing I feel I should be doing. How convenient. How (unusually) self-
consistent.

Assuaging, through writing, the drive to find a coherent narrative of my
life, personality, and decisions now feels important to me – fundamental,
even. Just as a newborn does not realise it has to eat,[1] I did not realise that
I needed to write. And, at times, it is elating to begin to learn, as Bradbury
would have it, 'how to tip ourselves over and let the beautiful stuff out'.[2]

My story may not be beautiful, but it was in there, and screaming to be tipped. A pyroclastic flow of it.

Feeling this way just now seems, *of necessity*, part of the story I am spinning myself. Have I found my *Tao* – 'the natural path, an unforced trajectory'?[3]

Reading a good story with a 'solid' central character, such as a *Sherlock Holmes* novel, will entertain and delight; 'being' the solid central character in a delightful story will entertain even more. Playing bit-parts in such stories is not so much fun – somebody else always seems to get the best lines. And the story gravitates towards the central character: Holmes gets to 'influence' the events in the story, whereas Lastrade's second cousin twice removed (if he 'exists') doesn't seem to cause even the merest of ripples in the narrative.

You might object that Holmes does not *know* that he is the central character, and so he does not, in any sense, *influence* the events: Holmes and all the situations in which he becomes involved were written by Conan-Doyle! I don't read *Sherlock Holmes* books (Terra loves them), but philosophers find him a useful 'tool' when trying to describe the concept of an entirely fictional character who is fleshed-out in such a way, and to such an extent, as to give us clear *expectations* about the *kinds* of things he would and would not do. Although he is not 'real', he has 'character traits'. He is, inasmuch as Conan-Doyle has written him that way, *self-consistent*. This self-consistency seems even to allow for time-shifting of his character to, for example, the Second World War era, or even to the present-day: He may write a blog and carry a smartphone, but he is still Sherlock Holmes.

I have not seen the RCS[4] production that depicts the most recent iteration of my previously-mentioned persistent playwright's cryonaut character. The protagonist, Adam, is, apparently, a diver. From the pictures I have seen, the production team seems to have taken advice along the lines of 'when painting sci-fi by numbers, use a lot of 1'. Do I smell bridges burning? White ones? My decision has inspired a new narrative, but the guts of me – the astringent mix of ideas, empathies, fears, and hatreds – are not required. Like Bradbury's alien vampire in 'The Man Upstairs', I am full of strange, colourful shapes.[5] Perhaps hollowed out and stuffed –

not with coins but with chaff – I am more *convincing*. Maybe *I* am not self-consistent enough to be believable. In truth, I have no idea how self-consistent a character Adam is. Perhaps if I had not had some small voice in my subconscious telling me that I would, some day, wish to write my own story in my own way, this would not matter to me.

And now for a different, more comfortably arm's-length, tack. When I was a teenager I had musical 'heroes'. John Lydon, of PiL, was one of them. I had never met John Lydon but enjoyed the music he had written (and has since written) and performed with PiL. I also admired him for what I had read about his rebellious, political, and provocative nature as a person. Seeing him, many years later, in an advertisement for butter did something to my perception of him – it changed it in a way that I did not like. It *jarred* with me. *This isn't what John Lydon is supposed to be all about*, I thought. *He's sold out!* This was a plot-twist, involving the central character, that just didn't seem to *work*.

John Lydon, unlike Sherlock Holmes (or Adam), is a real person. I am merely demonstrating that we seek, and indeed *expect*, a certain self-consistency from 'characters' with whom we feel familiar – be they real or fictional. When applied to real people, these expectations can be deeply unfair: I do not actually *know* John Lydon as a person; I don't know everything about his personal reasoning and motivations, so I should not make sweeping judgements about him on the basis of his appearance in a butter advertisement. He has obviously found some way of reconciling himself to these new events in his life. After all, he is well practised in the art of changing himself: He once thought that he would become a schoolteacher; he thought, before his hand was injured, that he would be a guitar player.[6]

Self-consistency also *seems* an intrinsic property of objects: We have faculties that allow us to point to a mug or a mountain and say, with some conviction, what it is; most of us don't concern ourselves that such objects are composed of many smaller parts, or that their appearance of consistency would not stand up to all levels of scrutiny. But we also talk with certainty, as if to imply *solidity*, about things such as clubs and nations – 'things' that are, in fact, *abstractions*.

Scotland sighs. She is restless in her relationship with England and wishes to take control of her own resources – to assert herself. The marriage was only ever one of convenience, and, while the turbulence present earlier in their relationship has abated, their ongoing differences seem irreconcilable. Perhaps it is time to walk away.

Although the above text was composed by me, it is the kind of emotive anthropomorphisation of nations that I have, from time to time, read in books and political literature. It is easy to slip into this mode of language when speaking of nations, as they each seem to have a *character*. In reality, nations are human constructs that only maintain self-consistency until the factors binding them together change beyond recognition or disappear altogether: the dictator is assassinated and the people take over; the populace decide, in a referendum, to secede from the crumbling empire. Some changes are gradual (e.g. the people begin to feel a 'cultural reawakening', and this sentiment grows steadily over time), and some are rapid and decisive (e.g. there is a military coup that topples the government).

Philosophers, such as Parfit, find this sort of 'nations' explanation useful because nations are abstractions that we can all understand: We are familiar with the 'component parts' that go to make up a nation, even if we do not fully realise that it is, as a whole, an abstraction.[7] We, effectively the *agents* (more so if we live in a democracy), are some of those component parts, but structures (governments, armies, police forces) and rules (laws, social norms) are also key binding factors maintaining the coherence of the whole. The source of *overall* control of the nation's identity as a nation is often difficult, if not impossible, to pin down: There is usually a leader, and there may even be a monarch, but it is not usually possible to say that *any one individual* is the *absolute* and *complete* source of the nation's identity.

And nations have their own type of *continuity*: The people gradually die and are gradually replaced, the form of governance changes, the 'cultural identity' evolves as peoples from different ethnic backgrounds bring their traditions into the mix; but a nation can continue to be regarded as the same nation if a 'critical mass' of characteristics (usually including its name) stay the same.

The above, you might think, is obvious sleight of hand: that I am trying to lull you into a false sense of insecurity with my transparent analogies. You might have strong objections: The nation doesn't *know* anything; it is not *conscious*; it cannot *speak*. A bit of time spent mulling over such objections will leave you wondering whether they are as clear-cut as they at first appear. There are certainly, for example, times when an 'indecisive', even 'mumbling', nation seems to speak clearly, 'as if with one voice'.

Mugs, mountains, nations, selves; in 'perceiving' such phenomena, what our sensory and cognitive faculties are actually doing is allowing (or forcing) us to privilege some relational qualities and levels of complexity over others. 'Thingness' is tricky, messy, relative, and conditional.

## THE MONSTER OF THE DEEP

Why do I feel driven to keep plying you with these irksome analogies? In the absence of feedback from you about whether or not I am succeeding, I persist in trying to break a habit (a habit that was, for me, broken by writers such as Parfit and Dennett): a habit of regarding the *self* as something very different from a highly-sophisticated abstraction.

Without the new awareness that flows from the opening of this particular dam, we are able to understand neither my decision to sign up for cryonic preservation nor a host of difficult truths that will come to hammer on the door of our intellects throughout the course of our lives. We all want answers, but I appreciate that not everyone wants *this* type of answer. Some of our most cherished notions are bound up in our sense of some special, undefinable essence that makes us what we are. For many people, letting go of that idea is painful or unthinkable. For me, it was like coming up for air.

Am I being destructive? Have I smashed the car to get the driver out? (Or have I, alternatively, smashed the *driver* to get the *car* out?) Do I break my clockwork self in seeing what makes it tick? I can only try to assure you that that is not my way. I am no nihilist. I am a builder, a borrower, a fixer, an *exapter*;[8] or at least that is the story I tell myself. Nothing is broken, but something has gone away: some useless stuff – a slew of bad concepts.

I am *not* claiming that *the self* is not a useful abstraction. On the contrary, selves are a *profoundly* useful result of human evolution. How would we interact without them? How would we even be able to pick ourselves out of the 'background'? Perhaps we could still interact with our environment using the sort of 'built in' self-boundary-definitions of, say, sea-slugs or red sea-urchins, but we wouldn't have any plans, dreams, hopes, fears, aspirations, friendships, *loves*.

In his 1970 book *I Seem to Be a Verb*, engineer and inventor Richard Buckminster Fuller (perhaps best known for his work on geodesic domes) wrote:

> I live on Earth at present, and I don't know what I am. I know that I am not a category. I am not a thing – a noun. I seem to be a verb, an evolutionary process – an integral function of the universe.[9]

Much as I love this quote from Buckminster Fuller, I still think that 'things' *can be* a useful springboard to understanding and self-recognition. We need to start *somewhere*. Isn't a thing the poorer for not knowing what it *is*? You might have expected to find things that don't know what they are 'outside' of you (grass, buttercups, rocks, fence posts), or even 'inside' of you (bones, blood cells, bacteria), but you probably didn't expect to find one sitting in your chair, or wherever you are located reading this right now. This, in my view, is a bad state of affairs, and one that cannot continue without, in various direct or indirect ways, causing you a deal of cognitive dissonance.

Persons instinctively object to being 'objectified' in this way. You may not want to be classed as a 'thing'. Nevertheless, you *are* – at *this* level – an object of *sorts*: A very special one – you happen to be a *sapient* object. Crucially, the emergent properties of specific collections of smaller 'things' endow you with the ability to transcend your 'thingness' to become an abstracted sapient self; Buckminster Fuller clearly recognised the higher-level process and wished to emphasise its importance. I remember a discussion with ethologist Professor Robert Hinde, and his wife, from some years ago, where they objected to my referring to humans as 'machines'. Perhaps if I had been less keen to shock, and had instead concentrated on the emergent, *virtual*-machine nature of consciousness, my choice of words might have met with less resistance.[10] Because the machine comparison appears to strike at the sacred-feeling, brimming, churning complexity of persisting selves – at their self-created wonder – perhaps it seems a reduction too far. In contrast, exalting the *emergent* properties affords us a certain amount of protection from the base 'nuts and bolts' of self-creation; abstracted human minds seem less objectionable than objectified human egos. Finding the right terminology in this tangled forest of abstraction is *difficult*.

Matter is itself an abstraction. In stressing, as I did at the beginning of this book, that everything is made of atoms, I did not mean to give false respite from the abstractive onslaught. Atoms are, after all, mostly empty space (the vast majority of their mass is contained in the tiny *nucleus*); and the 'sub-atomic particles' within them are better (though by no means completely) understood as 'packets' at various energies (measured in *electronvolts*), or even as *probability waves*,[11] than as bits of stuff in the macro sense that we experience through our bodily senses. It so happens that atoms – despite their 'immaterial' nature – persist in ways that allow them to bond together and become part of larger and vastly more complex structures such as human beings. Matter is an abstraction that persists at *our* 'level of being' in *our* umwelt.

So the normal distinction between material and immaterial is, in many ways, a false one. It is perhaps little wonder, then, that persons become confused about their constituent ingredients. *Mind*, for example, seems ripe with immaterial potential, providing non-reductionist opportunists with endless byzantine ways to conjure something (extra) from nothing. But we have no need of chicanery; we know that matter can produce what we call 'mind', and, crucially, we know that it is *all* that is available and necessary to create the phenomenon of mind. Emergence allows this. The false dichotomy of the *science*-based material/immaterial debate is one of the reasons why some in the fields of cognition and artificial intelligence have begun to avoid using terms like *materialism* and *reductionism*. But unfortunate associations with vitalism and crank medicine make integrationist terms like *holism* problematic.

I hope that you have a good idea of what you are made of and how, in evolutionary terms, you came to be. But we may have overlooked something important: what we are *not* made of; what is *not* an ingredient in the mix. Although matter is an abstraction, insofar as we know anything, we know there is nothing in this phenomenon that does not arise from it. It *seems* like there is something else, but that is only because the human brain has happened to evolve, through genetic and semiotic evolution, in such a way as to make this kind of seeming *seem* to be the truth. It isn't.[12]

Harth sees consciousness as 'the *joiner* of the countless bits and pieces in the world around us'. But as this 'joiner' *also* arises from 'countless bits and pieces', what could it mean, at the level of the various processes in our brains, to *be conscious* – to *know*? ' "Knowing" ', says Harth, 'here means nothing other than receiving the information and being able to pass it on.'[13]

There is an emergent 'unitarity' to this powerful, massively-parallel shuttling of information that, literally, pulls our *self* together. We need this. But we should recognise it for what it is. We *are* this.

You are a wonderful creation, and you have a certain high-abstract solidity. Like the *story* in a book, your *self* cannot be located; yet you are *there*, implicit in the words. You are a book that has come to realise that it is writing itself (and that its story will, some day, end). And, like a *centre of gravity*, you seem to pull in toward the middle of your head. Planets don't fly apart when physicists point out that centres of gravity are 'just' useful abstractions, and your mind will not fly apart when someone points out to you that 'centres of narrative gravity'[14] (selves) are also 'just' useful abstractions.

Perhaps there are 'centre-preserving' subsystems within our brains that try to lock us out of areas of introspection where dangerous, self-fracturing booby traps may lie in wait. Who seeks the potential truths hidden in such areas? You do. Who bars the way? You do. Who is in there behind the tripwire? Who do you think?

In trying to resolve equations, mathematicians and physicists some-times hit *infinities*.[15] Your self may appear, from your point of view, to be such an infinity – one that you cannot seem to resolve out without the *kludge* (a messy workaround) of the Cartesian ego. Perhaps we should here take a lesson from mathematics, in that when we hit an infinity it is usually time to clear down the whiteboard and start again with a new perspective.

Parfit refers to that 'special, undefinable essence' I mentioned as 'the deep further fact'.[16] He means that, for those who believe in it, this is a fact *beyond* everything we know from science and certain types of 'forensic' philosophy (I include his own in this definition) about what persons are. He concludes that there is *no* deep further fact. His dismissal of it is rigor-ous and complete, but it is also philosopher-calm. I, on the other hand, cannot remain calm while discussing the notion of the deep further fact, for I despise it. I regard it as a monster that blights lives. It confuses, it obfuscates, it terrorises, it maims, it *kills*. If you are looking for the worst type of memes you will find them here, rolled up in the belly of the deep further fact.

This 'deep further fact' was chewing up my self-story, spitting out thin characters and shaky plot-devices. There were parts of the 'narrative' in which I did not and *could not* believe. As in *The Matrix*, where Agent Smith describes 'the first Matrix' as having been rejected by the minds of the enslaved humans, the old version *would not take*.

To stand a chance of becoming the 'necessary hero' of my own story, I was left with no choice but to fight this monster.

## ANGSTROM AND THE HOLE-BOOK

Angstrom the gap-toothed wurzel lived in the House of Hands and believed himself to be real. He awoke each night and, taking care not to wake the other Inklings, wandered the dark rooms and corridors. The big mirror at the end of the hall troubled him, and he tried hard not to glance at it. Sometimes, though, he was drawn to it. When he looked, he saw, as expected, his own reflection, but there always seemed to be something else.

The something else lurked near the shoulder of his reflection, but was invisible. He peered hard into the mirror, and even tried talking to it. But the something else stayed hidden.

Angstrom did not know whom to ask about the something else. Most people he knew didn't talk about such things, except when somebody died. And, even then, they used strange language and symbols that made Angstrom feel dizzy and a bit sick.

He decided to read books about something elses. Most of them had only more, though different, strange language and symbols in them. But some did not. *Some* of the books had holes right through them.

Angstrom peered hard into the holes but saw nothing, not even glassy reflections, and certainly no something elses. He could not understand what such books were for.

One day, while peering particularly hard into one of the hole-books, he fell in. He landed, without a bump, right back where he started. He looked around. Everything looked just the same.

But, some time later, a curious thing happened – he found that he could understand the strange language and symbols. They were the words and placeholders of an old song, and the old song was just a singsong joke.

Angstrom picked up a new hole-book. He turned it to read the name on the spine, though he knew what it would say. It read: *Angstrom and the Hole-book* by Angstrom the Gap-toothed Wurzel.

The next time Angstrom looked into the big mirror, the something else was gone.

A fairyless fairy tale, or, if you like, a selfless self tale. I just can't seem to stop spinning, and neither can you. The epigraph at the beginning of this chapter is yet more spin: It is, as I mentioned, 'often attributed' to Hypatia of Alexandria, but it is, more than likely, yet another 'fable', written by some more-recent author with an axe to grind. Because it resonates with me, I am grateful to whoever wrote it, and I congratulate them (perhaps posthumously) on the success of their meme.

Some books and films, often of the science fiction genre, include 'characters' that discover that they are not real. Authors often insist on portraying these characters as distraught, or at least distressed, when they find this out. This sort of portrayal may make for good stories, but it also reveals a smug 'I am real and you are not' self-satisfaction, not only on the parts of the 'real' characters but also on the part of the author him, or her, self.

Isn't it enough that characters and their settings are self-consistent enough to allow persistence?

Curiously, fantasy literature sometimes does a better job of tackling the reality dichotomy. Neil Gaiman's character Fiddlers Green appears as a jovial old man, but he is also, it turns out, the personification of a *place*. At the time I first read it, I found this idea whimsically attractive, but I wasn't sure why. I was, back then, unfamiliar with, but evidently open to, the power of abstraction.

You need never come to regard yourself as unreal (even on those days when you feel decidedly sketchy). I am not, in any case, trying to make you feel that way – there's no reason to make this more unsettling than it actually is – or to leave you in any other state of existential distress. I *am* trying as hard as I can, however, to get across to you the fact that your *self* is an abstraction. I have also seen it called a 'user illusion',[17] but I no longer think that is the correct terminology. The word 'illusion' is suggestive of trickery, even pretence, but the emergent abstraction of self is, in and of itself, neither tricksy nor pretentious. It is, in the sense we have now established, real. Crucially, however, it is real only because it got that way: over time, through evolution, and through constantly-reinforced high-level self-reference. It is not immanent in you.

Certain philosophers and scientists, in attempting to explain this, are sometimes accused of trying to make persons think of themselves as automatons. Those so accused would be justified in retorting that non-reductionists try to convince people that they are under the internal 'remote-control' of their 'souls'. It's fair to call various brain processes 'zombie systems', in that they don't, in isolation, *know* anything.[18] But – viewed as a whole phenomenon – you are neither the zombie that appears to be implied by the former explanation nor the 'possessed' implied by the latter. It is clear, though, from superstition and from our traditions of ghost story and horror movie, that such fears are old and deep-rooted. And perhaps this is where they come from: I fear that there is *something* in me; I fear there is *nothing* in me.

It is understandable to feel fear upon setting out from the safe harbour of old certainties; the waters ahead are of unknown depth and temperament. With dwindling supplies of what Timothy Taylor calls 'visceral insulation',[19] we may begin to shiver in the chilly air. But at least we are on the move, and pushing out a bow wave. The ripples that follow it are sometimes strange and are certainly far-reaching. They can also be deeply fulfilling.

## RIPPLES

The philosopher, paleontologist, and Jesuit theologian Pierre Teilhard de Chardin envisioned a 'mind layer' of our planet – analogous to the atmosphere or the 'biosphere' – that he referred to as the 'noosphere'. (The *noos* part is derived from the Greek word for 'mind'.) At first glance, de Chardin's take on Vladimir Vernadsky's earlier concept looks like it may have been intended as a 'scientised' theological shorthand for the 'mass' of 'souls' inhabiting the planet. But it is rather more subtle than that: De Chardin is hypothesising about *consciousness*, and about what happens when many conscious minds overlap in an all-encompassing 'thoughtspace'.[20]

You will not be surprised to learn that the Catholic Church did not much care for his hypotheses. We can fairly assume that this was because of the strongly human-centric nature of his ideas: God got squeezed to the margins. A priest pushing ever-harder in this direction might end up a representative of something like Egan's 'Church of the God that makes no difference'.[21] The *noosphere* idea also sounds somewhat Buddhist: a shared consciousness, a shared understanding, a 'shared moment'.[22] It does *not* sound Christian – certainly not the glitzy, hero-worship version that is Catholicism.

Those with New-Age leanings eagerly adopt fuzzy-edged notions like the noosphere. Before you know it, however, they have become a kind of *telepathy* – a way in which we can all, if we 'free our minds', communicate with each other just by *thinking*. I will not pretend to understand exactly what de Chardin was really getting at; I don't think he exactly knew that himself. But there is nothing wrong with trying to find new and better ways of describing our sometimes-shared experience as human beings on our little planet in this indifferent universe.

My siblings and I overlap. We share certain inherited genetic traits including, as I mentioned earlier, the propensity to develop with certain areas of similarly-arranged neuronal configurations, leading to some similar, motor-cortex-driven physical mannerisms. We also overlap in other ways. We share at least some memories, and although these memories are encoded in the brains of different 'observers', it is likely that they will have certain features in common. There is at least a weak overlap of memory here. (In a stronger form, I could even come to regard my memory of crying at the circus as a *quasi-memory*,[23] rather than as a straightforwardly false one.) We have a partially-shared *history*: We have the same parents, and for the early parts of our lives we all lived together, at least for some of the time, in the same house. We continue to 'make' shared history by communicating, by seeing one another, and by getting together, *en masse*, for special occasions. We also make shared future by *making plans* to get together again for future events.

This kind of overlapping is ordinary but also profound, exquisite but exquisitely unstable. Without regular re-tying of the bonds that lash them in place, our sets begin to drift apart; the intersections of our Venn diagrams of shared experience wink out like closing eyes.

With our new understanding of the self, including a complete rejection of the *deep further fact*, we can envision, if we wish, a new kind of 'sphere. We could call it the *anattasphere*.[24] In *this* 'place', because none of us has a soul or even a *self*, the intellectual and emotional connections that we make with each other can go far beyond the linking together of individuals like nodes in a network; we become, overlapped and *en masse*, a larger 'entity'. This is not a super-organism. It is nebulous because it does not have a self any more than we do as persons – it is another abstraction. Actions occur in the anattasphere. Because it has no centre, these actions (events) bubble up like nonlinear noise in the system. There is no way of knowing how far the ripples of these actions will spread, nor with which other ripples they will interfere, perhaps forming larger waves or cancelling out entirely. It is a complex system.

I accept that this is a simplistic picture (of a complex system). We know that certain 'actors' – governments, specific individuals, nations, armies – are the centres out from which some of the largest and most destructive 'waves' radiate. But perhaps we should regard such actors as mere outliers: Because they have not accepted the selfless and soulless principles that 'govern' the anattasphere, they are not fully part of it. Ideas like this often begin to break down once we factor in the harsh realities of the worst of human behaviour. Outliers can become strange attractors. And what drives such behaviour? Religious dogma, fundamentalist political ideologies, uncontrolled greed, complacency: all the types of behaviour that spring from the mistaken belief that some people are *better* or *more entitled* than others are. *People* who display this type of behaviour neither realise what they are nor fully accept the repercussions of their actions.

Because we do not exist as individuals in the anattasphere, we are not summarily deleted from it when our brain and body die. Some traces of our network of actions remain imprinted on the system, for a time, after our death. These traces gradually fade out as the system forgets our actions. Some individual traces remain strongly imprinted, perhaps for hundreds of years, but even these eventually fade as the ripples from those actions weaken to irrelevance. Call this 'stage' a type of liminality, if you must; it is certainly no immortality. This abstraction, *as a whole*, changes over time in response to the combined effects of ripples from both the living and the dead; sometimes they combine in ways that push up great waves of change: We may call these paradigm shifts.

Memes are inherent in abstractions of this sort. We could even say that memes, not persons, are the key agents (and replicators) within this system.[25]

You may already be uncomfortable with my anattasphere abstraction. Perhaps it sounds silly and implausible, but bear with me a while. My steps may be faltering, but there is really nothing wrong with framing abstractions in new terms. We can all do it, provided we are clear (or at least as clear as we can be) about what we mean. In *Incomplete Nature*, Terrence Deacon coins the term *absential*, 'to denote phenomena whose existence is determined with respect to an essential absence.'[26] The term is unfamiliar and odd, but, importantly, it is not unnecessary. In Deacon's terms, your life is already full of absentials: 'longing, desire, passion, appetite', 'life', 'purpose', 'zero', 'money', 'information'[27] ('death' must also qualify, as must many apparently-trivial phenomena such as 'a patch of blue sky'). 'Anattasphere' is intended as such an absential.

When Richard Dawkins posited *memes* in *The Selfish Gene*, as a way of explaining a key point about replicators,[28] he could not have been sure that the idea of memes would itself become a successful meme. He certainly could not have guessed that a new environment (the internet) would arrive in which his nascent meme would come to flourish. In contrast, we only need *my* absential meme for the duration of this book. I have created a shorthand abstraction and have used it, in terms consistent with that abstraction, to describe what happens in *this* system while persons live, and when they die. Its absential status is, I think, justified by the sense in which we *feel* this place that isn't a place. Every time we express an opinion – whether face-to-face, on the phone, in our writings, or typed on some online forum – we are pushing out to flex the wider 'network' around us. It can be intoxicating. It can be addictive. It seems to matter greatly to us that our existence is registered, and indeed, many *suffer* greatly when theirs is not.[29]

The anattasphere can 'benefit' from the arrival of new persons but also from the survival of existing ones. The longer persons survive, the more strongly their traces become imprinted on the system. You might consider this a bad thing: 'Bad' long-lived individuals might lay deep tracks in the system, 'dominating' it and 'steering' it in directions that are 'bad' for the anattasphere as a whole. That is possible. However, such reactions ignore the effects, both probable and speculative, of dramatically increased lifespan. Long-living, healthy individuals will continue to contribute to the total 'pool of intellect', or, as I would prefer to think of it, *the wisdom pool*, for the durations of their extended lives. In this perspective, there is an 'evolutionary pressure' toward ever-greater wisdom in the system: Wise traces survive both *inside* and *outside* individuals as memes because they increase the survival-chances of the next 'generation' of wise traces, and so on.[30]

Perhaps the preservation of individuals is a normal part of the evolution not only of the anattasphere but also of our species. In our evolutionary past, it mattered only that our genetic payload was copied forward in time: As long as our genes survived in our offspring, our *continued* existence as individuals did not matter. This blind process eventually produced brains that could see the benefits of individual survival, creating the potential for individuals to break free of the 'dumb' gene-transcription process of which they were a result. In this sense, because of its evolved intelligence, the brain can bootstrap itself to find potential forms of survival that are novel but which are *still a result* of the evolutionary genetic instructions that created it; 'the transfer of function to progressively more suitable substrates'[31] may no longer need biological-evolutionary timescales to play out. 'End-directed' behaviour – the kind of specifically-constrained self-organisation that Deacon calls *teleodynamic*[32] – seems an obvious marker of intelligence, but it seems obvious only because we have a well-worn word for it. What we call 'intelligence' has emerged via simpler subsystems of much less 'selfy' self-reference to become a dynamic tool of ever-unfolding, loop-amplified, gain-adjusted, goal-directed impact. 'Intelligence', as Grand puts it, 'is a mechanism for enhancing persistence'.[33] Those of us who seek long-term survival see nothing 'unnatural' about applying the technological fruits of human intelligence to our ongoing quest to persist as phenomena in the universe.

This individual survival instinct is, then, a result of both genetic and memetic evolution: The *idea* of individual survival is persistent and will replicate itself forward in time *if* it is a fit replicator. The 'rightness' of this, in any moral sense, is largely irrelevant; only the survivability and evolvability of the idea matters in this context. Only if it leads to the extinction, for whatever reason, of the idea can we say that it has been an unsuccessful meme.[34] If the meme persists in individual survivors who propagate it within the anattasphere, then it is a fit replicator.

This is the proper context within which to place my choice to be cryonically preserved. That choice can then be viewed as a teleodynamically-generated feature on the continuum of the memetic natural selection of ideas about death. The pool of human *knowledge* continues to grow, but it may be the case that certain types of *wisdom* will always be excluded from the wisdom pool if the individuals who carry that wisdom, or who may come to carry it, cannot persist to spread it. 'Wise memes' are by no means the preserve of highly-persistent (or recurrent) individuals alone, but wisdom takes time to develop, and many of us will perish long before we get a chance to contribute memes that could be considered wise.

It does take a deal of optimistic speculation to assert that wise memes, rather than memes in general, will become the most efficient replicators. But to do otherwise assumes that the motivations of healthy, long-living individuals would be *just the same* as our current motivations. It makes little sense, given the complexity, to assume that their motivations would be just the same. And to take a deeply pessimistic view – that their motivations would be *worse* – can only lead one to conclude that the decline of the system is inevitable. There is nothing constructive in assuming that human wisdom (and therefore behaviour) must inevitably decline; such sour-grape dystopian views are, I think, merely conservatism in disguise.

It is useful, before we go to the lengths of radical continuity measures, to figure out in what senses we are actually *alive* and to what extent we *already* survive our own deaths. This is neither obscurantist nor theological; these questions arise as a logical consequence of the new understanding of the self.

In his 1992 paper The Technical Feasibility of Cryonics, engineer Ralph Merkle titled one of his sections 'The information theoretic criterion of death'. Through his involvement with Alcor and his understanding of information theory, Merkle had identified a pressing need to find a better, scientifically accurate, definition of what 'dead' means. His terminology, now shortened to information-theoretic death, has come into wide usage in scientific discourse. Of the structures of the brain he stated:

If they are sufficiently intact that inference of the state of memory and personality are feasible in principle, and therefore restoration to an appropriate functional state is likewise feasible in principle, then the person is not dead.[35]

Unlike earlier definitions, Merkle's recognised the prime importance of the data. By his definition, other elements of what we think of as death (so-called clinical death) are mere precursors to true death – the loss of the data. His version of information-theoretic death focuses solely on the brain of the 'deceased'; I think it accurate, but not quite complete.

Information is physical, and vice versa.[36] If we survive as healthy individuals, the information we carry in our brains (including wise traces laid down over time and through experience) is, largely, preserved. If we have made no provision for the preservation of this *dataset* – in our writings, in computer files, in songs, in recorded speech and video, in our vitrified brains – it will be lost to entropy when we die. Only others who remember us or whose thought/action we have influenced indirectly will then carry our continuing contribution to the anattasphere. This is *some* kind of 'transference' of data, but it is weak and inefficient. The remaining dataset will certainly be *small*, but not in the sense that it has been efficiently *compressed*.[37] It will be small because the majority of the data has been 'infected' by entropy, and has thus become unrecordable and irrecoverable. By our failure to preserve such information, we are *accelerating* the pace of entropy in the system; we are *encouraging* 'the spread of ignorance'.[38]

Within the above description of datasets (and their retention or loss), you can find one answer to the question of the extent to which you are alive and 'the dead' are not. You are alive to the extent that the *information* you carry within your connectome, and which is carried by other means outside of you, is intact. You are, then, on a continuum – 'death' to life to death – with fuzzy, but ultimately truncated, beginning and end. At first you do not exist (although you might exist, pre-conception, as an idea in the minds of your parents); as you develop in the womb, your connectome gradually accrues structure and complexity; this complexity grows rapidly after you are born and begin to learn; after *synaptic pruning*, your connectome reaches some peak of structured complexity in adulthood;[39] that structured complexity begins to be lost to entropy as you age and start to forget things (or worse still, develop a structure-wrecking *neurodegenerative* illness such as Alzheimer's); you die, and if your brain is not preserved, your information-bearing structures disintegrate (you now meet Merkle's information theoretic criterion of death); you remain only as memetic traces in the anattasphere; your traces fade to chaotic irretrievability. This is *not* a cycle: There are no means of reversing the process once you have dissipated. Remember the arrow of time.

I, too, am on this continuum. But, for me, there is a 'pause' of indeterminate length after my clinical death. We cannot call this point of *my* death 'truncation', because much structure will remain intact within my vitrified brain. We do not know the point at which my continuum *will* be truncated. Still, we will then know everything we can know about my situation: My brain is vitrified; I have not yet been (apart from in the most literal sense) truncated. Alive/dead status questions here tend towards emptiness.

With the continuum idea in mind, we could come, in future, to regard any persons revived from cryostasis as *recurring* persons. Given the likelihood of *some* memory-damage, their personalities might be rather different from their pre-stasis state. Such persons could be 're-introduced' to themselves by means of archived information about their lives. Alternatively, they might wish to take on new identities more in keeping with their post-stasis personalities. In a similar sense, I now try to regard myself as a *series-person*. This concept, created by Nagel and exapted by Parfit,[40] has become a useful shorthand for me to describe the way that I can change – gradually over time or even dramatically as a result of substrate-change or stasis-induced memory-loss – and still be the same series-person. Persons do not seem to be allowed such attributes, so I prefer to be a series-person.

The sense in which dead persons might be re-established in future is similar to Parfit's example of clubs and the way they can be re-established after periods in abeyance.[41] The new version of the club may have different members, but it can take on the identity of the original club should the members choose to grant it that identity. I can hope, then, that though I will die, I will be reconvened at some future time.

My self *currently* happens to reconvene each morning in a way that provides psychological continuity. This type of living is what you and I would regard to be what Parfit calls 'ordinary survival'. (He also points out that this ordinary type of survival repeatedly robs us of consciousness, and that it is, in this and other ways, '*about as bad as being destroyed and Replicated*';[42] in other words – in mine – it is a copied-Jackman type of survival.) If my self were to reconvene one morning in a way that, while not indicating death, did *not* provide psychological continuity, others might say that I had amnesia. If this continued for a long time, they might have questions about whether I was still the same person as prior to the amnesia. If I were to fail to wake up at all, they would say that I was dead. While not – at least in the minds of those who knew me – *completely* stripping me of my identity, it would be assumed, by most, that my death would mean that I could not be reconvened. It would certainly be true, in that case, that the normal 'automatic' reconvening of me had failed, but it does not follow that it *must* be true that under *all* circumstances any and all future attempts by others to reconvene me must also fail.

A common argument against cryonics hinges, oddly, not on the assumption that it will fail but on the idea that it might succeed in a point-defeating way. In this commonly-raised scenario, a person is revived after many years in stasis but has absolutely no memory of her life prior to revival. This, on a face-value reading of it, would be about as bad as permanent death – there has been *no* psychological continuity or connectedness, and so all identity has been lost. This objection is, literally, extreme: It raises a case at one extreme end of what might be, in reality, a *spectrum* of cases where psychological continuity, in the form of memory, holds to one degree or another. At the other extreme of this spectrum, the person remembers everything; in the middle of the spectrum, she still has half of her memories; near to the objection end, she recalls nothing but a beautiful arrangement of tiger lilies in a red vase with a gold rim; at the objection end, she remembers nothing.[43]

We can do other things apart from preserving our brains after death that might help us to survive longer in the anattasphere. By the measures I outlined above, we can become better archivists of our lives: We can use the new technologies that have become, and will become, available to record as many aspects of our unique personalities as we possibly can. And we can learn to 'carry each other': We can distribute our selves among as many persons as possible, and welcome opportunities to carry strong traces of others. There are already wonderful, empathetic people in the world who strive always to do this, but recognition of the 'fuzzy-ended-ness' of life might encourage more to behave in this truly selfless way. I carry my loved ones more since I began writing this book than I did previously, simply because I have *thought* about them more, and in so doing have imprinted them more strongly onto my own pattern.

Whether or not we choose to accept the abstracted non-place that I have called the anattasphere, we *should* accept that new means of representation of the traces of dead persons are becoming available. *Avatars* are such a means of representation. The word derives from a Hindustani one meaning 'descent' (as in a god descending to earth) and implies 'manifestation'. Avatars are commonly used in computer gaming, where they represent an individual in the context of a computer-generated environment. Some players choose to create avatars that look similar to themselves, while others choose radically different types: animals, androids, objects, different genders, and so on. But we can equally apply the term 'avatar' to representations of individuals who are no longer *alive*.

Advertising agencies are well used to the buying and selling of 'image rights' for famous dead persons, but this territory has changed. It is becoming increasingly straightforward to represent a dead individual, on screen, *as if* he/she were still alive – think Einstein in that bread advertisement, or Gene Kelly's twisting body in the car-ad electro remix of 'Singin' in the Rain'. The Gene Kelly example sprang to mind because of the uncanny way 'he' moves: The computer animation/overlaying of his body has rendered him a floppy meat-marionette, being jerked around the flooded set on invisible strings. The dead have been re-animated.

Like our ancestors, we want the dead to speak back to us. When they will not, we sometimes puppet them back to life.

Would Gene Kelly have *wanted* this? Would Einstein? They are dead, so they no longer have any say in these matters, but here anattasphere-like concepts show their true relevance: In terms of the anattasphere, these persons are not utterly dead, and are having their opinions ignored simply because they no longer have a voice. In *Sum: Tales from the Afterlives*, Eagleman touches on moral-philosophical areas that we can relate to such avatar scenarios: 'Since we live in the heads of those who remember us, we lose control of our lives and become who they want us to be.'[44]

I note that Adam Yauch of the Beastie Boys, who died of cancer in 2012, added a clause to his will preventing any future use of his image, name, or music for advertising purposes.[45] Perhaps it was his understanding of the concept of *anatta* (he had become a Buddhist) rather than a more straightforward desire to protect his work from commercial interests that prompted this decision.

When we grieve after a death, a large part of that grief has to do with the gut-wrenching irretrievability of the deceased person. In 'coming to terms' with the loss of a person, we use various cognitive and symbolic methods to try to mitigate this. Ultimately, because the irretrievability cannot be mitigated, there seems to be no such coming to terms. Writing about the death of his friend the physicist Heinz Pagels, Seth Lloyd quotes from a Kenzaburo Oe novel: 'You can't make the absoluteness of death relative, no matter what psychological tricks you use.'[46]

I once read an online article that criticised Douglas Hofstadter's attempts to construct his own framework for internalising (or relativising) a death. With a sickening lurch, we find out part-way through *I am a Strange Loop* that Hofstadter's beloved wife, Carol, had died suddenly of a brain tumour, aged just 42.[47] While he employs complex and convoluted 'tricks' to explain the persistence of his wife in objects such as photographs and videos – and, indeed, in his own brain – it is harsh to say that he is, in doing this, trying to make her 'not dead'. Hofstadter is, to me, consistent, in that the subject-matter of *Loop* is an extension of his earlier work on self-referencing phenomena to be found in books such as *Gödel, Escher, Bach: An Eternal Golden Braid*.[48] Given this, he would be inconsistent if he had not applied his deep understanding of the distributed, looping nature of self to a bereavement of such devastating closeness.

What Hofstadter *is* positing is something like an anattasphere, in which traces of Carol survive in his own brain, in the brains of friends and relatives, and in 'triggering' objects – those which either meant a great deal to her or which, in the forms of photos and videos, hold a recorded representation of her alive. It is painfully clear that even a person of the powerful counterintuitive capabilities of Hofstadter cannot sustain this precariously multi-faceted view in mind for long periods at a time; indeed, it often collapses into the grief-laden rubble of the 'classical' view. Nevertheless, his efforts to construct and maintain this majestic arc of 'Carolness' are, to my mind, poignant, valiant, and (for him at least) essential.

We hold on, but, over time, we forget them. Their traces in our brains weaken. In this regard, asking that we continually re-remember our feelings about loved ones who have died can be like asking us to constantly revisit our grief. This is painful for us. Often, however, the *clarity* of memories and the *pain of loss* weaken at different rates: We learn to carry the memory traces somewhat separately from the remaining grief.

There is no worship here, and no spirits – only memories in the minds of others and consequences of past actions. Theology corrupts this simple truth.

I have noticed a temptation among those who have abandoned *religious* concepts, perhaps in favour of a humanist standpoint, to replace one kind of death-related mysticism with another (sometimes to the extent of talking about physicists as if those people were modern-day shamans or druids). It *is* possible that physics will provide, for *some*, a type of solace. But it is not to the most basic levels of the often-cited law of conservation of energy that we should appeal in seeking consolatory signs of persistence after death. For in doing this, we ignore fundamental second law facts about entropy and chaos; although the *energy* is conserved, persons cannot and *do not* persist in the kinds of ways that matter to us within such disorder. We should appeal, instead, to facts about the order-*involving* structures upon which now-dead persons became partially imprinted while alive – structures such as the brains of other still-living persons.

The physics-as-mysticism angle plays to an apparent need to believe – because we tend to think that some now lost 'subject of experiences' (some *I*) was involved during life – that there *can be* no relevant order-involving ways to persist after death. This is fatalistic and false. Furthermore, we are inclined to believe that the ways that remain after a death – shared mannerisms, memories, ripples from actions – are not just *weaker* than, but fundamentally *different* from, the order-involving ways that pertained while the person was still alive. This, too, is incorrect.

A water wave is a fragile thing: persistent in open ocean, but quickly shoaled then dashed to non-existence on contact with the non-propagating substrate of land. There you were, a wave travelling hopefully through your spacetime medium, when suddenly there was nowhere to go. Death reared up – scythe-blades of jagged rock in front of you. This was the end. But did it *have* to be? Could a pause, or a change of substrate, have allowed the continued propagation of you and those with whom you travel?

Your name may be 'writ in water',[49] but for now, you have the great advantage of being a wave that can *think*. And even though you might never be able to think your way around the rocky finality of this problem, there will perhaps come a time when the suspicion that there *might* be a solution will crest in *your* mind as it did in mine.

Half asleep. Dulled senses and slow registration. Stubborn right eyelid droops in faint supplication to the fading undertow of dreamless slumber. At this time of day I have peace to write, but the strong inclination just to curl up again within the warm ambit of dozing Terra.

We have been away for a few days. Wine-fogged, I had watched her sleep in our room at the faded Victorian-era hotel. Twenty-two years together. How others face the terror of their mayfly existence I do not know. Time tears at us. Alcohol dissolves the glue holding together the pieces of my learned coping-structure; internal incantations fail. The love and pain well up to bursting.

'An empty box, a dry ocean, a lost monster, nothing, nothing, nothing...' At this lowest ebb, these words of the Georges Méliès character in *The Invention of Hugo Cabret*[50] could aptly describe the way I feel.

Slow registration. Guitar's portrait of me, gifted to me nearly twenty years earlier, is finally complete, and hangs in plain sight. I do not notice until our third day at his home. He has painted out the background. Appropriately context-free, my coal-bucket-adorned head can be made, from certain viewing angles, to float disembodied in *chiaroscuro* void.[51]

By the fire, under the rippling canopy, we talk of transcendence. I use the word as a placeholder. Guitar and his wife are artists; they understand this striving, stretching for something *further*. Without religion, this becomes more poignant because we recognise that it exists, for now, only in our minds; we must *get it down* while we can – it is fleeting. For Guitar's wife, now ten years ill, normality is the current aspiration – to walk in the haar-shrouded forest, to step onto a train without assistance – and transcendence is greedy.

My parents' anniversary party – golden balloons and rosy good-cheer – lies behind us in the arc of this special summer. I have learned more about them now and have come to see some of the errors I have made in judging their characters and relationship. My father can put *some* words, though faltering ones, to his past depression: 'I'm not sure what happened. I just, well, I wasn't interested in...' He is a poet with repressed cantos of his own to attend to. My mother wanted more from life. And she struggles with the irretrievability of moments, such as those of the waning and passing of her father, when the voicing of hard truths might have brought different outcomes and, perhaps, resolutions. They love us but are disappointed in us. We are not the vehicles they would have chosen. This is a familiar story. We are fortunate still to have them both here – to learn from, and to teach.

Gamma, Epsilon, Zeta, and I take these rare opportunities to reinforce, in determined revelry, our brotherly bonds. Those bonds are incoherent, ramshackle; we are all so different. But strong. Epsilon has set 'House of

Hands' to a new acoustic-guitar arrangement he has written. I manage to sing a verse or two with him before losing my voice to the rawness of my emotions.

I now feel less need, when I speak to Epsilon, to dig for significance in his personality traits. My disappointment from the night he told me that perhaps there was just nothing there, has gone. He was right. It seems that he has not been monstered by the deep further fact; he has no need to hide in riddles.[52] I think, though, that he would enjoy learning more about the specialness of the nothingness that he, and everybody else, weaves into such vibrant patterns day by day. He has clarity but also an ebullient fatalism. I hope that he can sustain the former but, along with the cigarettes, ditch the latter. It's a killer.

Beta seems reticent about hurling herself into the emotional whirlpool. She is there and not there; she frequently goes away to smoke. We drag her back to a past that she longs for and does not, remembers and does not. These smarting areas become, like us, greyer by the day.

Alpha feels unwell and takes to bed early. We miss his musical input. I struggle to converse with him sometimes: His recourse to the language of faith and ambiguity threatens to form a new barrier – one through which it may become increasingly hard to make out his true shape. He disapproves of my philosophy and finds my preoccupations grisly. Our intersections are fragile. Maybe the saying is true – the older we get, the more like ourselves we become. Perhaps becoming too much like ourselves leaves no room for others. I must try to bear this in mind.

His twin daughters are now doting mothers, each also carrying another new life inside her, due to unfold into this unfolding world a few days hence. His two younger daughters play around the marquee. His teenage son hides-out, along with Gamma's, in the kitchen, close to the beer supply. Epsilon's stepdaughter joins us by the roaring fire; his son kicks balloons around the twilight lawn.

Terra and I meet Thor and his wife on the pier one sunny July day. We abandon our unproductive mackerel fishing, to visit their new home. Thor's wife's multiple sclerosis has worsened since we last saw her – the *demyelination* of her nerve-fibres continues apace.[53] She takes delight in her new voice-recognition-equipped iPad. She is brittle thin. Thor has been commissioned to take photographs for a coffee-table book about the mountains of the West Highlands. He is modest, kind, solid. He ages like his father – worn but never care-worn; weathered, never beaten. Their eldest son is now a tousle-headed teenager, replete with mumbling disinterest.

The billowy morning I read of Bradbury's Martian 'fire balloons'[54]: white-tendrilled cirrus sky dappled with fluffy cumulus; a cool east wind carrying salt-scent and sound up from the shore and through the wide-open roof window. Things are good. I imagine the fire balloons, blue-glowing, above my mountains: the 'old ones', gone out of their bodies and become eternal drifting mind.

In this vein, I reach out from my captive mind to sense the landscape beyond my home; the idea of connecting with my environment makes some sense now, for I have realised that I, too, am part of it. I am, however, paralysed in the same way that we all are – from the tips of my extremities out: I cannot feel my ocean depths or my distant mountains; my clouds are entirely numb.

New memories – some glittering and cleanly darrow-hooked, some gaffed. I prize them, but will they keep well, freeze well? In reflecting on these new memories from within my hall of mirrors, I feel empathy. I feel it in retrospect but also in the moment. Despite my scientific and philosophical investigations, I still have no idea what level of empathy is appropriate for one's own mental health. Great surges of emotion triggered by particular events and situations, and by certain books and films, spring, of course, from the sometimes-harrowing human stories, but also, I think, from their echoes of that transcendent yearning for greater comprehension and interconnectedness. Is this too much for us? Should I leave it be, or risk dissolving into 'a pool of bubbles and tears' like the 'Squonk' from a Genesis song that Alpha used to play on his old Decca when I was small?

My mind drifts now to the woman from the party, living with the crushing unanswered question of whether she will develop Huntington's;[55] now to the friendly retired couple along the road, beset by the spectre of *metastasis* of his pancreatic cancer; to the vitrified girl in the dewar; to the bo(d)y in the river. Too much, too much. Breathe. Find equilibrium.

I am not a Buddhist, and will not become one: There is no convenient badge for the state of mind I aspire to hold. The balancing is, for each of us, sometimes subtle and sometimes desperate. It has a wonder of its own. Branding seems a defilement.

Optimism is a trait that I feel strongly within myself but which, I think, others often fail to see in me. It is a strange, stretched-out optimism – one within which complexity and pragmatism admonish child-like hope. I have fallen in love with the stories of our lives, our everyday mythologies. How will they unfold? What will become of us?

Promise and peril tango in the half-light. Ceaseless. Breathless.

## TO THE POWER OF YET

In Chapter 1, I wrote of the damned inscrutability of time. Most of what we encounter during our lives is, at many levels, inscrutable to us. Some philosophers (and even some scientists) think that to understand the 'hard problems' of science, such as consciousness and making conscious machines, the human race may need to cross some critical threshold of intelligence. We could call such a threshold *intellectual escape velocity* (IEV), as the idea is not dissimilar to de Grey's *longevity escape velocity* (LEV). It is not possible to define clearly what, beyond IEV, would be the upper limit of human (and machine) intelligence. The universe would open up to us, and many of the hard problems would, over time, be solved.

There is much debate around the question of whether we are actually getting smarter or not. James Flynn, a professor of political studies who also investigates IQ and intelligence, has claimed that IQ scores have improved year-on-year because the ability of humans to *abstract* has increased.[56] This change, the 'Flynn effect', can explain why we may find it difficult to get our grandparents to understand modern concepts such as, for example, sexual freedom or cyberspace; because they had little or no experience of such abstractions during their formative years, the modes required to navigate them may not be present in their minds. While it is true that we cannot pin down an absolute definition of what 'more intelligent' actually means, we may be able to agree that improved abilities to abstract enrich our experience of the world. It is reasonable to infer an 'upward' or 'outward' trajectory in the progress of our ability to absorb and navigate abstractions (in part, because there are more of them to deal with than in the past), which we may choose to call 'increasing intelligence'.

The intelligence pessimists, in contrast, may see no such increase or improvement, and no particular reason why we should *ever* be able to cross the threshold. Some even choose to ignore the potential for intelligence enhancement via 'artificial' methods such as nootropics, BCIs, and nanotech. Perhaps, then, we are like poor pre-op Charlie Gordons (Charlie is the central character in *Flowers for Algernon* by Daniel Keyes[57]), but stuck forever below that critical threshold above which the answers to the hard problems begin to become apparent. Unlike Charlie, we never get to taste the rarefied air of life in the ultra-high-IQ (whatever that might mean) stratum. Many people think that would be the best situation for humanity, and, if they know the story, might point to Charlie's tribulations after his IQ is made to rocket to 185: He becomes unable to relate to former friends and colleagues; in some sense, he has left them behind. Better, then, that he loses the intelligence he has gained and returns to 'normal'.

The paucity of vision in this kind of thinking is breathtaking. It reminds me of the Aesop fable about the fox who, finding that the potentially sweet and thirst-quenching grapes are out of his reach, concludes that they must be sour anyway. Does that mindset sound familiar? Do you, then, see the striking similarity between the small-minded closing of the book of intelligence and the conclusion that cryonic preservation *must* be futile? Here lies the territory of preachers and dogmatists: Be happy with your lot; there is *virtue* and *grace* in ignorance and death, and only bitterness in striving beyond them.

There is, of course, an unavoidable 'preachiness' in making the counter-arguments. And how the religious love to use their own barren terminology to attack their enemies: They say that *our* 'faith' is in science and technology; they say we are 'evangelical' about those things. The stark difference is that *we* have closed nothing – no books, no reasonable avenues of investigation, no real possibilities.

We stretch for the grapes – sour as they may be; and for the olives – bitter as we *know* them to be; there are abundant *potentialities* here, and, given time, we might gather them in, process them, and acquire a taste for them.

A poster on the wall of the room where Terra teaches her class of seven-to-ten year-olds tells her pupils of the importance of 'yet'. It is intended to point out to them that when they feel stuck and unable to progress they should, as well as seeking help, remind themselves that the problem at hand is only something they cannot resolve *yet*. Children, with their wide-open minds, can understand this. And there is a level beyond the obvious upon which they realise that they are not 'the finished article', that they are in a process of *becoming* – a process of which some feel bright hints, some dark. Perhaps, in this particular regard, it would do us *all* good to think of ourselves as children. This might help us to internalise the power of yet. In doing this, we would have to avoid all the religious connotations of seeing ourselves as childlike. For we are not helpless; we have no divine 'father'. As neatly encapsulated in an internet meme I read recently, 'The Lord is not my shepherd, for I am not a sheep'.

But it is not only about the capacity of the human race to reach a higher plane of intelligence that we make such mistakes; we also make fundamental mistakes about how far we have progressed as a species. Despite the fact that we have no yardstick by which to measure our potential, we often assume that we can say something useful about it solely based on the progress we have made since our emergence as a separate species. *I* think that we live at the beginning of human history – that our most important metamorphoses are still ahead of us. From our chill vantage point, it sometimes seems that we can do little but look backwards and shudder, while trying desperately to learn something from what we see from there. What is yet more chilling is the fact that we are still living that past. Our era remains *feudal* in that monarchs, dictators, and monopolistic capitalist corporations may still control our lives to a frightening degree. Despite the presence of a solid framework for understanding the universe and our place in it, we still turn to the magical thinking of superstition/religion for grand solace; we bow and scrape before entities both real and imagined, all of them undeserving of our piteous obeisance. And far too many still engage in the diabolical actions that lead to the sufferings and deaths of others, ignorant of the stark sense in which they are also, in doing so, hurting and killing themselves.

Though we are, in these ways and in many others, primitive, we cannot *deserve* to suffer for this.[58] (Who, or what, could possibly be the arbiter of this suffering?) Our dead past amounts to not even a particle in the ocean of deep time, yet we still seem too much trapped in this our Age of Mourning, in this our Dead Zone.

We can escape this place. We can *soar* away from it. But we're not out yet.

## THE STATE OF THE ARTEFACT

When I began writing this book, I thought that my decision to sign up for cryonic preservation would be the cold foundation on top of which all the other elements would be built. But the foundation has shifted and resettled as something else – something that was in me long before I ever made that decision. It was the inkling that a terrible lie was being promulgated.

*Naturally* – and, yes, I also mean that in the evolutionary and biological sense – I came to think that the lie was something to do with death. My 'gut instincts' told me, from an early age, that people were making a mistake in believing that they would go to some other place 'after' it. But that part of it is a simple, even obvious, conclusion to reach. The lie was *outside* in the world of superstitious belief but it was also *in me*: It got into *this* brain having been relayed up through my ancestry in an unbroken chain from the first hominin that ever thought a thought that felt a bit like I.

Perhaps we need to forgive others whom we may blame for inculcating various bad forms of the lie in us. Some of them did not know any better.

We certainly need to forgive *ourselves* for having believed in that lie. It is so deeply rooted in our memes and genes that even deep contemplation of the truth will not always succeed in shocking it away. But we can't forgive until we begin to address the addiction. It isn't easy; so much time spent believing we had ghosts in our machines has accustomed us to haunting ourselves.

Maybe the best we can hope for is a see-saw of realisation: In the absence of a counterweight, our sense of personal identity will dominate; it may just hang around like an over-familiar house-guest,[59] but we should recognise it as one who may try, when our guard is down, to foist on us the weird and wrong idea that we have some further and undefinable entity within us. We should oppose this brute with the weight of our growing understanding that we have no such thing, that we are not such things. We have no Cartesian egos. Our *Is* are abstractions.

If you have ever wondered what a Buddhist monk is 'doing' while sitting silently in deep contemplation, then what I have described above is it. It is, at least, what he or she is (according to the most intelligible interpretations I have read) *supposed* to be doing. Try to imagine the training it would take to be able to maintain a balanced state so mindful of the non-existence of *I* that even one's own death would not matter all that much: a trained understanding of the indeterminacy of personhood that renders us purely *santana* (stream or flow); a 'state of grace' not merely self-effacing but self-*erasing*. This is not a special form of fatalism, it's just an upshot (and not necessarily the most important one) of being able to hold that particular state of mind.

The part of a person that kicks out the soul, finding no meaning or use for the concept, is the same part of a person that can decide that death, being a 'mechanical' process, can be defied. But that, while appearing for some to be a massive shift of paradigm, is only a small step.

I now know that my own mortality should not matter as much to me as it apparently does. Equally, though, I find no reason to ignore the *wonder* and *possibility* brought to light by the fact of the self abstraction. A lingering self-interest is present in this, but I also now strongly wish others to feel such freedom: freedom to ditch old irrationalities and to begin to view life as a beautiful shared experiment in connectedness, continuity, and overlap.

The current state of affairs cannot, and will not, continue. The rise of this realisation will shape the future of humanity; the welcome inception of the soulless era is now upon us. You may balk at the idea of a post-spiritual world, but you can look at it another way: Can the notion that we are *not* part of the fabric of the universe we inhabit – a prerequisite assumption at the core of non-reductionist beliefs – persist? And if you think it can, do you also believe that will be for the best? I have been clear from the outset that I regard that arcane meme to be an extremely harmful one. I want to live in a world where we are honest about what we are, and where we get to find out what new levels of cooperation and mutual understanding spring from that. Parfit says of this realisation:

> While we are not Reductionists, the further fact seems like the sun, blazing in our mental sky. The continuities are, in comparison, merely like a day-time moon. But when we become Reductionists, the sun sets. The moon may now be brighter than everything else. It may dominate the sky.[60]

I have found my place and my voice. I belong here, but I also belong in *that* future. I am strung taut between the body of here and the machine head of 'there': a standing wave, vibrating. I fully accept that I am part, along with everyone and everything else, of what Lloyd calls 'a shared computation'.[61] The criterion of a spectrum of surviving *memories* is far too narrow. We must take into account all levels of our distributed connectedness. If I should go down into the liquid nitrogen and return with *all* of my memories – my 'wavefronts' of psychological continuity – intact, I would still not be at the positive extreme of this true spectrum. The overlaps I share with others will be missing if those others are not there. And if they *are* there then we will be missing *other* overlaps that were previously present in the lives of those loved others, who themselves previously overlapped with people I may not even know.

Alone we will be a shard of what was once that larger interconnected structure: an inkling,[62] but also (to borrow a word from Alastair Reynolds) a *shatterling*.[63] We may be incomplete, but, in any scenario, some connections and continuities will persist. 'Any scenario' includes the scenario where we all die. But, in that case, the connections and continuities will be orders of magnitude weaker and, I think, much the poorer for that.

While I can bear this in mind, I have less fear of the dark. My way is moonlit.

If you were not expecting a book about cryopreservation to turn so philosophical, then that makes two of us. But it was necessary. An evolving heterogeneous view takes root then fractals outward as we first *consider*, then *reach*, then *live with* such a decision. Science tells us something about the mechanics – both current and speculative – of the processes involved, but seldom says much about the wider context within which people may come to choose such radical means. The context is non-trivial. Philosophy – the modern kind formulated in the light of scientific evidence – is required for the proper expression of the semantic content of far-reaching debates that rage in and between the minds of thinkers, and of the objective reasons why the answers *matter* so much. Religion, in hard contrast, tells us less than nothing – it only ever obfuscates.

The dark glee has been allowed vent. But what was it? Those precious moments, perhaps, when my young mind rejected the old program and touched the transcendent, chaotic joy of becoming. It danced in the dancer void; in the unfolding spaces in between; in its own blind spot; and found a swirling, complicated peace there. It knew little, yet sensed that hidden even in the apparent vacuum of its own callow self were worlds and multitudes to explore.

As I grew up, I tried to let it go. I tried not to be a thinker. I failed.

And does my claim that I am a practical person stand up to detailed scrutiny? At times, I have been racked by terrible anxieties and nightmares. Those fears have not always had rational bases. I, like most, fear my own death and the deaths of my loved ones. So I have tried to do a practical thing – one that, on some level, helps to mitigate my fear. In and of itself it is not a solution, only a certain type of action – one that contrasts sharply with the practiced *in*action of standard death conventions. Epsilon once insisted that it must be the case that I *believe* in cryonics. I still disagree. As I said at the time, and to Epsilon's exasperation, my *belief* is not required: Should my dead body be buried or cremated, my chance of persisting would be zero; should it be cryonically preserved, that chance becomes non-zero. Cryonic preservation is, therefore, a rational choice. Belief belongs with the zero-chance scenarios.

My decision may be deemed self-interested, but it is not immoral. Nor, though, could it be described as fully moral: It does not seem to have Kantian 'moral worth', in that it would be hard to describe it as a *duty*.[64] I think it less entropy-promoting – and so less *imprudent* – than allowing my brain to rot in the ground or burn to ashes. And, in the ways that matter, you too lack the relevant duty: You are not *duty-bound* to *dissipate* in those traditional ways. So there are ways to see the *comparative* moral worth of cryonic preservation as non-zero. Here, for the sake of avoiding further argument about ethics, I will claim only that it can be fairly deemed morally *neutral*, and that it is morally *permitted* though not morally required.

Of those scientists and philosophers who, despite all their curiosity and vision, still refuse to regard cryonics as anything other than the province of cranks, I ask this: What did you expect us to do? We *are* practical people. *You* taught us to look at life in radically different ways, and we welcomed that. You swept away old gods, evils, and certainties, and we welcomed that. Would you really now have us concede, despite the knowledge of our patternistic and potentially-recordable architecture that *you* helped to provide, that when the time comes we should just lie down and die in the *old* way? Do not waste your time and ours insisting that cryonic preservation must be no more than a modern-day form of Pascal's wager.[65] We value your creativity and intellect; we don't deserve your scorn.

Instead, think of cryonics as an ultra-dark 'Black Swan' strategy: A revival would be an unprecedented extreme event allowed by a low-probability / high pay-off experiment in deep temporal randomness. As Black

Swan coiner and rider Nassim Nicholas Taleb says (though in a different context), 'absence of randomness equals guaranteed death'.[66]

I have learned something of what a brain is, what consciousness is, what a mind is, and what persons are. The things I have learned support my suspicion that it might be possible to pause a recently-deceased person and, later, restore them to *some form* of state that we could call living. This is becoming, in certain scientific quarters, a working hypothesis – but one for which, for obvious reasons, the research is fraught with difficulty.[67] Some scientists allow it, while many others deny it. Few have attempted refutation of it.

I did *not* realise, at the time I made my decision, that I was still clutching a false premise about the nature of *myself* and other selves. This journey has laid it bare. Recognising and letting go of the false premise has been liberating: It removes barriers between persons like you and persons like me; it makes understandable the true nature of persons and their relationship to all other things in the universe; it cuts away the tumour of false belief, allowing us the longed-for chance to heal. Used skilfully it is a sharp and practical tool, like Occam's razor[68] – a scalpel, perhaps.

It is high tide. I take to the sea, just beyond the rope boom holding back the slurry of dead-crab carapace and rotting seaweed. Terra takes a photograph. My head is turned away from the shore. I am cold, a little out of my depth, suspended...

...And a new moment begins. I have no soul. I persist. There is no deep further fact. I am free.

# NOTES

## INTRODUCTION – INTO THE WHITE

1 'Planck length: The smallest possible size for anything, the *quantum* of size. In centimetres, roughly a decimal point followed by 32 zeroes and a 1.' From: Gribbin, *In Search of the Multiverse*, 128, 214.

2 Mikulecky, Gilman, and Brutlag, *AP Chemistry For Dummies*, 34–35; Lippincott, *Anatomy and Physiology*, 18. The atoms in our bodies are present as *elements*, most of which are made up of more than one *isotope*. By way of example, carbon accounts for some 18.5 percent of the mass of the human body. Most of this is carbon-12, a *stable* isotope (one that does not undergo *radioactive decay*), and around one percent is carbon-13, also stable. Tiny amounts of the unstable carbon *radioisotope* carbon-14 are also present.

3 Roach, *Stiff*, 264–266. On Wiigh-Masak process for shattering and composting dead bodies, allowing their use as plant fertiliser.

4 Feynman, BBC Horizon: The Pleasure Of Finding Things Out.

5 Wachowski and Wachowski, *The Matrix*.

## 1 THE BLOODY ARROW OF TIME

1 Abbott, *Flatland*.

2 Atkins, BBC Horizon: What Is One Degree? Not a direct quote from Professor Atkins, but close to what he said in this interview, and to what he has said about the nature of entropy in other talks and interviews.

3 Byrne and Harrison, *Heaven*.

4 For an accessible summary, see: Ferguson, *Prisons of Light*, 32.

5 Lloyd, *Programming the Universe*, 67. Provides further clarification of Clausius' definition and the origins of the entropy concept.

6 Lambert, "Disorder – A Cracked Crutch for Supporting Entropy Discussions"; Lloyd, *Programming the Universe*, 42.

7 Parfit, *Reasons and Persons*, 178–179. As Parfit points out, this answer to the question of how we fast we move through time may be unsatisfying, but it is the only possible answer.

8 Cox and Forshaw, *Why Does E=mc2?*, 92–96.

9 Ibid., 96–101.

10   Dawkins, *The Selfish Gene*, 19.

11   Diamond, *Guns, Germs and Steel*.

12   Atkins, "Entropy."

13   See e.g., Mlodinow, *The Drunkard's Walk*.

14   Parfit, *Reasons and Persons*, 160.

15   Ibid., 159–160. Parfit also talks of such biases as 'discount rates': 'We have a *discount rate* with respect to time, and we discount the *nearer* future at a *greater* rate.'

16   Gilbert, *Stumbling on Happiness*, 6–9.

17   Rilling, "Human and Nonhuman Primate Brains."

18   Adams, *The Restaurant at the End of the Universe*, chap. 8. In the story, Zaphod Beeblebrox is forced to go into the Vortex. It turns out, however, that he is in an artificial universe created for his benefit, so the Vortex does not have the intended effect. Instead of causing madness, it merely boosts his already enormous ego.

19   Parfit, *Reasons and Persons*, 174–177.

20   See e.g., Pearce, "Wirehead Hedonism versus Paradise-Engineering."

21   Edwards, "The Dreaming: A Question of Time"; Willis, *World Mythology*, 279.

22   Hinde, *Why Gods Persist*, 12. 'These involve an entity or entities which ... are usually independent of time.'

23   Hubbard. (Often attributed to Frank Ward O'Malley.)

## 2   INKLINGS OF THE DEAD ZONE

1   Lloyd, *Programming the Universe*, 185–186.

2   "Paraquat," s.v. a toxic fast–acting herbicide, which becomes deactivated in the soil.

3   Activated charcoal is now used in preference to fuller's earth; see e.g., Okonek et al., "Activated Charcoal Is as Effective as Fuller's Earth or Bentonite in Paraquat Poisoning." The use of syrup of ipecac as an emetic is now discouraged; see e.g., Höjer et al., "Ipecac Syrup for Gastrointestinal Decontamination."

4   "Creosote Oil (note: Derived from Any Source) – Identification, Toxicity, Use, Water Pollution Potential, Ecological Toxicity and Regulatory Information."

5   *Seanair* is a Gaelic word for 'grandfather'. From what I can remember, though, everyone called him Seanair.

6   In the West Highlands of Scotland, we call the young of saithe (our name for coalfish) 'cuddies'.

## 3   DEAD ROOTS

1   Taylor, *The Buried Soul*, 22.

2   Dickson, *The Dawn of Belief: Religion in the Upper Paleolithic of Southwestern Europe*, chap. 4.

3   The finds at the site in question, Sima de los Huesos in Spain, are of animal and *Homo heidelbergensis* bones, and may even date back to as early as 500,000 years ago. However, the burials hypothesis is disputed. See e.g., Rabada, "Were There Ritual Burials in the Sima de Los Huesos Outcrop? (Atapuerca Range, Burgos, Spain)."

4    Pettitt, *The Palaeolithic Origins of Human Burial*, chap. 4.

5    Nishida and Kawanaka, "Within-Group Cannibalism by Adult Male Chimpanzees"; Barley, "How Chimps Mourn Their Dead."

6    Sommer, "The Shanidar IV 'Flower Burial.'"

7    Knight, "Ochre in Prehistory."

8    Taylor, *The Buried Soul*, 214.

9    Pinker, *The Language Instinct*, 363–364.

10   See e.g., Halverson, "Art for Art's Sake in the Paleolithic."

11   Kerrigan, *The History of Death*, 46–47.

12   Removal of the brain (excerebration) before embalming may not have been common practice in ancient Egypt. Recent studies suggest 'a high degree of variability' in brain treatment. See: Wade, Nelson, and Garvin, "Another Hole in the Head?"

13   Taylor, *The Buried Soul*, 15, 28.

14   Ibid., 128, 183.

15   Ibid., chap. 7. An initiate woman of the tribe presides over the Viking Rus ceremony referred to here. The horrific details of the ceremony are recounted by the Arab traveller Ibn Fādlan, who uses the term *Malak al-Maut* to describe the woman. The Angel of Death (and/or Retribution) appears in the Qur'an and other lores, and is known in English as *Azrael*.

16   Dawkins, *The God Delusion*, 242–243.

17   Kearney, "Myths and Scapegoats." On the 'victimage (scapegoat) mechanism' concept originally put forward by René Girard.

18   Kerrigan, *The History of Death*, 136.

19   Taylor, *The Buried Soul*, 7–11; Perlmutter, "The Semiotics of Honor Killing & Ritual Murder"; McCauley, "Understanding the 9/11 Perpetrators: Crazy, Lost in Hate, or Martyred."

20   Kerrigan, *The History of Death*, 30.

21   Ibid., 135–136; Willis, *World Mythology*, 234–235; Taylor, *The Buried Soul*, 101.

22   MacCulloch, *The Religion of the Ancient Celts*, 427.

23   Roach, *Stiff*, 40.

24   Taylor, *The Buried Soul*, chap. 6.

25   Parker Pearson et al., "Further Evidence for Mummification in Bronze Age Britain"; Lobell and Patel, "Cladh Hallan."

26   van Gennep, *The Rites of Passage*.

27   MacDougall, "Hypothesis Concerning Soul Substance Together with Experimental Evidence of The Existence of Such Substance"; Letter to the Editor, American Medicine, "Hypothesis Concerning Soul Substance."

28   John Calvin and John Knox were contemporaries. For further information about the relationship between them and their theologies, see e.g., Hazlett, "Contacts with Scotland," 124–125.

29   Porter, *Blood and Guts*, 9–11.

30   Jackson, *The Lovely Bones*.

31  "Immanent," s.v. existing or operating within; inherent; (of God) permanently pervading and sustaining the universe. Often contrasted with TRANSCENDENT. I have used this word several times throughout the book to flag up the religious or quasi-religious sense/idea that certain complex phenomena have 'always just been that way'.

32  Taylor, "What Questions Have Disappeared?"; Taylor, *The Buried Soul*, fig. 2: absorption–isolation spectrum. In his diagram, Taylor puts both mummification and 'cryogenics' at the 'individualized immortality' extreme of his scale.

## 4  MONKEY IN THE MACHINERY

1  My wife once told me about a maths teacher who wanted to name his offspring using consecutive letters of the Greek alphabet. From this point on, I will adopt this naming scheme for my siblings; hence, oldest sibling will become Alpha, second-oldest Beta, and so on.

2  Quote Investigator, "Writing About Music Is Like Dancing About Architecture." An article about the origins of this quote, which has been attributed to various people, including Frank Zappa. The article concludes that it originally came from the American actor and musician Martin Mull.

3  Banks, *Walking on Glass*.

4  Banks, *The Bridge*.

5  Gray, *Lanark*. I did not read *Lanark* until long after I read Banks' *The Bridge*. While it felt familiar, I didn't sense the connection to *The Bridge* until I began writing this book. A quick bit of online research confirmed my hunch: Wilson, "Iain Banks Interview."

6  Carpenter, "On the Influence of Suggestion in Modifying and Directing Muscular Movement, Independently of Volition."

## 6  CONTINUITY

1  I will later discuss the Buddhist view. Some would argue that it is so radically different from both (a) and (b) that it requires a category of its own; I will argue that it is a special form of (b).

2  Dennett, *Darwin's Dangerous Idea*, 81–83.

3  Hameroff and Penrose, "Orchestrated Reduction of Quantum Coherence in Brain Microtubules"; Hameroff and Penrose, "Consciousness in the Universe."

4  Tegmark, "Importance of Quantum Decoherence in Brain Processes."

5  Davies, "Quantum Fluctuations and Life"; Sahu et al., "Multi-Level Memory-Switching Properties of a Single Brain Microtubule."

6  Huang, "The Scale of the Universe 2." An elegant interactive animation that starts at our human scale, and is zoomable down to the quantum foam and up to the edge of the observable universe.

7  Simmons, *The Hyperion Omnibus*; Simmons, *The Endymion Omnibus*.

8  Thera, "Buddhism in a Nutshell – Is Buddhism a Religion"; Firth, *Religion: A Humanist Interpretation*, 32. Firth points out that the faith-based *Lotus Sutra* had 'no direct connexion with the Buddha', was 'composed long, perhaps centuries, after his death', and that 'Its

essential teaching, salvation by faith, is in flat opposition to ... the older orthodox 'Hinayana' (Theravada) teaching.'

9 Thera, "Buddhism in a Nutshell – Anatta."

10 Kerrigan, *The History of Death*, 12.

11 See e.g., Parfit, *Reasons and Persons*, 210; also, Hofstadter, *I Am a Strange Loop*, chap. 21: A Brief Brush with Cartesian Egos.

12 Dennett, *Consciousness Explained*, sec. Why Dualism is Forlorn. A clear (and quite amusing) take on Descartes' notion of a 'seat of selfhood', and the many problems with it.

13 Porter, *Blood and Guts*, 67–68.

14 Kulstad and Carlin, "Leibniz's Philosophy of Mind."

15 Hájek, "Pascal's Wager." Though Pascal's wager may seem purely theological, it is connected to the field of mathematics known as *probability theory* (known in other areas of study as *decision theory* and *game theory*). While Pascal's *framework* for rational decision-making is still of value, in many ways the wager highlights the dangers of allowing untestable beliefs and assumptions to become theoretical variables.

16 Kant, *The Metaphysics of Morals*, 232.

17 Hume, *A Treatise of Human Nature*, vol. 1, sec. Of Personal Identity; Thera, "Buddhism in a Nutshell – Anatta."

18 Hofstadter, *I Am a Strange Loop*, 16–23. Hofstadter's choice of 'huneker' as his 'souledness' unit alludes to James Huneker's comment that 'small-souled men' should not attempt to play a particular piece by Chopin; see: *Chopin: The Man and His Music*.

19 Hofstadter, *I Am a Strange Loop*, 51, 194.

20 Recorded by Plutarch (45–120 CE) in *The Life of Theseus*, 1st century.

21 See e.g., Villazon, "What Percentage of My Body Is the Same as Five Years Ago?"; also, Spalding et al., "Dynamics of Hippocampal Neurogenesis in Adult Humans."

22 Hughes, "Compassionate AI and Selfless Robots: A Buddhist Approach."

23 Parfit, *Reasons and Persons*, chap. 10: What we believe ourselves to be. Parfit's thought experiment was later dramatised by neuropsychologist Paul Broks as a sci-fi short story – "To Be Two or Not to Be."

24 Nolan, *The Prestige*, n. Based on the book by Christopher Priest; Hood, *Supersense*, 228.

25 The bundles of nerve fibres (axons) that connect the hemispheres are known as *cerebral commisures*; an operation involving cutting through these is known as a *commisurotomy*. Many commisurotomies involve cutting through only the corpus callosum (the largest bundle), so these are sometimes called *corpus callosotomies*. In other commisurotomies, *all* the commisures are cut.

26 Gazzaniga, "The Split Brain in Man"; Bear, Connors, and Paradiso, *Neuroscience*, 628–632.

27 Ramachandran, "Split Brain with One Half Atheist and One Half Theist."

28 Mackay, "Divided Brains – Divided Minds?" Dennett is also critical of exaggerated 'two spheres' views of the outcomes of split-brain procedures. However, he thinks that such oversimplification is often a non-deliberate result of difficulties in describing these outcomes. And, where Mackay stresses the physical ways in which a largely-divided brain

might still be a cohesive *mind*, Dennett challenges the ingrained dualism inherent in both the single 'monolithic mind' and the 'two hemispheres, two selves' dogmas. See: Dennett, *Consciousness Explained*, 423–426.

29  See e.g., Devlin et al., "Clinical Outcomes of Hemispherectomy for Epilepsy in Childhood and Adolescence."

## 7  WHERE'S THE FIRE?

1  From Book VII of *The Republic*, written around 380–360 BCE.

2  Bostrom, *Superintelligence*, 126.

## 8  MAKING IT TO THE BRIDGE

1  Rohr, *Sundials*, 127. Rohr's book gives the direct translation: 'Each *one* wounds, the last one kills'. Because of the context (on sundials), the phrase is often translated from the Latin as 'Each *hour* wounds; the last one kills', though the Latin for hour/time (*hora*) does not actually appear in it.

2  For a comprehensive look at the origins and controversies of the 'fat hypothesis', see: Taubes, *The Diet Delusion*, pt. 1: The Fat–Cholesterol Hypothesis.

3  de Grey, "'We Will Be Able to Live to 1,000.'"

4  de Grey, *A Maintenance Approach to Combat Aging-Related Diseases*.

5  Kuhn, *The Structure of Scientific Revolutions*, 10. In this book, Thomas Kuhn coined the now-popular (and often abused) phrase 'paradigm shift'.

6  Wise, "About That Overpopulation Problem."

7  Dennett, *Darwin's Dangerous Idea*, sec. The Tools for R and D: Skyhooks or Cranes?

8  Hughes, *Citizen Cyborg*, 151.

9  Burns, "A Man's A Man For A' That." Burns would have been familiar with the work of Kant, Hume, and other important Enlightenment thinkers. A recurrent theme in Kant's work is that all people, whether good or bad, have 'moral status' (dignity, supreme value); Parfit, *On What Matters*, 1:244. Such human-universal moral status represents a kind of equality that applies regardless of abilities and physical differences. The theme of striving for a common ground of human solidarity – and the sense that the trajectory of humankind was *towards* such an enlightened moral understanding – features strongly in many of Burns' works.

10  Shubin, *Your Inner Fish*.

11  Varki and Altheide, "Comparing the Human and Chimpanzee Genomes."

12  Gaarder, *The Ringmaster's Daughter*, 106–111.

13  Parfit, *Reasons and Persons*, 227–228.

14  de Condorcet, *Sketch for a Historical Picture of the Progress of the Human Mind*.

15  Many scientific papers published over the last twenty years or so have examined the links between telomeres, telomerase, cancer, and ageing. See e.g., Shay, "Aging and Cancer"; Donate and Blasco, "Telomeres in Cancer and Ageing."

16  Curtis, *The Way of All Flesh*.

17  Skloot, *The Immortal Life of Henrietta Lacks*, 2, 392.

18  Cohen et al., "Calorie Restriction Promotes Mammalian Cell Survival by Inducing the SIRT1 Deacetylase."

19  Mark Mattson provides a succinct definition of biological hormesis, which refers to the work of Masoro and others on caloric restriction: Mattson, "Hormesis Defined."

20  Parfit examines some of the philosophy of 'causing to exist' in *Reasons and Persons*, chap. 16: The Non–identity Problem. He asks, 'If we cause someone to exist, who will have a life worth living, do we thereby benefit this person?'

21  Moore, *Enhancing Me*, chap. 8: Adding Technology.

22  Porter, *Blood and Guts*, 134.

23  More, "Technological Self-Transformation."

## 9  SHADOWS, REFLECTIONS, AND TRANSPARENCIES

1  'The "little brain," as the *enteric division* of the ANS is sometimes called, is a unique neural system embedded in an unlikely place: the lining of the esophagus, stomach, intestines, pancreas, and gallbladder.' From: Bear, Connors, and Paradiso, *Neuroscience*, 495.

2  Kerrigan, *The History of Death*, 15. Kerrigan states that 'Sigmund Freud ... yoked together the principles of *eros*, or sexual love, and *thanatos*, 'the death instinct'.'

3  Porter, *Blood and Guts*, 25–30.

4  In: Vesalius, *De Humani Corporis Fabrica*.

5  Bear, Connors, and Paradiso, *Neuroscience*, 505–506.

6  Ibid., 505.

7  Reynolds, *Revelation Space*, chap. 1.

8  Grand, *Creation*, sec. Persistence is a virtue.

9  Hood, *Supersense*, 139.

10  Dennett, *Consciousness Explained*, 14, 261.

11  Minsky, *The Society of Mind*; Minsky, *The Emotion Machine*, 4.

12  Dennett, *Consciousness Explained*, 261. Dennett states: 'Anyone whose skeptical back is arched at the first mention of homunculi simply doesn't understand how neutral the concept can be, and how widely applicable.'

13  Ibid.

14  Dennett, The Normal Well-tempered Mind. Terrence Deacon also allows neurons agency, in the form of a type of cellular-level 'vegetative sentience'; see: Deacon, *Incomplete Nature*, 509.

15  Minsky, *The Emotion Machine*, 17, 109–112.

16  Moore, "Interview with Alan Moore."

17  Hofstadter, *Gödel, Escher, Bach*, 17. Hofstadter mentions the 'liar paradox' (also known as 'Epimenides paradox') in *GEB*. He develops further his ideas on the relationship between consciousness, the liar paradox, and other self-referential phenomena in *I Am a Strange Loop*.

18  Yeats, "Sailing to Byzantium."

19  McLuhan, *Understanding Media*. McLuhan's well-known phrase encapsulates his sense of the symbiosis of medium and message in the realm of advertising and the media. It is also

an apt phrase to describe the type of apparent symbiosis I am discussing.

20  Yeats, "Among School Children."

21  In Buddhism, *mindstream* is the flow of conscious awareness.

22  One Planck time equals approximately $5 \times 10^{-44}$ seconds. It is considered the *quantum* of time, in the same way that the Planck length (mentioned in Introduction) is considered the quantum of length. Planck units are employed in physical theories such as *loop quantum gravity*.

23  Harth, *The Creative Loop*, 48–49; Dennett, *Consciousness Explained*, sec. How the brain represents time.

24  This is why, for example, *awareness* of a painful experience – and even a certain amount of cognitive reflection upon that experience – arrives 'in consciousness' *before* the pain. As Dennett says, 'When you hit your finger with a hammer, the fast (myelin-sheathed) nerve fibers send a message to the brain in about 20msec; the slow, unmyelinated C-fibers send pain signals that take much longer – around 500msec – to cover the same distance. Dennett, *Consciousness Explained*, 103.

25  Kant, "Of the Ground of the Division of All Objects into Phenomena and Noumena."

26  Minsky, *The Society of Mind*, 155; Minsky, *The Emotion Machine*, 117–119.

27  Dawkins, *The God Delusion*, sec. The mother of all burkas.

28  Eagleman, *Incognito*, 77. This term was first introduced by Jakob von Uexküll, in *Umwelt und Innenwelt der Tiere*. Dennett uses the same terminology: 'Their *Umwelt und Innenwelt*, their Surroundworld and Innerworld', in *Consciousness Explained*, 446.

29  Pariser, *The Filter Bubble*.

30  MacLennan, "Conditioned Existence."

31  Egan, *Permutation City*.

32  Dyson, "Time without End"; Lloyd, *Programming the Universe*, 206–208. Lloyd suggests that under such circumstances 'eventually, it would take billions of years to have a single thought.'

33  Heinlein, "By His Bootstraps."

34  Harth, *The Creative Loop*, 80.

35  University of Cambridge, "Surprising Solution to Fly Eye Mystery." The article quotes the lead authors of the study in question, Professor Roger Hardie and Dr Kristian Franze of the University of Cambridge. It also states: 'A fly's vision is so fast that it is capable of tracking movements up to five times faster than our own eyes.'

36  Dennett, *Consciousness Explained*, 346.

37  Dennett, "Intuition Pumps."

38  Cox and Forshaw, *The Quantum Universe*, 52. Terrence Deacon helps to clarify the issue of teleology by bringing in related *telos* (end/purpose/goal) terms for phenomena that *appear* to be directed towards pre-specified outcomes. The most minimal level he calls *teleonomic* (Pittendrigh, 1958) and the next level – taking in 'automatically achieved' ends/states such as thermostatic control of temperature – *teleomatic* (Mayr, 1974). Deacon, *Incomplete Nature*, sec. Pseudopurpose.

39  Harth cites the emergence of toolmaking as the beginning of purposiveness: 'I will therefore

call the emergence of purpose and function among the objects of the physical world the *creative animation* of matter. The creators are almost exclusively humans, and the creations are called machines.' Harth, *The Creative Loop*, 25–28. And Deacon stresses how different pre-purpose lifeforms were from those that emerged with the advent of brains capable of symbolic thinking: 'There was nothing that was considered right or wrong, valuable or worthless, good or evil on our planet until the first human ancestors began thinking in symbols.' Deacon, *Incomplete Nature*, 144.

40 Eacott and Easton, "Mental Time Travel in the Rat." This paper discusses the extent to which rats can, like us, perform 'mental time travel' in planning ahead to complete tasks. The results of the experiments discussed suggest that so-called *episodic memory* is not unique to humans. It would be quite a leap, however, to conclude from this that rat brains are capable of generating from episodic memory the type of forward-thinking and temporal flow perceived by humans.

41 Kant, *Groundwork of the Metaphysics of Morals*, 40; Parfit, *On What Matters*, 1:177. This 'formula' is at the heart of Kantian philosophy. Parfit puts it this way: '*The Formula of Humanity*: We must treat all rational beings, or persons, never merely as means, but always as ends.'

42 Rawls, *A Theory of Justice*, 432.

43 Parfit, *On What Matters*, 1:103–107. Also, Parfit rejects '*the State-Given Theory*', which suggests that we should adopt some belief or desire when we are in 'states' where that belief or desire would make things go better. He concludes that, 'Whenever it would be better if we had certain beliefs or desires, we have reasons to want to have these beliefs or desires, and to make ourselves have them, if we can. But we do not, I suggest, have *state*-given reasons to have beliefs or desires.' Ibid., vol. 1, sec. Appendix A: State–Given Reasons.

44 Glover, *What Sort of People Should There Be?*, 136.

45 Parfit, *Reasons and Persons*, 281.

46 Ibid., 215.

47 Geim et al., "Microfabricated Adhesive Mimicking Gecko Foot-Hair"; Geim et al., "Electric Field Effect in Atomically Thin Carbon Films." Graphene is an atom-thin form of *carbon*, as discussed in Chapter 1, *Outrageous Fortune*

48 Thuras, "Sokushinbutsu of Dainichi Temple." Such a monk would starve himself to death by a method (involving much fluid-loss) intended to limit decomposition of his body. Finally, he would be sealed into a dry tomb to die.

49 "Automatones: Animate Statues." 'THE AUTOMATONES were metallic statues of animals, men and monsters crafted and made animate by the divine smith Hephaistos. Automatones were also manufactured by the great Athenian craftsman Daidalos.'

50 The hairpin appeared in a dream I had around the time I was completing this book. Unfortunately, it manifested as a sharp, and in the dream fatal, bend in the road.

51 Harth uses an illustration of *The Thinker*, with the caption 'The isolated brain. (Nothing comes in, nothing goes out.)'. Harth, *The Creative Loop*, 54.

52 A state of focussed, effortless concentration. Sports psychologists might explain it to their clients as being 'in the zone'.

53  See e.g., Raichle and Snyder, "A Default Mode of Brain Function." The *default mode network* (DMN) is also known as the *task-negative network* (TNN).

54  Seung, *Connectome*, chap. 15: Save As...

55  Abbott, *Flatland*. Abbott mentions 'Spaceland' as the 3-D place to which his 2-D Square character is taken by the Sphere. Later in the book, the Square imagines a dimension above the Third, which he calls 'Thoughtland'.

56  Bear, Connors, and Paradiso, *Neuroscience*, 728b. Provides the following definition of synaesthesia: 'Synesthesia is a phenomenon in which sensory stimuli evoke sensations usually associated with different stimuli.' This definition is given in the context of a section about a synaesthetic patient of the Russian psychologist Alexandr Luria.

57  Edson, "The Floor."

58  Eagleman, *Incognito*, 70.

## 10  BITS OF SELVES

1   Einstein, *The Ultimate Quotable Einstein*, 475. This collection lists the quote, with potential source information and various qualifications, as *possibly* or *probably* by Einstein. It may have much earlier origins.

2   See: HBP, "The Human Brain Project."

3   Bradley, "Neuroscientists Count on Technology Evolution."

4   Bear, Connors, and Paradiso, *Neuroscience*, 77.

5   Seung, *Connectome*, chap. 3: No Neuron Is an Island. Seung states that, 'In many neurons, the electrical signals of dendrites are continuously graded, unlike the all-or-none spikes of the axon.'

6   Rosenblatt, "The Perceptron."

7   UCL, "Human Brain Project Wins Major EU Funding."

8   Waldrop, "Computer Modelling." See also Markram's spat with Dr Dharmendra Modha about Modha's 'cat brain simulation' project for IBM: Adee, "Cat Fight Brews Over Cat Brain."

9   Seung, *Connectome*, chap. 15: Save As... Seung defines Peters' Rule as 'a theoretical principle stating that connectivity is random.' For original sources of Peters' Rule, see: Peters and Feldman, "The Projection of the Lateral Geniculate Nucleus to Area 17 of the Rat Cerebral Cortex. I. General Description"; Braitenberg and Schüz, "Peters' Rule and White's Exceptions."

10  Seung, *Connectome*, chap. Introduction. 'Neurons adjust, or "reweight," their connections by strengthening or weakening them.'

11  Hofstader discusses video feedback loops in some detail, in *I Am a Strange Loop*, chap. 5: On Video Feedback.

12  Though some reviewers had suggested that the *Mindflex* might not contain a proper EEG sensor, 'teardowns' of its headband by hardware hackers have proved that this toy incorporates an EEG-processing chip also found in more-advanced devices made by NeuroSky. "MindFlex Teardown: A Look inside the Mindflex"; Mika, Vidich, and Yuditskaya, "How to Hack Toy EEGs."

13 My gaming headset picks up generalised *biopotentials* – secondary signals caused by electrical changes in the nerves and skin – and other signals such as muscle movement. It detects neuronal signals only vaguely and indirectly, so it is not a true EEG measurement device. Nevertheless, once amplified by the headset, these weak signals are enough to allow the software algorithm to detect whether my brain is generating, on average, alpha, beta, or theta waves.

14 The involvement in conscious thought of cortical areas such as the *medial temporal lobe* (MTL), which contains the *hippocampus*, an area known to be important in memory processing, cannot be ruled out. See e.g., Zeithamova, Schlichting, and Preston, "The Hippocampus and Inferential Reasoning." However, the *prefrontal cortex* of the *frontal lobe* is thought to be the main integrative 'higher-order' (abstract thought) *association area* in the human brain. See e.g., Martini, *Anatomy and Physiology*, sec. Integrative Centers.

15 Hawkins and Blakeslee, *On Intelligence*, 42.

16 Herculano-Houzel, "The Remarkable, yet Not Extraordinary, Human Brain as a Scaled-up Primate Brain and Its Associated Cost."

17 Sironi, "Origin and Evolution of Deep Brain Stimulation." A comprehensive paper on the history and evolution of DBS techniques. See also, Parkinson's UK, "Deep Brain Stimulation."

18 Nishimoto et al., "Reconstructing Visual Experiences from Brain Activity Evoked by Natural Movies." See also, Anwar, "Scientists Use Brain Imaging to Reveal the Movies in Our Mind."

19 Neumann, "Some Aspects of Phenomenal Consciousness and Their Possible Functional Correlates"; Dennett, *Consciousness Explained*, 180. Dennett cites Odmar Neumann's paper, stating that it 'suggests that orienting responses are the biological counterpart to the shipboard alarm "All hands on deck!"'

20 Dennett, *Consciousness Explained*, 144. Dennett states that, 'The brain's task is to guide the body it controls through a world of shifting conditions and sudden surprises, so it must gather information from that world and use it *swiftly* to "produce future" ... .' See also this fascinating talk by neuroscientist/engineer Daniel Wolpert, in which he claims that the brain evolved to control movement: "The Real Reason for Brains."

21 Bach-y-Rita et al., "Form Perception with a 49-Point Electrotactile Stimulus Array on the Tongue"; Tyler, Danilov, and Bach-Y-Rita, "Closing an Open-Loop Control System." For a summary, see this news article discussing successful substitution of *vestibular system* function in one patient, and 'sight' in another, using different versions of the BrainPort: Blakeslee, "New Tools to Help Patients Reclaim Damaged Senses."

22 Maturana and Varela, *Autopoiesis and Cognition*.

23 Dennett, *Consciousness Explained*, 111–143.

24 Herculano-Houzel and Lent, "Isotropic Fractionator"; Lent et al., "How Many Neurons Do You Have?"

25 See e.g., Jonas and Buzsaki, "Neural Inhibition." 'The specific firing patterns of principal cells in a network will depend largely on the temporal and spatial distribution of inhibition.'

26 See e.g., Bear, Connors, and Paradiso, *Neuroscience*, 43b.

27  See e.g., Kolb, Gibb, and Robinson, "Brain Plasticity and Behavior."

28  Cajal, "The Structure and Connexions of Neurons." Cajal calls it 'the inextricable forest of the brain'. Seung draws upon this analogy in *Connectome*, chap. Introduction.

29  Strictly speaking, neurons do not touch each other. As mentioned before, the 'connection' between axon and dendrite is actually a tiny gap called a synapse. However, neurites are held together at the synaptic cleft by *adhesion molecules*.

30  Harth, *The Creative Loop*, 54.

31  Though we feel that our decision-making happens fully consciously, scientific evidence indicates that unconscious neural activity precedes conscious awareness of decision-making. Our 'free decisions' are not really free. See: Soon et al., "Predicting Free Choices for Abstract Intentions."

32  Following: Seung, *Connectome*, chap. 3: No Neuron Is an Island. As Seung puts it: 'Just as a redwood "wants" to be struck by light, a neuron "wants" to be touched by other neurons.' Donald Hebb's theory of neuron adaptation in learning is often summarised as, 'Cells that fire together, wire together'; this phrase is attributed to neuroscientist Carla Shatz. Hebb, *The Organization of Behavior*; Shatz and Goodman, "Developmental Mechanisms That Generate Precise Patterns of Neuronal Connectivity"; Doidge, *The Brain That Changes Itself*, 63.

33  Edelman, *Neural Darwinism*.

34  Goldstein, "Emergence as a Construct."

35  Gell-Mann, "Beauty, Truth and ... Physics?"

36  Deacon, *Incomplete Nature*, 203.

37  Ibid., 228.

38  Reynolds, *Boids*; Ball, *Critical Mass*, 152–156.

39  Conway, *Life*; Deacon, *Incomplete Nature*, 170. According to Deacon, 'He [Conway] called it "life" because the algorithm determined whether cells were on (alive) or off (dead) by virtue of being activated by proximity to live neighbours or inactivated due to lack of live neighbours.'

40  This saying has been in use since at least the early 19th century, and may derive from earlier related expressions conveying a similar sense of a ludicrous/ridiculous/impossible task involving circularity of self-action. The term *bootstrapping* as used in computing may derive more directly from Robert Heinlein's sci-fi story "By His Bootstraps." In the story, a man enters a series of time loops containing various versions of himself by leading 'himself' through 'time gates'. The paradoxical, 'self-instantiating' nature of the concept makes it an appropriate term for describing the recursive nature of certain computer algorithms.

41  Seung, *Connectome*.

42  Sporns, Tononi, and Kotter, "The Human Connectome"; Hagmann, Meuli, and Thiran, "From Diffusion MRI to Brain Connectomics."

43  It is difficult to find a consistent estimate of the number of neurons and synapses per cubic millimetre of adult human cortex. Some estimates are as high as one hundred thousand neurons and one billion synapses, while others talk of neuron densities of around 40,000 per cubic millimetre. I have gone with a common 'guesstimate' of 50,000 neurons, with up

to 10,000 synapses per neuron. Drachman, "Do We Have Brain to Spare?"; Leuba and Garey, "Evolution of Neuronal Numerical Density in the Developing and Aging Human Visual Cortex"; Voytek and King, "How Do We Know That There Are 100 Billion Neurons in the Brain?"

44  Seung, *Connectome*, chap. 8, 9. Based on *some* estimates of neuronal and synaptic packing densities.

45  Woodruff and Pitts, "Monozygotic Twins with Obsessional Illness." An early paper on the subject that discusses a study of monozygotic twins and suggests that 'obsessional illness' – extreme mannerisms, if you like – can be inherited.

46  Seung, *Connectome*, chap. 6: The Forestry of the Genes. Seung mentions autism and schizophrenia as potential connectopathies: 'disorders of neural connectivity'. Given that we can genetically inherit predispositions to develop such disorders, we should accept that inheriting predispositions to develop other connectomic configurations may lead not only to shared mannerisms but also to shared modes of thought and behaviour.

47  Turkheimer, "Three Laws of Behavior Genetics and What They Mean."

48  See e.g., Doidge, *The Brain That Changes Itself.*

49  Premack and Woodruff, "Does the Chimpanzee Have a Theory of Mind?"; Saxe and Kanwisher, "People Thinking about Thinking People. The Role of the Temporo-Parietal Junction in 'Theory of Mind'"; Saxe and Wexler, "Making Sense of Another Mind."

50  Hölzel et al., "Mindfulness Practice Leads to Increases in Regional Brain Gray Matter Density."

51  'Everyone is necessarily the hero of his own life story.' From: Barth, *End of the Road,* 71. I use this quote and the word 'necessarily' in the sense that it is hard to imagine how any person could have 'his own life story' at all *without* himself or herself as at least the protagonist, if not the 'hero'.

52  'Why not use connectomics to critically examine the claims of cryonics?' Quoted from: Seung, *Connectome*, chap. 14: To Freeze or to Pickle?

53  The 'God of the gaps' concept is attributed to evangelist Henry Drummond, who used the phrase 'gaps which they will fill up with God' in *The Lowell Lectures on the Ascent of Man,* 333.

54  Lloyd, *Programming the Universe,* 32.

55  Semon, *The Mneme,* 55.

56  Craddock, Tuszynski, and Hameroff, "Cytoskeletal Signaling"; ScienceDaily, "Scientists Claim Brain Memory Code Cracked."

57  Grand, "Email Correspondence: AI Wisdom." This observation by Grand reminds us that we should be wary of shrill, media-hyped interpretations of the M.O. of single neurons in storing and triggering memories. Some overblown reports of results included in a 2005 paper on the subject claimed that scientists had discovered a 'Jennifer Aniston neuron'. Original paper: Quiroga et al., "Invariant Visual Representation by Single Neurons in the Human Brain."

58  Throughout the history of neuroscience, there have been various competing methods of defining brain 'regions' or 'areas'. Perhaps the most widely used is the Brodmann map,

developed by the German anatomist Korbinian Brodmann and published in 1909. Brodmann showed his areas as numbered subdivisions of larger brain areas such as *primary motor cortex* and *primary somatosensory cortex*. For example, *Brodmann area 17* refers to the *primary visual cortex* (also commonly called *V1* – visual 1). Because Brodmann's maps were based upon his rigorous studies of neuronal organisation (including their organisation within the cortical *layers*) they are still considered to be of value in neuroscience today, although in a much-refined form.

59  Cajal attempted to classify neurons into different types, primarily based upon their appearance: A *pyramidal neuron* was one with a roughly pyramid-shaped cell body, and also differentiated by having an especially thick dendrite at its apex; a *stellate cell* was one with neurites radiating out, roughly equally, in all directions from the cell body, thus lending it a 'star-shaped' appearance.

60  Somewhat mischievously, I have based this word upon a Greek-derived one, *egeiro*, which usually only appears in biblical texts. See: "Strong's Greek: 1453. Ἐγείρω (egeiró) – to Waken, to Raise up." It is sometimes used to mean 'raise up' (as in necromancy-style 'raising from the dead'), but it can also mean 'awakening'. I use it here mainly in its other sense of 'cause to appear' (or even 'cause to appear to cause to appear') because it seems appropriate to the computer bootstrapping process in which algorithms seed other algorithms, eventually causing a (non-heavenly) host of useful applications to appear.

61  The terms 'sapient' and 'sentient' are often confused. The etymological roots of 'sapient' (as used in *Homo sapiens*, for example) are to do with the concept of wisdom, and this term is usually applied – albeit questionably – only to human beings. The term 'sentient' is more connected with the concepts of sensing, sensation, and perception, although it also touches upon consciousness. We tend to use 'sentient' about animals that are capable of flexible learning in response to what they perceive, and that also includes us, of course. The ability to *abstract* at a high level – to conceive of and then seek abstracts such as *meaning* – is perhaps the main factor that differentiates a sapient being from a sentient one.

62  Ettinger, *The Prospect of Immortality*, 113.

63  Grand, *Growing up with Lucy*, 129.

64  Harth, *The Creative Loop*, 64.

65  Grand, *Growing up with Lucy*, 129.

66  Ibid., 112.

67  Deacon, *Incomplete Nature*, 394.

68  A common perception of the field of cybernetics is that it is all about robots; it is actually about communication, control, and constraint in systems. The term was coined by mathematician Norbert Wiener, in *Cybernetics; Or, Control and Communication in the Animal and the Machine*.

69  Eagleman, *Incognito*, 107.

70  Dennett, *Consciousness Explained*, 333, 335.

71  Hawkins and Blakeslee, *On Intelligence*, 87.

72  Bernard, *Introduction à l'étude de la médecine expérimentale*.

73  'In mammals, the requirements for life include a narrow range of body temperatures and

blood compositions. The hypothalamus regulates these levels in response to a changing external environment. This regulatory process is called *homeostasis*, the maintenance of the body's internal environment within a narrow physiological range.' From: Bear, Connors, and Paradiso, *Neuroscience*, 484.

74 Deacon, *Incomplete Nature*, 178.

75 Hofstader coined the term 'strange loops', and emphasised their 'tangled hierarchy', in *Gödel, Escher, Bach*, 10. In discussing 'temporal transition', Deacon refers to Hofstadter's 'strange loops' and restates this tangling of what he calls their 'hierarchic ontological dependency'; see *Incomplete Nature*, 178. This looping and tangling of sequence in thought is a challenge for those who seek to explain the importance of temporal dependencies in the brain and to incorporate them into their computer models; Jeff Hawkins' hierarchical temporal memory (HTM) is an ambitious example a 'neocortical algorithm' that attempts to emulate the apparent hierarchical and time-dependent features of human memory and intelligence. See: Hawkins, "Why Can't a Computer Be More Like a Brain?" .

76 Nagel, "What Is It Like to Be a Bat?"

77 Dennett, *Consciousness Explained*, 441–448.

78 Eagleman, *Sum*, sec. Descent of Species.

79 Turing, "On Computable Numbers, with an Application to the Entscheidungsproblem."

80 Costandi, "How to Build a Brain." The article quotes Ton Engberson of the SyNAPSE neuromorphic computing project. The project is led by Dharmendra Modha, whose neuron modelling approach is very different from that of Henry Markram. This has caused some friction between them; see Notes to Chapter 10, note 8.

81 Indiveri et al., "Integration of Nanoscale Memristor Synapses in Neuromorphic Computing Architectures."

82 Turing, "Computing Machinery and Intelligence."

83 Using advanced 'chatbot' systems, it is relatively easy to fool a human into believing they are speaking to another person and not to a machine. So the ability to pass the Turing test would not alone be a sufficient criterion for identifying true intelligence. Claims that a chatbot known as 'Eugene Goostman' had passed the test were widely dismissed on the grounds that the bot was designed to simulate the responses of a 13-year-old Ukrainian boy with a poor grasp of English, and not those of an intellectual and linguistic equal of the judges. See e.g., Sample and Hern, "Scientists Dispute Whether Computer 'Eugene Goostman' Passed Turing Test." Stringent tests more in keeping with Turing's 'imitation game' have been proposed by Kapor and Kurzweil, among others. See: Kurzweil and Kapor, "A Wager on the Turing Test: The Rules."

84 A term coined by AI-researcher Hugo de Garis, who believes that a war between supporters and opponents of strong AI is inevitable. See: de Garis, *The Artilect War*.

85 Science fiction author Karl Schroeder coined this term to describe, in the 'speculative realism' context of his books, a kind of consciousness-like object-to-object communication that does not relate, on any level, to human beings. See: *Ventus*.

86 Tononi, "Consciousness as Integrated Information"; Tegmark, "Consciousness as a State of Matter."

87   Deacon, *Incomplete Nature*, 392.

88   Lloyd, *Programming the Universe*, 3.

89   'Decoherence: The way *quantum* interference effects get spread out among larger groups of particles in the outside environment, so that quantum information is spread out more widely and lost in the noise of everything else that is going on.' Definition from: Gribbin, *In Search of the Multiverse*, 209.

90   Though not necessarily *intelligent* matter *per se*, others have referred to a hypothetical universal computing substrate as *computronium*. See: Amato, "Speculating in Precious Computronium." Amato cites Margolis and Toffoli of MIT as originators of the hypothesis.

## 11   HIGH ON EXTROPY

1   It turned out that the word already existed, but with a different meaning: 'To show the introvolutions, **extravolutions** of which the animal frame is capable... .' Lamb, *Literary Sketches and Letters*, 167.

2   Dennett, *Consciousness Explained*, 211. In a footnote, Dennett mentions his borrowing of the term 'exapt' (Gould, 1980): 'To put to an extended use ... .' Gould coined the term 'exaptation' as an aid to discussion of adaptation of traits, in evolutionary biology. I use the term in the spirit of Dennett's 'extended use': to put across the sense in which an old term can be adapted (where appropriate) to encapsulate a somewhat different/novel meaning.

3   Barker, *The Great and Secret Show*.

4   "Moore's Law (computer Science)," s.v. Prediction made by American engineer Gordon Moore in 1965 that the number of transistors per silicon chip doubles every year. The true doubling time is approximately 18 months.

5   Kurzweil, *The Singularity Is Near*; Vinge, "The Coming Technological Singularity."

6   Doctorow and Stross, *The Rapture of the Nerds*.

7   Drexler, *Engines of Creation*.

8   Feynman, BBC Horizon: The Pleasure Of Finding Things Out.

9   An acronym for *molecular nanotechnology* (MNT): Broderick, *The Spike*, 18.

10   DeLillo, *White Noise*.

11   Though rare, cases of revival after accidental deep hypothermia – profound hypothermia with prolonged cardiac arrest – have been documented for decades. Most cases involve cardiac arrest either in snow (e.g., avalanche victims) or in very cold water. See e.g., Oberhammer et al., "Full Recovery of an Avalanche Victim with Profound Hypothermia and Prolonged Cardiac Arrest Treated by Extracorporeal Re-Warming." Recent studies of their long-term medical outcomes suggest that 'neurologic deficits' (brain damage) are usually minimal or absent in such patients. See e.g., Walpoth et al., "Outcome of Survivors of Accidental Deep Hypothermia and Circulatory Arrest Treated with Extracorporeal Blood Warming." It's somewhat confusing that similar terminology – profound hypothermic circulatory arrest (PHCA) – is sometimes used to describe therapeutic non-accidental deep hypothermia (as discussed later in the book).

12   Donald Hebb's 'dual-trace' hypothesis suggests that it is like both RAM *and* hard drive.

See: Hebb, *The Organization of Behavior*. This hypothesis incorporates the idea that stability of personality and memory is maintained by the structure of synaptic connections within the brain (the 'hard drive'), while the more 'volatile' thoughts and very short-term memories are carried as transient electrochemical signals (the 'RAM'). There is constant exchange/dumping of information between these two 'systems', just as there is between the RAM and hard drive in a computer. Thus we should expect that people revived following PHCA would lose some short-term memories, but would not lose their longer-term memories or their personalities.

13 See: "Von Neumann Machine (computer Science)." From this entry: 'Instructions should be encoded so as to be modifiable by other instructions ... it meant that one program could be treated as data by another program.' See also: "Turing Machine (computing Device)."

14 Dennett, *Consciousness Explained*, 211. Dennett has this phrase in inverted commas, so I am not sure if he is quoting someone else (there is no reference) or just (more likely) making it clear that he is speaking figuratively.

15 While referencing a term coined by Dennett, I discovered that he, too, had heard the fantasy of 'universal acid' while he was at school; Dennett, *Darwin's Dangerous Idea*, 63. Indeed, the idea of a universal-solvent-like substance, known as the 'alkahest', has been around since at least the time of the alchemists. See e.g., Wedekind and Helbigk, *Dissertatio inauguralis medica de alkahest ...*

16 Despite his scepticism about the progress of artificial intelligence, Harth agrees: 'This gives the brain the capacity for bootstrap processes, to make something out of virtually nothing using the ever-present noise as the source of its unpredictability.' Harth, *The Creative Loop*, 170.

17 Gardner, *The Mind's New Science*, 385.

18 I had watched some documentaries about plasma fusion, and found the images of the unstable, ultra-hot plasma in its magnetic confinement system – known as a *tokamak* – very beautiful. Unlike thought, plasma fusion is not yet self-sustaining. But a self-sustaining, sun-like nuclear reaction is the goal of plasma- and laser-fusion scientists.

19 Taylor, *My Stroke of Insight*, 152.

20 de Wolf and O'Neal, "The Case for Whole Body Cryopreservation."

21 I have since found out that few cryonicists choose to make their names publicly available.

22 A word often used synonymously with *cryonics, cryonic preservation, cryostasis, suspended animation*.

23 My version of 'futurology' concentrates on what might be the true and ultimate (though somewhat hidden) aims of human beings, and how our technologies might develop as a result of striving towards those ultimate ends. If, for example, one of our ultimate aims is to transcend our physical bodies, then we can view the evolution of various technologies – such as virtual-reality gaming – as part of the 'trajectory' towards that end. Understanding *ends* will provide us with a better guide to the future than will concentrating solely on the 'bells and whistles' of what might well be, in reality, only faddish technological *means* toward greater ends.

24 As quoted in: Ellis, *Teaching and Learning Elementary Social Studies*, 431. Bohr probably

got the phrase from Danish cartoonist Robert Storm Petersen, but it may have a much earlier source.

25   See Notes to Chapter 10, note 13.

26   Robbins, *A Symphony in the Brain.*

## 12   I SING THE BODY ECCENTRIC

1   More et al., "Transhumanist FAQ."

2   Hughes, *Citizen Cyborg*, xv, 221–227, 234–240.

3   Hood, *Supersense*, 205.

4   Diamond, *Collapse*, 167–168.

5   Hughes, *Citizen Cyborg*, sec. Siding with the X–Men.

6   Reynolds, *Revelation Space*. In the book, a 'beta-level' simulation is based only upon computer models of the *behaviour* of the simulated person, and so is not considered to be a full representations of its 'original'.

7   See e.g., Eagleman, *Incognito*, sec. The Shift from Blame to Biology.

8   Kershaw, *A History of the Guillotine*, 20.

9   Roach, *Stiff*, 215.

10   Taylor, *The Buried Soul*, 266.

11   Thomson, "First Human Head Transplant Could Happen in Two Years"; Moscow Times, "Russian Programmer Volunteers for 'Frankenstein' Head Transplant."

12   Dreger, *One of Us*, chap. 2: Split Decisions.

13   Dworschak, "'My Second Birth.'"

14   See Notes to Chapter 4, note 6.

15   Increased blood flow in the brain is tightly correlated with increased brain activity, but it is important to understand that we are not directly 'seeing' brain activity in fMRI scans.

16   Stein, "'Vegetative' Woman's Brain Shows Surprising Activity."

17   "Outlier," s.v. a person or thing situated away or detached from the main body or system. This term is also used in the statistical mapping of distributions. In this particular instance, our Buddhist outliers 'detached from the main body or system' are the likes of the self-mummifying Sokushinbutsu monks (see Notes to Chapter 9, note 48). Monks who take their own lives by self-immolation are also radical outliers.

18   Hayward and Varela, *Gentle Bridges*, 152.

19   Dalai Lama, *The Universe in a Single Atom*. In his book, he states: 'If scientific analysis were conclusively to demonstrate certain claims in Buddhism to be false, then we must accept the findings of science and abandon those claims.'

20   Koene, "The Case for Substrate-Independent Minds and Whole Brain Emulation."

21   These futuristic enhancements – and many others – are standard genetic issue in the citizens of Iain Banks' fictional *Culture* civilisation.

22   Harth talks of the counterpart of AI being the '*indigenous android (IA)*': the conformist, dehumanised human who behaves like a machine with relentless, destructive tendencies. Harth, *The Creative Loop*, 31–32.

23   Hughes, *Citizen Cyborg*, 195.

24  Ibid., 210.

25  Using knowledge gleaned from her experience as a palliative nurse, Bronnie Ware discusses the most common regrets of the dying in her book *The Top Five Regrets of the Dying*. A summary of her main findings is given in: Steiner, "Top Five Regrets of the Dying."

26  Dennett, *Consciousness Explained*, 81–82.

27  The source of this phrase is unclear. It likely first arose in theological discussions, and it also appears in scholarly papers. See e.g., Price, "Fairly Bland." Greg Egan uses it in *Permutation City*, 34.

28  Again, the source of this word is unclear. Though still considered a neologism, it is in now widely used among life-extensionists.

29  Alloy and Abramson, "Judgment of Contingency in Depressed and Nondepressed Students."

30  Wheeler, "Hyperactivity in Brain May Explain Multiple Symptoms of Depression." This press release quotes study author Dr Andrew Leuchter.

31  Dawkins, *The Selfish Gene*, chap. 11. Memes: the new replicators.

32  Moran, "What Is a Gene?" Moran here provides a concise definition of a gene as 'a DNA sequence that is transcribed to produce a functional product.' However, he also mentions the abstract/philosophical 'Gene-P' definition: 'The "P" stands for "phenotype" indicating that this gene concept defines a gene by its phenotypic effects and not its physical structure.' While we may readily take on board the notion of memes as abstracts defined by their 'phenotypic effects', we may be markedly less keen to think of them as real 'sequences' producing real 'functional products' (thoughts, attitudes, behaviours, actions). To highlight this difficulty, Dennett asks us to consider whether *words* are real: Dennett, "Dangerous Memes."

33  "Consequentialism," s.v. the doctrine that the morality of an action is to be judged solely by its consequences.

34  The exact origin of the term 'negentropy' is unclear. Some sources attribute it to American Physicist Robert Lindsay (c. 1942). It was certainly used by Leon Brillouin, in the early 1950s, in "The Negentropy Principle of Information." More commonly, however, it is associated with Erwin Schrödinger's concept of 'negative entropy', as discussed in his book *What Is Life?*. Lloyd explains it this way: 'The opposite of entropy is called "negentropy." Negentropy consists of known, structured bits. A system's negentropy is a measure of how far away that system is from its maximum possible entropy. A living, breathing human being has lots of negentropy, as opposed to, say, a gas of helium atoms at uniform temperature, which has *no* negentropy.' Lloyd, *Programming the Universe*, 191.

35  See references to work the work of Iain M. Banks in Chapter 4. See also this article I wrote about him, shortly after his death in 2013: MacLennan, "The Massive, Magnanimous Liberty of The Culture."

36  Hughes, *Citizen Cyborg*, xiii.

37  A phrase coined by Brian Eno, one of the founders of The Long Now Foundation. 'The long now' spans ten thousand years into the past and ten thousand into the future. See: Long Now Foundation, "About The Long Now."

38  Ettinger, *The Prospect of Immortality*, 110.

## 13  ABEYANCE

1   Bender, *The Sandman Companion*, 104–105.

2   Gaiman, *The Sandman: Season of Mists*.

3   The source of the *butterfly effect* trope is unclear. Sources posited include Bradbury (1952), Lorenz (1961), and Merilees (1972). It seems likely that some combination of these sources ultimately led to the generation of what has now become an established meme.

4   Orrell, *The Future of Everything*, 137–145.

5   Lloyd, *Programming the Universe*, 88. According to Lloyd, 'In a chaotic system, the invisible information in the microscopic bits infects the macroscopic bits, causing the observable characteristics to wander in an uncertain fashion – like the butterfly's effect on the course of a hurricane.'

6   Grand, *Creation*, sec. Running the ridge. And in a similar vein, Deacon refers to attractors as 'a "warping" of the space of probable configurations.' Deacon, *Incomplete Nature*, 230.

7   People often confuse *weather* with *climate*. This NASA article succinctly explains the difference: Gutro, "What's the Difference Between Weather and Climate?"

8   Orrell, *The Future of Everything*, 172.

9   Harth, *The Creative Loop*, 146.

10  Hofstadter, *I Am a Strange Loop*, 45–51.

11  Bostrom, *Superintelligence*, 33.

12  Quoted in: Costandi, "How to Build a Brain."

13  Baudrillard, *Simulacra and Simulation*, 1–3, 123.

14  Deacon, *Incomplete Nature*, 473.

15  Perhaps we live within a simulation (complete with simulated rain). Nick Bostrom makes a well-argued case for the logic of concluding that we do: If (as is likely) people of the future design sophisticated 'ancestor-simulations', and if (as is likely) they run many such simulations, then what are the chances that we live in the one real world as opposed to one of the many simulations? See: Bostrom, "Are We Living in a Computer Simulation?"

16  Noise can be fascinating. In the short time I spent as a sound engineer, I learned that there are different 'colours' of noise, such as white, pink, and blue. Colours of noise are also used in physics to define the different types of noise present in various physical systems.

17  Dennett, *Consciousness Explained*, 135.

18  See Chapter 9, *Tick...followed...tock*, for references to *Permutation City*.

19  Kerrigan, *The History of Death*, 12.

20  I am referring here to *myocardial infarction*, commonly known as heart attack.

21  Borjigin et al., "Surge of Neurophysiological Coherence and Connectivity in the Dying Brain." According to this paper, based on data collected from studies of brain activity in dying rats: 'These data demonstrate that the mammalian brain can, albeit paradoxically, generate neural correlates of heightened conscious processing at near-death.'

22  Bear, Connors, and Paradiso, *Neuroscience*, sec. Cell Death. 'The expression of cell death genes causes neurons to die by a process called *apoptosis*, the systematic disassembly of the

neuron. Apoptosis differs from *necrosis*, which is the accidental cell death resulting from injury to cells.'

23 Martin et al., "Neurodegeneration in Excitotoxicity, Global Cerebral Ischemia, and Target Deprivation." Though often involving free glutamate, the mechanisms of excitoxic cell death are complex and difficult to categorise, as this paper shows: 'We conclude that cell death in the CNS following injury can coexist as apoptosis, necrosis, and hybrid forms along an apoptosis-necrosis continuum. These different forms of cell death have varying contributions to the neuropathology resulting from excitotoxicity, cerebral ischemia, and target deprivation/axotomy.'

24 "Formaldehyde." 'Formaldehyde preserves or fixes tissue or cells by ... cross-linking of primary amino groups in proteins ... .'

25 Graham-Rowe, "Rabbits Kept Alive by Oxygen Injections."

26 Freitas, "Exploratory Design in Medical Nanotechnology."

27 It would be immensely useful to be able to greatly lower body temperature while a patient is conscious, but this is not possible because of the self-regulatory *homeostatic* systems of the human body: Below a certain temperature a *conscious* person will begin to shiver, goose bumps will form on the skin, etc., making their body temperature return to normal.

28 Thomson, "Suspended between Life and Death."

29 See e.g., Arrich, "Clinical Application of Mild Therapeutic Hypothermia after Cardiac Arrest"; also, Olsen, Weber, and Kammersgaard, "Therapeutic Hypothermia for Acute Stroke."

30 Bradford, "Cap 'Cools' Brain for Stroke Trial." This article quotes Professor Malcolm Macleod, head of experimental neuroscience at the University of Edinburgh's Centre for Clinical Brain Sciences: 'We think that it may be that within all of our cells there are really quite primitive protective mechanisms that protect us from rapid changes in temperature. We think that what we might be able to do is reawaken those processes in the brains of patients who have had strokes to protect their brain cells until such time as the body's own repair mechanisms can kick in and have their effect.'

31 Stix, "Cool Aid"; Taylor et al., "A New Solution for Life Without Blood."

32 Gater, "NASA's Craziest Ideas: Astronaut Hibernation." The article quotes Dr John E. Bradford of SpaceWorks Engineering.

33 Witelson, Kigar, and Harvey, "The Exceptional Brain of Albert Einstein."

34 Diamond et al., "On the Brain of a Scientist"; Galaburda, "Albert Einstein's Brain"; Falk, Lepore, and Noe, "The Cerebral Cortex of Albert Einstein"; Men et al., "The Corpus Callosum of Albert Einstein's Brain."

35 Sample, "Quest for the Connectome." Jeff Lichtman, quoted in this article, discusses his automated tape-collecting lathe ultramicrotome (Atlum), which, as the name suggests, does away with the painstaking task of manually transferring the slivers onto slides. There also now exists a method (SBFSEM) of high-resolution scanning the serially-cut faces of a sample, instead of the slices: Denk and Horstmann, "Serial Block-Face Scanning Electron Microscopy to Reconstruct Three-Dimensional Tissue Nanostructure."

36 de Wolf, "Chemical Preservation and Human Suspended Animation," sec. Limits of conn-

ectome preservation.

37  Michael Shermer, founding publisher of *Skeptic* magazine, chose the example of 'a can of frozen strawberries' in this often-cited but rather lacklustre sceptical take on cryonics: Shermer, "Nano Nonsense & Cryonics."

38  Meryman, "Osmotic Stress as a Mechanism of Freezing Injury"; Fuller, Lane, and Benson, "Mechanisms of Cryoinjury."

39  Skloot, *The Immortal Life of Henrietta Lacks*, 114.

40  Alcor, "Alcor: Procedures."

41  Hickey, "The Brain in a Vat Argument." For a different take on the standard version – one involving a brain in a vat still remotely attached to its 'host' body – see: Dennett, "Where Am I?" My own summarised take on it makes the Scarecrow from *The Wizard of Oz* the host in question: "If I Only Had a Brain."

42  Fahy et al., "Cryopreservation of Organs by Vitrification"; Chen et al., "Neonatal Outcomes after the Transfer of Vitrified Blastocysts"; Zhang et al., "Cryopreservation of Whole Ovaries with Vascular Pedicles."

43  Merkle, "The Allocation of Long Term Care Costs at Alcor." As well as providing detailed information about 'Bigfoot' dewar capacities, this article suggests that promoting neuropreservation over whole-body preservation – leading eventually to more-efficient use of space and resources – could make neuropreservation very cheap, or even free.

44  de Wolf, "Cryopreservation of the Brain: An Update," 22.

45  *Cryogenics* is the study of what happens to materials – including biological materials – at very low temperatures. The distinction between *cryonics* and *cryogenics* is, obviously, an important one, but one that is often confused, causing some consternation for cryogenicists. See e.g., Cryogenic Society, "Cryonics Is NOT the Same as Cryogenics."

46  This saying has been attributed to Nick Bostrom, and a variation of it to Ben Best. However, it is a common saying among cryonicists. We use it to point out that – while obviously an awful outcome – it seems illogical to view cryopreservation as a worse outcome than assuredly-permanent death. Nevertheless, I accept there may be people who have entirely conquered their fear of death who might either prefer its permanence or see no difference.

47  Grand, "Email Correspondence: AI Wisdom."

48  E.g., Harth, *The Creative Loop*, chap. 15. In the final chapter of his book, Harth is rather dismissive of the idea that there could be a way of copying the patterns that constitute selves. He is no dualist, but he sees the existence of the pattern as being too dependent upon its substrate to be copyable – and, in effect, *separable* – from it.

49  Lloyd, *Programming the Universe*, 194. Lloyd is here discussing physical systems in general, so the principle also applies to brains. While much less bit-centric, Grand's holograph analogy provides a way to understand why it might be possible to 'reconstruct the person' if we could extract those regularities from the randomness and transform them in the correct ways. He goes on to describe the holograph analogy as 'a broad mathematical concept – the idea that information is blurred through convolution and perhaps even something like a Fourier transform into a distributed form quite unlike the discrete bits of a digital

computer.' Grand, "Email Correspondence: AI Wisdom."

50 Drexler, *Engines of Creation*, 133–134. Drexler discusses fixation using *glutaraldehyde* molecules. He also states, 'Fixation and vitrification together seem adequate to ensure long-term biostasis.'

51 Saenz, "Brain Preservation Technology Prize." You may be familiar with work of Gunther von Hagens, who invented a plastination technique and uses it in his educational/artistic work on cadavers. See: von Hagens, "Plastination."

52 Reynolds, *Revelation Space*. In the story, the early 'alpha-level' (complete) simulations are created via a (not always successful) destructive scanning process. Alpha-levels are considered true simulations, unlike the 'beta-levels', which I mentioned in Chapter 12.

53 Pfister, *Transcendence*.

54 Seung, *Connectome*, chap. 14: To Freeze or to Pickle?

55 Drexler, *Radical Abundance*, x.

56 Porotto et al., "Synthetic Protocells Interact with Viral Nanomachinery and Inactivate Pathogenic Human Virus."

57 Reynolds, *Chasm City*.

58 de Wolf, "A Skeptic's Guide to Cryonics," para. 2. According to de Wolf: 'backed up by histological research where it has been established that the neuroanatomical basis of identity does not just implode within 5 minutes of circulatory arrest.'

59 "Thomas K. Donaldson." Thomas K. Donaldson died of a brain tumour in January 2006, and was cryopreserved at Alcor.

60 Donaldson, "He's Dead, Jim: The Irreversibility of Death as a Circular Argument."

61 Society for Venturism, "Kim Suozzi Charity." For a full case report, see: de Wolf, "Cryopreservation of Kim Suozzi (Patient A-2643)."

## 14 THE WAKE

1 Grand, *Creation*, 213.

2 Bradbury, *Zen in the Art of Writing*, 112.

3 Deacon, *Incomplete Nature*, 213.

4 Royal Conservatoire of Scotland. Formerly known as the Royal Scottish Academy of Music and Drama (RSAMD).

5 Bradbury, "The Man Upstairs."

6 Lydon, Zimmerman, and Zimmerman, *Rotten: No Irish, No Blacks, No Dogs*.

7 Parfit, *Reasons and Persons*, 211–212.

8 See Notes to Chapter 11, note 2.

9 Fuller, Agel, and Fiore, *I Seem to Be a Verb*, 1.

10 Many people (some academics included) get hot under the collar about use of the word 'machine' in reference to human beings. It can certainly be useful, used in this way, for making people sit up and take notice. Paul Davies and John Gribbin at least get hot under the collar about it in an interesting way, writing, 'we can see that Ryle was right to dismiss the notion of the ghost in the machine – not because there is no ghost but because there is no machine.'; in: *The Matter Myth*. Nice, but still, I feel, a non sequitur. If there is no

machine then there can be no ghost either. 'Ghost' is far more loaded – and with crass spirituality at that – than 'machine' has ever been.

11 Put forward by physicist Max Born (1882–1970) as a better way of describing what were previously thought of as 'matter waves'. See e.g., Ponomarev and Kurchatov, *The Quantum Dice*, sec. Probability Waves.

12 There may always be unknowable things, and unknowable ways of knowing (eternal versions, perhaps, of Donald Rumsfeld's 'unknown unknowns'). Human brains can parse into information what is available to them from their umwelt, and pass it along in particular ways. But as our processing capacity is not infinite, we cannot know *all* things in *all* ways. Strictly in that sense, then, there *is* a 'something else'. However, by this definition it is *everything else* that we do not yet know and may not ever know. In a strange way, non-reductionists seek to reduce this 'everything else' down to a 'something else' that it is possible to know right now (specifically, to certain kind of plaything universe scenario). This is a fallacy. To quote J. B. S. Haldane, 'My own suspicion is that the universe is not only queerer than we suppose, but queerer than we *can* suppose.' Haldane, *Possible Worlds and Other Essays*.

13 Harth, *The Creative Loop*, 140, 66; Dennett, *Consciousness Explained*, 235. Harth is here (p66) discussing groups of neurons in the visual pathway, but the same principle could apply to *any* group of neurons, or indeed, to any brain structure. Similarly, Dennett writes of 'passive, rubber-stamping functionaries', but he uses this semi-pejorative phrase mainly as a device to show that the idea of a 'hierarchy' in the brain passing information along to some place where all the 'knowing' happens, is wrong.

14 Dennett, *Consciousness Explained*, 418.

15 "Infinity (mathematics)." When they create/use an equation, mathematicians and physicists are usually trying to get a clear – *finite* – result. In this context, equations resulting in infinities are only useful where they are trying to prove that something is infinite (i.e. not very often).

16 Parfit, *Reasons and Persons*, 242.

17 Dennett, *Consciousness Explained*, 216, 220, 311–312.

18 Eagleman, *Incognito*, 132.

19 Taylor, *The Buried Soul*, 277, 288. 'Visceral insulation is a recoil from corporeality, as if we feel that, by coming too close to what is bodily, our inevitable mortality will somehow make itself too painfully known.'

20 Teilhard de Chardin, *The Phenomenon of Man*; Hughes, *Citizen Cyborg*, 174.

21 Egan, *Permutation City*, 18.

22 Simmons, *The Endymion Omnibus*. Inspired by the work of Teilhard de Chardin, Simmons provides an interesting sci-fi-meets-religion take on the 'shared moment' concept. In this context, the 'moment' happens when the minds of all sapient lifeforms simultaneously experience the agony and death of one particular individual.

23 Parfit, *Reasons and Persons*, 220.

24 See *anatta* in Chapter 6, *What is Moral Philosophy?*

25 Dennett, *Consciousness Explained*, 200–210.

26  Deacon, *Incomplete Nature*, 3.

27  Ibid., 3, 8–11, 28–29, 373.

28  Dawkins, *The Selfish Gene*, chap. 11. Memes: the new replicators.

29  Leader and Corfield, *Why Do People Get Ill?*, 181.

30  It could be argued that 'wisdom' is merely a contextual value judgement. In part, I agree. What can be judged 'wise' will, of course, change over time: We know from historical documents that many brutal rulers were once considered wise. However, I have chosen to use this word to put across my sense of optimism about the potential of the human race (in whatever future form it may come to take). I use it in the sense of taking prudent, balanced account of relevant variables before and during action. Because I think that those 'relevant variables' will, in 'the long now', come to include what will make things go best for all sentient, sapient, and thalient beings, I am optimistic about the capacity of 'wisdom' to evolve, and thus to improve lives.

31  Deacon, *Incomplete Nature*, 425.

32  Ibid., chap. 9.

33  Grand, *Creation*, 162.

34  Dennett, "Dangerous Memes." In this fascinating talk, Dennett gives the example of the evidently self-defeating celibacy meme of the religious sect known as the Shakers. He states: 'The meme for Shakerdom was a sterilising parasite.'

35  Merkle, "The Technical Feasibility of Cryonics."

36  As discussed, we should try to bear in mind that *everything* that exists is information-bearing, so things we normally think of as physical (such as rocks and persons) are 'made of' information. And vice versa, what we normally think of as mere data is really structured information and is, therefore, also physical (even if that physicality does not manifest at the same 'level' as the one where we normally perceive solidity/physicality).

37  'Lossy' types of compression are designed to retain the most important features of datasets at the expense of the comparatively homogeneous. For example, in a digital image of a person standing on a mountain against a background of blue sky, the algorithm would retain much less information about the area of sky (as it can be simplified down to a more general mathematical description indicating an area of fairly-homogeneous blue) than about the pixels describing the person's face.

38  Lloyd, *Programming the Universe*, 79–85. I am assuming that we consider human 'information states' to be of value, and that preservation of those states would not lead, somehow, to an increase in entropy at some other informational level. Relatedly, one could make a case that the counter-entropic richness of human information states would be best captured by preserving them at peak of connectomic complexity. While this might be more 'efficient' in terms of information preservation, the value I place on my current existence outweighs the value I *currently* place upon achieving optimal preservation fidelity.

39  See e.g., Petanjek et al., "Extraordinary Neoteny of Synaptic Spines in the Human Prefrontal Cortex."

40  Parfit, *Reasons and Persons*, 289–290. According to Parfit, 'Nagel describes the concept of a *series-person*.'

41 Ibid., 213–214.

42 Ibid., 280.

43 Following: Ibid., sec. The Psychological Spectrum.

44 Eagleman, *Sum*, 24. Some objections to cryonics centre on a related loss-of-control aspect: The ongoing welfare of the cryonically-preserved may end up in the hands of people who do not remember them and who would not, therefore – according to this line of reasoning – have their best interests at heart.

45 Cubbarubia, "Adam Yauch's Will Prohibits Use of His Music in Ads."

46 Lloyd, *Programming the Universe*, 216. Lloyd quotes Oe, *A Personal Matter*.

47 Hofstadter, *I Am a Strange Loop*, 227.

48 Hofstadter, *Gödel, Escher, Bach*.

49 The inscription upon John Keats' gravestone includes the words 'Here Lies One Whose Name was writ in Water'. It does not bear his name.

50 Selznick, *The Invention of Hugo Cabret*, 299.

51 The painting is now entitled *Donald Quixote*.

52 I only recently found out from Epsilon that he *did* go through a difficult period of battling with invasive supernaturalistic thoughts.

53 See e.g., Bear, Connors, and Paradiso, *Neuroscience*, 96b. Further information about demyelination in MS.

54 Bradbury, *The Martian Chronicles*, 118.

55 Bear, Connors, and Paradiso, *Neuroscience*, 467. 'Huntington's disease is a hereditary, progressive, inevitably fatal syndrome … . The disease is particulary insidious because its symptoms usually do not appear until well into adulthood.'

56 Flynn, "Why Our IQ Levels Are Higher than Our Grandparents'."

57 Keyes, *Flowers for Algernon*,.

58 Parfit, *On What Matters*, 1:269–272. From the index entry (p527) pointing to Parfit's full explanation: 'Since our acts *are* merely events in time, we cannot deserve to suffer'.

59 This brings to mind the strange creature in Edward Gorey's *The Doubtful Guest*.

60 Parfit, *Reasons and Persons*, 309.

61 Lloyd, *Programming the Universe*, 54, 210. Lloyd states (p54) that 'the universe is indistinguishable from a quantum computer.'

62 A group of famous authors, including C.S. Lewis, J.R.R. Tolkien, and Charles Williams, referred to themselves as 'The Inklings' for the purposes of their extra-curricular gatherings. In that sense, the word can be used to mean, if you like, 'child of ink'. I use it here in a similar sense: the narrative self; the stories we tell (write for) ourselves that make us 'selves'. Are we inklings written in our own ink?

63 Reynolds, *House of Suns*.

64 Parfit, *On What Matters*, 1:299–300. Parfit's book makes much reference to the work of Kant. At various points, it examines what Kant means by 'moral worth'.

65 Seung, *Connectome*, chap. 14: To Freeze or to Pickle? Cryonicists reject the type of Pascal's wager comparison made by Seung and others. (Accepting, however, the utility of modern *probabilistic* interpretations of the wager, Ettinger allowed this term to used in one of the

prefaces to *The Prospect of Immortality*, pt. Preface by Jean Rostand.) In Pascal's terms, achieving 'immortality' requires only the exercise of faith-based belief and ritual. It would never be possible to test the efficacy of exercising such precepts. Cryonics is not faith-based, and is potentially testable. Seung knows this, and admits as much earlier in his book (see Chapter 13, *Ultrastructures*), so his Pascal's wager comparison is a curious and incongruous departure.

66  Taleb, *Antifragile*, 102; Taleb, *The Black Swan*. Taleb highlights the importance of recognising '*asymmetric outcomes*', but says that it is an error to call this Pascal's wager. Ne notes that 'Pascal's argument is severely flawed theologically: one has to be naïve enough to believe that God would not penalize us for false beliefs.' Ibid., 210.

67  Cryonics organisations, such as Alcor, invest heavily in this type of research. See, for example, Graber, "Research and Development Update." However, even using the most advanced current scanning techniques, it is not possible to know to what extent the connectome is intact within a cryopreserved brain. With cryonics patients, there is the additional limitation that intactness cannot be checked via ultramictrotome slicing. The kind of investigative collaboration between connectomics experts and cryonicists suggested by Seung (see Chapter 13, *Ultrastructures*) would be welcomed by the cryonics community.

68  Deacon, *Incomplete Nature*, 186. On William of Occam/Ockham (1287–1347): 'his famous exhortation to eliminate redundant hypotheses in favour of the simplest or fewest … .'

# BIBLIOGRAPHY

Aainsqatsi, K. *World_line.svg*, 2007. http://en.wikipedia.org/wiki/File:World_line.svg.

Abbott, Edwin A. *Flatland: A Romance of Many Dimensions*. First published in 1884. Amherst, NY: Prometheus Books, 2005.

Adams, Douglas. *The Restaurant at the End of the Universe*. Pan Books, 1980.

Adee, Sally. "Cat Fight Brews Over Cat Brain." *IEEE.org*, November 23, 2009. http://spectrum.ieee.org/tech-talk/semiconductors/devices/blue-brain-project-leader-angry-about-cat-brain.

Alcor. "Alcor: Procedures." *Alcor.org*, January 12, 2004. http://www.alcor.org/procedures.html.

Alloy, Lauren B., and Lyn Y. Abramson. "Judgment of Contingency in Depressed and Nondepressed Students: Sadder but Wiser?" *Journal of Experimental Psychology: General* 108, no. 4 (1979): 441–85. doi:10.1037/0096-3445.108.4.441.

Alpha. "MindFlex Teardown: A Look inside the Mindflex." *Bigmech.com*, November 10, 2009. http://www.bigmech.com/misc/mindflex/.

Amato, Ivan. "Speculating in Precious Computronium." *Science* 253, no. 5022 (August 23, 1991): 856–57. doi:10.1126/science.253.5022.856.

American Medicine. "Hypothesis Concerning Soul Substance." *American Medicine* 2, no. 7 (July 1907): 395–97.

Anwar, Yasmin. "Scientists Use Brain Imaging to Reveal the Movies in Our Mind." *UC Berkeley News Center*, September 22, 2011. https://newscenter.berkeley.edu/2011/09/22/brain-movies/.

Arrich, Jasmin. "Clinical Application of Mild Therapeutic Hypothermia after Cardiac Arrest." *Critical Care Medicine* 35, no. 4 (April 2007): 1041–47. doi:10.1097/01.CCM.0000259383.48324.35.

Atkins, Peter. BBC Horizon: What Is One Degree? Interview for TV documentary, January 10, 2011.

———. "Entropy." Excerpt of speech presented at the Beyond Belief conference, 2007. http://youtu.be/E8660-s6Oy8.

"Automatones: Animate Statues." *Theoi.com*, October 14, 2007.

http://www.theoi.com/Ther/Automotones.html.

Bach-y-Rita, P, K A Kaczmarek, M E Tyler, and J Garcia-Lara. "Form Perception with a 49-Point Electrotactile Stimulus Array on the Tongue: A Technical Note." *Journal of Rehabilitation Research and Development* 35, no. 4 (October 1998): 427–30.

Ball, Philip. *Critical Mass: How One Thing Leads to Another*. London: Arrow, 2005.

Banks, Iain. *The Bridge*. London: Abacus, 1990.

———. *The Wasp Factory*. London: Futura, 1985.

———. *Walking on Glass*. London: Futura, 1988.

Barker, Clive. *The Great and Secret Show: The First Book of the Art*. New York: Harper & Row, 1989.

Barley, Shanta. "How Chimps Mourn Their Dead." *Newscientist.com*, April 26, 2010. http://www.newscientist.com/article/dn18818-how-chimps-mourn-their-dead.html.

Barth, John. *End of the Road*. New York: Avon Books, 1958.

Baudrillard, Jean. *Simulacra and Simulation*. Translated by Sheila Glaser. First published in the original French in 1981. Ann Arbor: University of Michigan Press, 1994.

Bear, Mark F., Barry W. Connors, and Michael A. Paradiso. *Neuroscience: Exploring the Brain*. 3rd ed. Philadelphia, PA: Lippincott Williams & Wilkins, 2007.

Bender, Hy. *The Sandman Companion*. London: Titan, 2000.

BenFrantzDale. *Necker_cube.svg*, 2007. http://en.wikipedia.org/wiki/File:Necker_cube.svg.

Bernard, Claude. *Introduction à l'étude de la médecine expérimentale*. Paris: J. B. Baillière et fils, 1865.

Blakeslee, Sandra. "New Tools to Help Patients Reclaim Damaged Senses." *The New York Times*, November 23, 2004, sec. Science. http://www.nytimes.com/2004/11/23/science/23sens.html.

Borjigin, Jimo, UnCheol Lee, Tiecheng Liu, Dinesh Pal, Sean Huff, Daniel Klarr, Jennifer Sloboda, Jason Hernandez, Michael M. Wang, and George A. Mashour. "Surge of Neurophysiological Coherence and Connectivity in the Dying Brain." *Proceedings of the National Academy of Sciences*, August 12, 2013, 201308285. doi:10.1073/pnas.1308285110.

Bostrom, Nick. "Are We Living in a Computer Simulation?" *The Philosophical Quarterly* 53, no. 211 (2003): 243–55. doi:10.1111/1467-9213.00309.

———. *Superintelligence: Paths, Dangers, Strategies*. First edition. Oxford: Oxford University Press, 2014.

Bradbury, Ray. "The Man Upstairs." In *The Small Assassin*. London: Grafton, 1976.

———. *The Martian Chronicles*. St. Albans: Panther, 1977.

———. *Zen in the Art of Writing*. Santa Barbara: Joshua Odell Editions, 1994.

Bradford, Eleanor. "Cap 'Cools' Brain for Stroke Trial." *BBC*, May 1, 2013, sec. Scotland. http://www.bbc.co.uk/news/uk-scotland-22368408.

Bradley, Simon. "Neuroscientists Count on Technology Evolution." *Swissinfo.ch*, October 28, 2013. http://www.swissinfo.ch/eng/science_technology/Neuroscientists_count_on_technology_evolution.html?cid=37190572.

Braitenberg, V., and A. Schüz. "Peters' Rule and White's Exceptions." In *Anatomy of the Cortex*, 109–12. Studies of Brain Function 18. Springer Berlin Heidelberg, 1991. http://link.springer.com/chapter/10.1007/978-3-662-02728-8_21.

Brillouin, L. "The Negentropy Principle of Information." *Journal of Applied Physics* 24, no. 9 (1953): 1152–63. doi:10.1063/1.1721463.

Broderick, Damien. *The Spike: How Our Lives Are Being Transformed By Rapidly Advancing Technologies*. New York: Forge, 2001.

Broks, Paul. "To Be Two or Not to Be." In *Into the Silent Land: Travels in Neuropsychology*. Atlantic, 2003.

Burns, Robert. "A Man's A Man For A' That." Glasgow: Stewart and Meikle, 1799.

Burton, Tim. *The Melancholy Death of Oyster Boy & Other Stories*. London: Faber and Faber, 2005.

Byrne, David, and Jerry Harrison. *Heaven*. Fear of Music. Sire, 1979.

Cajal, Santiago Ramón y. "The Structure and Connexions of Neurons," 1906. http://www.nobelprize.org/nobel_prizes/medicine/laureates/1906/cajal-lecture.pdf.

Canavero, Sergio. "HEAVEN: The Head Anastomosis Venture Project Outline for the First Human Head Transplantation with Spinal Linkage (GEMINI)." *Surgical Neurology International* 4, no. Suppl 1 (June 13, 2013): S335–42. doi:10.4103/2152-7806.113444.

Carpenter, William B. "On the Influence of Suggestion in Modifying and Directing Muscular Movement, Independently of Volition." *Proceedings of the Royal Institution of Great Britain* 1 (1852): 147–53.

Chen, Yuan, Xiaoying Zheng, Jie Yan, Jie Qiao, and Ping Liu. "Neonatal Outcomes after the Transfer of Vitrified Blastocysts: Closed versus Open Vitrification System." *Reproductive Biology and Endocrinology: RB&E* 11 (2013): 107. doi:10.1186/1477-7827-11-107.

Cohen, Haim Y., Christine Miller, Kevin J. Bitterman, Nathan R. Wall, Brian Hekking, Benedikt Kessler, Konrad T. Howitz, Myriam Gorospe, Rafael de Cabo, and David A. Sinclair. "Calorie Restriction Promotes Mammalian Cell Survival by Inducing the SIRT1 Deacetylase." *Science* 305, no. 5682 (July 16, 2004): 390–92. doi:10.1126/science.1099196.

"Consequentialism." *Oxford Dictionaries*. Oxford University Press, 2014. http://www.oxforddictionaries.com/definition/english/consequentialism.

Conway, John. *Game of Life*, 1970.

Costandi, Moheb. "How to Build a Brain." *BBC Focus*, July 2012.

Cox, Brian, and J. R. Forshaw. *The Quantum Universe: Everything That Can Happen Does Happen*. London: Penguin, 2012.

———. *Why Does E=mc2? (and Why Should We Care?)*. Cambridge, MA: Da Capo Press, 2009.

Craddock, Travis J. A., Jack A. Tuszynski, and Stuart Hameroff. "Cytoskeletal Signaling: Is Memory Encoded in Microtubule Lattices by CaMKII Phosphorylation?" *PLoS Comput Biol* 8, no. 3 (March 8, 2012): e1002421. doi:10.1371/journal.pcbi.1002421.

"Creosote Oil (note: Derived from Any Source) – Identification, Toxicity, Use, Water Pollution Potential, Ecological Toxicity and Regulatory Information." *Pesticideinfo.org*,

March 18, 1999. http://www.pesticideinfo.org/Detail_Chemical.jsp?Rec_Id=PC36578.

Cryogenic Society. "Cryonics Is NOT the Same as Cryogenics." *Cryogenicsociety.org*, June 13, 2008. http://www.cryogenicsociety.org/cryonics/.

Cubbarubia, R. J. "Adam Yauch's Will Prohibits Use of His Music in Ads." *Rolling Stone*, August 9, 2012. http://www.rollingstone.com/music/news/adam-yauchs-will-prohibits-use-of-his-music-in-ads-20120809.

Curtis, Adam. *The Way of All Flesh*. Documentary, 1997.

Dalai Lama. *The Universe in a Single Atom: The Convergence of Science and Spirituality*. Crown Publishing Group, 2005.

Davies, P. C. W. "Quantum Fluctuations and Life." *arXiv:quant-ph/0403017*, May 25, 2004, 1–10. doi:10.1117/12.561309.

Davies, P. C. W., and John Gribbin. *The Matter Myth: Dramatic Discoveries That Challenge Our Understanding of Physical Reality*. New York: Simon & Schuster, 1992.

Dawkins, Richard. *A Devil's Chaplain: Selected Essays*. Edited by Latha Menon. London: Phoenix, 2004.

———. *The God Delusion*. London: Transworld, 2006.

———. *The Selfish Gene*. 30th anniversary edition. Oxford; New York: Oxford University Press, 2006.

Deacon, Terrence William. *Incomplete Nature: How Mind Emerged from Matter*. 1st ed. New York: W.W. Norton & Co, 2012.

De Condorcet, Antoine-Nicolas. *Sketch for a Historical Picture of the Progress of the Human Mind*. Translated by June Barraclough. First published in the original French in 1795. New York: Noonday Press, 1955.

De Garis, Hugo. *The Artilect War: Cosmists vs. Terrans: A Bitter Controversy Concerning Whether Humanity Should Build Godlike Massively Intelligent Machines*. Palm Springs, CA: ETC Publications, 2005.

De Grey, Aubrey. *A Maintenance Approach to Combat Aging-Related Diseases*, 2013. http://youtu.be/WEwlOjRVQIk.

———. "'We Will Be Able to Live to 1,000.'" *BBC News*, December 3, 2004, sec. UK. http://news.bbc.co.uk/1/hi/uk/4003063.stm.

DeLillo, Don. *White Noise*. New York: Viking, 1985.

Denk, Winfried, and Heinz Horstmann. "Serial Block-Face Scanning Electron Microscopy to Reconstruct Three-Dimensional Tissue Nanostructure." *PLoS Biol* 2, no. 11 (October 19, 2004): e329. doi:10.1371/journal.pbio.0020329.

Dennett, Daniel C. *Breaking the Spell: Religion as a Natural Phenomenon*. London: Penguin Books, 2006.

———. *Consciousness Explained*. London: Penguin, 1993.

———. "Dangerous Memes." Speech presented at the TED conference, February 2002. http://www.ted.com/talks/dan_dennett_on_dangerous_memes.html.

———. *Darwin's Dangerous Idea: Evolution and the Meanings of Life*. New York: Touchstone, 1996.

———. "Intuition Pumps." In *Third Culture: Beyond the Scientific Revolution*, edited by John

Brockman, 181–97. New York: Touchstone, 1996.

———. The Normal Well-tempered Mind, January 8, 2013.
http://edge.org/conversation/the-normal-well-tempered-mind.

———. "Where Am I?" In *Brainstorms: Philosophical Essays on Mind and Psychology*.
Cambridge, Mass.: MIT Press, 1981.

Devlin, A. M., J. H. Cross, W. Harkness, W. K. Chong, B. Harding, F. Vargha-Khadem, and
B. G. R. Neville. "Clinical Outcomes of Hemispherectomy for Epilepsy in Childhood and
Adolescence." *Brain* 126, no. 3 (March 1, 2003): 556–66. doi:10.1093/brain/awg052.

De Wolf, Aschwin. "A Skeptic's Guide to Cryonics." *Cryonics*, September 2013.

———. "Chemical Preservation and Human Suspended Animation." *Cryonics*, January 2013.

De Wolf, Aschwin, and Mike O'Neal. "The Case for Whole Body Cryopreservation." *Cryonics*,
February 2014.

De Wolf, Chana. "Cryopreservation of Kim Suozzi (Patient A-2643)." *Cryonics*, March 2014.

———. "Cryopreservation of the Brain: An Update." *Cryonics*, September 2013.

Diamond, Jared M. *Collapse: How Societies Choose to Fail or Survive*. New edition. London:
Penguin Books, 2011.

———. *Guns, Germs and Steel: A Short History of Everybody for the Last 13,000 Years*. London:
Vintage, 1998.

Diamond, Marian C., Arnold B. Scheibel, Greer M. Murphy Jr., and Thomas Harvey. "On the
Brain of a Scientist: Albert Einstein." *Experimental Neurology* 88, no. 1 (April 1985): 198–
204. doi:10.1016/0014-4886(85)90123-2.

Dickens, Charles. *A Christmas Carol*. London: Chapman & Hall, 1843.

Dickinson, Emily. "Forever Is Composed of Nows." In *Further Poems of Emily Dickinson:
Withheld from Publication by Her Sister Lavinia*, edited by Martha Dickinson Bianchi and
Alfred Leete Hampson. Boston, Mass.: Little, Brown, and Company, 1929.

Dickson, D. Bruce. *The Dawn of Belief: Religion in the Upper Paleolithic of Southwestern Europe*.
Arizona: University of Arizona Press, 1990.

Doctorow, Cory, and Charles Stross. *The Rapture of the Nerds*. London: Titan, 2013.

Doidge, Norman. *The Brain That Changes Itself: Stories of Personal Triumph from the Frontiers
of Brain Science*. London: Penguin Books, 2008.

Donaldson, Thomas K. "He's Dead, Jim: The Irreversibility of Death as a Circular Argument."
Alcor.org, August 12, 2004. http://www.alcor.org/Library/html/hesdeadjim.htm.

Donate, L. E., and M. A. Blasco. "Telomeres in Cancer and Ageing." *Philosophical Transactions
of the Royal Society B: Biological Sciences* 366, no. 1561 (November 29, 2010): 76–84.
doi:10.1098/rstb.2010.0291.

Drachman, David A. "Do We Have Brain to Spare?" *Neurology* 64, no. 12 (June 28, 2005):
2004–5. doi:10.1212/01.WNL.0000166914.38327.BB.

Dreger, Alice Domurat. *One of Us: Conjoined Twins and the Future of Normal*. Harvard
University Press, 2004.

Drexler, K. Eric. *Engines of Creation: The Coming Era of Nanotechnology*. New York: Anchor,
1986.

———. *Radical Abundance: How a Revolution in Nanotechnology Will Change Civilization*. First edition. New York: BBS PublicAffairs, 2013.

Drummond, Henry. *The Lowell Lectures on the Ascent of Man*. New York: J. Pott & co., 1908. http://archive.org/details/lowelllecture00drumiala.

Dschwen. *Lorenz_attractor.svg*, 2006. http://en.wikipedia.org/wiki/File:Lorenz_attractor.svg.

Dworschak, Manfred. "'My Second Birth': Discovering Life in Vegetative Patients." *Spiegel Online*, November 25, 2009. http://www.spiegel.de/international/spiegel/my-second-birth-discovering-life-in-vegetative-patients-a-663022.html.

Dyson, Freeman J. "Time without End: Physics and Biology in an Open Universe." *Reviews of Modern Physics* 51, no. 3 (July 1, 1979): 447–60. doi:10.1103/RevModPhys.51.447.

Eacott, Madeline J., and Alexander Easton. "Mental Time Travel in the Rat: Dissociation of Recall and Familiarity." *Behavioral and Brain Sciences* 30, no. 03 (2007): 322–23. doi:10.1017/S0140525X07002075.

Eagleman, David. *Incognito: The Secret Lives of the Brain*. Kindle Edition. Edinburgh: Canongate, 2011.

———. *Sum: Tales from the Afterlives*. Kindle Edition. Edinburgh: Canongate, 2009.

Edelman, Gerald M. *Neural Darwinism: The Theory of Neuronal Group Selection*. New York: Basic Books, 1987.

Edson, Russell. "The Floor." In *The Clam Theater*. Middletown, Conn.: Wesleyan University Press, 1973.

Edwards, William Howell. "The Dreaming: A Question of Time." In *An Introduction to Aboriginal Societies*. Victoria: Social Science Press, 1988.

Egan, Greg. *Permutation City*. London: Millennium, 1998.

Einstein, Albert. *The Ultimate Quotable Einstein*. Edited by Alice Calaprice. Princeton University Press, 2010.

Ellis, Arthur K. *Teaching and Learning Elementary Social Studies*. Boston: Allyn and Bacon, 1977.

Ettinger, Robert C. W. *The Prospect of Immortality*. 2nd edition. New York: Macfadden Books, 1966.

Fahy, Gregory M, Brian Wowk, Jun Wu, John Phan, Chris Rasch, Alice Chang, and Eric Zendejas. "Cryopreservation of Organs by Vitrification: Perspectives and Recent Advances." *Cryobiology* 48, no. 2 (April 2004): 157–78. doi:10.1016/j.cryobiol.2004.02.002.

Falk, Dean, Frederick E Lepore, and Adrianne Noe. "The Cerebral Cortex of Albert Einstein: A Description and Preliminary Analysis of Unpublished Photographs." *Brain: A Journal of Neurology* 136, no. Pt 4 (April 2013): 1304–27. doi:10.1093/brain/aws295.

Ferguson, Kitty. *Prisons of Light: Black Holes*. Cambridge: Cambridge University Press, 1996.

Feynman, Richard P. BBC Horizon: The Pleasure Of Finding Things Out. Documentary, 1981. http://www.bbc.co.uk/programmes/p018dvyg.

———. *The Pleasure of Finding Things Out: The Best Short Works of Richard P. Feynman*. Edited by Jeffrey Robbins. New York: Perseus Books, 1999.

———. "What Is and What Should Be the Role of Scientific Culture in Modern Society." Speech presented at the Galileo Symposium, Italy, 1964.

Feynman, Richard P., and Ralph Leighton. *Surely You're Joking, Mr. Feynman!: Adventures of a Curious Character*. Edited by Edward Hutchings. London: Vintage books, 1992.

Firth, Raymond. *Religion: A Humanist Interpretation*. London: Routledge, 1996.

Flynn, James. "Why Our IQ Levels Are Higher than Our Grandparents'." Speech presented at the TED conference, March 2013.
http://www.ted.com/talks/james_flynn_why_our_iq_levels_are_higher_than_our_grand parents.html.

"Formaldehyde." *Wikipedia, the Free Encyclopedia*, January 27, 2014.
http://en.wikipedia.org/w/index.php?title=Formaldehyde.

Freitas, R. A., Jr. "Exploratory Design in Medical Nanotechnology: A Mechanical Artificial Red Cell." *Artificial Cells, Blood Substitutes, and Immobilization Biotechnology* 26, no. 4 (July 1998): 411–30.

Fuller, Barry J., Nick Lane, and Erica E. Benson. "Mechanisms of Cryoinjury." In *Life in the Frozen State*. Florida: CRC Press, 2004.

Fuller, Richard Buckminster, Jerome Agel, and Quentin Fiore. *I Seem to Be a Verb*. Bantam Books, 1970.

Gaarder, Jostein. *The Ringmaster's Daughter*. Translated by James Anderson. London: Weidenfeld & Nicolson, 2002.

Gaiman, Neil. *The Sandman: Season of Mists*. London: Titan, 1992.

Galaburda, Am. "Albert Einstein's Brain." *The Lancet* 354, no. 9192 (November 1999): 1821. doi:10.1016/S0140-6736(05)70590-0.

Gardner, Howard. *The Mind's New Science*. New York: Basic Books, 1985.

Gater, Will. "NASA's Craziest Ideas: Astronaut Hibernation." *BBC Focus*, January 2014.

Gazzaniga, Michael S. "The Split Brain in Man." *Scientific American* 217, no. 2 (August 1967): 24–29. doi:10.1038/scientificamerican0867-24.

Geim, A. K., S. V. Dubonos, I. V. Grigorieva, K. S. Novoselov, A. A. Zhukov, and S. Yu. Shapoval. "Microfabricated Adhesive Mimicking Gecko Foot-Hair." *Nature Materials* 2, no. 7 (July 2003): 461–63. doi:10.1038/nmat917.

Geim, A. K., K. S. Novoselov, S. V. Morozov, D. Jiang, Y. Zhang, S. V. Dubonos, I. V. Grigorieva, and A. A. Firsov. "Electric Field Effect in Atomically Thin Carbon Films." *Science* 306, no. 5696 (October 22, 2004): 666–69. doi:10.1126/science.1102896.

Gell-Mann, Murray. "Beauty, Truth and ... Physics?" Speech presented at the TED conference, March 2007.
http://www.ted.com/talks/murray_gell_mann_on_beauty_and_truth_in_physics.html.

Gilbert, Daniel Todd. *Stumbling on Happiness*. London: Harper Perennial, 2007.

Glover, Jonathan. *What Sort of People Should There Be?*. Harmondsworth; New York: Penguin, 1984.

Golding, William. *Pincher Martin*. London: Faber and Faber, 1956.

Goldstein, Jeffrey. "Emergence as a Construct: History and Issues." *Emergence: Complexity and Organization* 1, no. 1 (March 1999): 49–72. doi:10.1207/s15327000em0101_4.

Gorey, Edward. *The Doubtful Guest*. London: Bloomsbury, 1998.

Graber, Steve. "Research and Development Update." *Cryonics*, January 2013.

Graham-Rowe, Duncan. "Rabbits Kept Alive by Oxygen Injections." *Nature*, June 27, 2012. doi:10.1038/nature.2012.10899.

Grand, Steve. *Creation: Life and How to Make It*. London: Phoenix, 2001.

———. "Email Correspondence: AI Wisdom," March 29, 2012.

———. *Grandroids* (version in progress), 2011.

———. *Growing up with Lucy: How to Build an Android in Twenty Easy Steps*. London: Weidenfeld & Nicolson, 2004.

Gray, Alasdair. *Lanark: A Life in Four Books*. London: Macmillan, 1994.

Gribbin, John. *In Search of the Multiverse*. London; New York: Allen Lane, 2009.

Gutro, Rob. "What's the Difference Between Weather and Climate?" *Nasa.gov*, February 1, 2005. http://www.nasa.gov/mission_pages/noaa-n/climate/climate_weather.html.

Hagmann, Patric, Reto Meuli, and Jean-Philippe Thiran. "From Diffusion MRI to Brain Connectomics," 2005. http://doc.rero.ch/record/7576.

Hájek, Alan. "Pascal's Wager." In *The Stanford Encyclopedia of Philosophy*, edited by Edward N. Zalta, Winter 2012., 2012. http://plato.stanford.edu/archives/win2012/entries/pascal-wager/.

Haldane, John Burdon Sanderson. *Possible Worlds and Other Essays*. London: Chatto and Windus, 1928.

Hall, J. Storrs. *Beyond AI: Creating the Conscience of the Machine*. Amherst, N.Y: Prometheus Books, 2007.

Halverson, John. "Art for Art's Sake in the Paleolithic." *Current Anthropology* 28, no. 1 (February 1987): 63–89.

Hameroff, Stuart, and Roger Penrose. "Consciousness in the Universe: A Review of the 'Orch OR' Theory." *Physics of Life Reviews*, August 20, 2013. doi:10.1016/j.plrev.2013.08.002.

———. "Orchestrated Reduction of Quantum Coherence in Brain Microtubules: A Model for Consciousness." *Math. Comput. Simul.* 40, no. 3–4 (April 1996): 453–80. doi:10.1016/0378-4754(96)80476-9.

Harth, Erich. *The Creative Loop: How the Brain Makes a Mind*. London: Penguin, 1995.

Hawkins, Jeff. "Why Can't a Computer Be More Like a Brain?" *IEEE Spectrum* 44, no. 4 (April 2007): 21–26. doi:10.1109/MSPEC.2007.339647.

Hawkins, Jeff, and Sandra Blakeslee. *On Intelligence*. New York: Owl Books, 2005.

Hayward, Jeremy W., and Francisco J. Varela, eds. *Gentle Bridges: Conversations with the Dalai Lama on the Sciences of Mind*. Boston, Mass.: Shambhala Publications, 1992.

Hazlett, Iain. "Contacts with Scotland." In *The Calvin Handbook*, edited by H. J. Selderhuis, 124–25. Cambridge: Wm. B. Eerdmans Publishing, 2009.

HBP. "The Human Brain Project." *Humanbrainproject.eu*, April 15, 2012. https://www.humanbrainproject.eu/en_GB.

Hebb, Donald Olding. *The Organization of Behavior: A Neuropsychological Theory*. Wiley, 1949.

Heinlein, Robert A. "By His Bootstraps." *Astounding Science Fiction*, October 1941.

Herculano-Houzel, S. "The Remarkable, yet Not Extraordinary, Human Brain as a Scaled-up Primate Brain and Its Associated Cost." *Proceedings of the National Academy of Sciences*

109, no. Supplement_1 (June 20, 2012): 10661–68. doi:10.1073/pnas.1201895109.

Herculano-Houzel, Suzana, and Roberto Lent. "Isotropic Fractionator: A Simple, Rapid Method for the Quantification of Total Cell and Neuron Numbers in the Brain." *The Journal of Neuroscience: The Official Journal of the Society for Neuroscience* 25, no. 10 (March 9, 2005): 2518–21. doi:10.1523/JNEUROSCI.4526-04.2005.

Hickey, Lance P. "The Brain in a Vat Argument." *Internet Encyclopedia of Philosophy*, 2005. http://www.iep.utm.edu/brainvat/.

Hinde, Robert A. *Why Gods Persist: A Scientific Approach to Religion*. London; New York: Routledge, 1999.

Hofstadter, Douglas R. *Gödel, Escher, Bach: An Eternal Golden Braid*. London: Penguin, 2000.

———. *I Am a Strange Loop*. New York: Basic Books, 2007.

Höjer, J, W G Troutman, K Hoppu, A Erdman, B E Benson, B Mégarbane, R Thanacoody, et al. "Position Paper Update: Ipecac Syrup for Gastrointestinal Decontamination." *Clinical Toxicology (Philadelphia, Pa.)* 51, no. 3 (March 2013): 134–39. doi:10.3109/15563650.2013.770153.

Hölzel, Britta K., James Carmody, Mark Vangel, Christina Congleton, Sita M. Yerramsetti, Tim Gard, and Sara W. Lazar. "Mindfulness Practice Leads to Increases in Regional Brain Gray Matter Density." *Psychiatry Research: Neuroimaging* 191, no. 1 (January 30, 2011): 36–43. doi:10.1016/j.pscychresns.2010.08.006.

Hood, Bruce M. *Supersense: From Superstition to Religion – The Brain Science of Belief*. London: Constable, 2009.

Huang, Cary. "The Scale of the Universe 2." *Htwins.net*, 2012. http://htwins.net/scale2/.

Hubbard, Elbert. *The Philistine Magazine*, December 1909.

Hughes, James. *Citizen Cyborg: Why Democratic Societies Must Respond to the Redesigned Human of the Future*. Cambridge, MA: Westview Press, 2004.

———. "Compassionate AI and Selfless Robots: A Buddhist Approach." In *Robot Ethics: The Ethical and Social Implications of Robotics*, edited by Patrick Lin, Keith Abney, and George A. Bekey. Cambridge, Mass: MIT Press, 2012.

Hume, David. *An Enquiry Concerning Human Understanding*. Reprinted from the Posthumous edition of 1777. Oxford: University College, 1902.

———. *A Treatise of Human Nature*. Edited by David Fate Norton and Mary J Norton. First published anonymously in 1739. Vol. 1. 2 vols. Oxford: Oxford University Press, 2007.

Huneker, James. *Chopin: The Man and His Music*. New York: Scribner's, 1918.

"Immanent." *Oxford Dictionaries*. Oxford University Press, 2013. http://www.oxforddictionaries.com/definition/english/immanent.

Indiveri, Giacomo, Bernabe Linares-Barranco, Robert Legenstein, George Deligeorgis, and Themistoklis Prodromakis. "Integration of Nanoscale Memristor Synapses in Neuromorphic Computing Architectures." *Nanotechnology* 24, no. 38 (September 27, 2013): 384010. doi:10.1088/0957-4484/24/38/384010.

"Infinity (mathematics)." *Encyclopedia Britannica Online*. Encyclopedia Britannica, 2014. http://www.britannica.com/EBchecked/topic/287662/infinity.

Jackson, Peter. *The Lovely Bones*. Drama, Fantasy, 2010.

Jonas, Peter, and Gyorgy Buzsaki. "Neural Inhibition." *Scholarpedia* 2, no. 9 (2007): 3286. doi:10.4249/scholarpedia.3286.

Kaku, Michio. *Parallel Worlds: The Science of Alternative Universes and Our Future in the Cosmos*. London: Penguin, 2006.

Kant, Immanuel. *Groundwork of the Metaphysics of Morals*. Edited by Mary Gregor and Jens Timmermann. First published in the original German in 1785. Cambridge: Cambridge University Press, 2012.

———. "Of the Ground of the Division of All Objects into Phenomena and Noumena." In *The Critique of Pure Reason*, translated by J. M. D. Meiklejohn, 1781.

———. *The Metaphysics of Morals*. Edited by Mary J. Gregor. First published in the original German in 1797. Cambridge University Press, 1996.

Kearney, Richard. "Myths and Scapegoats: The Case of René Girard." *Theory, Culture & Society* 12, no. 4 (November 1, 1995): 1–14. doi:10.1177/026327695012004002.

Kerrigan, Michael. *The History of Death: Burial Customs and Funeral Rites, from the Ancient World to Modern Times*. Guilford, Conn.: Lyons Press, 2007.

Kershaw, Alister. *A History of the Guillotine*. London: John Calder, 1958.

Keyes, Daniel. *Flowers for Algernon,*. New York: Harcourt, Brace & World, 1966.

Knight, Chris. "Ochre in Prehistory." In *Blood Relations: Menstruation and the Origins of Culture*. New Haven: Yale University Press, 1995.

Koene, Randal A. "Embracing Competitive Balance: The Case for Substrate-Independent Minds and Whole Brain Emulation." In *Singularity Hypotheses*, edited by Amnon H. Eden, James H. Moor, Johnny H. Søraker, and Eric Steinhart, 241–67. The Frontiers Collection. Springer Berlin Heidelberg, 2012. http://link.springer.com/chapter/10.1007/978-3-642-32560-1_12.

Kolb, Bryan, Robbin Gibb, and Terry E. Robinson. "Brain Plasticity and Behavior." *Current Directions in Psychological Science* 12, no. 1 (February 1, 2003): 1–5. doi:10.1111/1467-8721.01210.

Kuhn, Thomas S. *The Structure of Scientific Revolutions*. Chicago: University of Chicago Press, 1962.

Kulstad, Mark, and Laurence Carlin. "Leibniz's Philosophy of Mind." In *The Stanford Encyclopedia of Philosophy*, edited by Edward N. Zalta, Winter 2013., 2013. http://plato.stanford.edu/archives/win2013/entries/leibniz-mind/.

Kurzweil, Ray. *The Singularity Is Near: When Humans Transcend Biology*. London: Gerald Duckworth, 2005.

Kurzweil, Ray, and Mitch Kapor. "A Wager on the Turing Test: The Rules." *KurzweilAI.net*, April 9, 2002. http://www.kurzweilai.net/a-wager-on-the-turing-test-the-rules.

Lamb, Charles. *Literary Sketches and Letters: Being the Final Memorials of Charles Lamb, Never Before Published*. New York: D. Appleton, 1849.

Lambert, Frank L. "Disorder – A Cracked Crutch for Supporting Entropy Discussions." *Journal of Chemical Education* 79, no. 2 (February 1, 2002): 187. doi:10.1021/ed079p187.

Leader, Darian, and David Corfield. *Why Do People Get Ill?*. London: Penguin, 2008.

Lent, Roberto, Frederico A. C. Azevedo, Carlos H. Andrade-Moraes, and Ana V. O. Pinto.

"How Many Neurons Do You Have? Some Dogmas of Quantitative Neuroscience under Revision." *European Journal of Neuroscience* 35, no. 1 (2012): 1–9. doi:10.1111/j.1460-9568.2011.07923.x.

Leuba, G, and L J Garey. "Evolution of Neuronal Numerical Density in the Developing and Aging Human Visual Cortex." *Human Neurobiology* 6, no. 1 (1987): 11–18.

Lippincott. *Anatomy and Physiology*. Lippincott Williams & Wilkins, 2002.

Lloyd, Seth. *Programming the Universe: A Quantum Computer Scientist Takes on the Cosmos*. London: Vintage, 2007.

Lobell, Jarrett, and Samir Patel. "Cladh Hallan." *Archaeology Magazine*, May 2010. http://archive.archaeology.org/1005/bogbodies/cladh_hallan.html.

Long Now Foundation. "About The Long Now." *Longnow.org*, 1996. http://longnow.org/about/.

Lydon, John, Keith Zimmerman, and Kent Zimmerman. *Rotten: No Irish, No Blacks, No Dogs: The Authorised Autobiography, Johnny Rotten of the Sex Pistols*. London: Hodder & Stoughton, 1994.

MacCulloch, J. A. *The Religion of the Ancient Celts*. First published in 1911. The Floating Press, 2009.

MacDougall, Duncan. "Hypothesis Concerning Soul Substance Together with Experimental Evidence of The Existence of Such Substance." *American Medicine* 2, no. 4 (April 1907): 240–43.

Mackay, Donald M. "Divided Brains – Divided Minds?" In *Mindwaves: Thoughts on Intelligence, Identity and Consciousness*, edited by Colin Blakemore and Susan Greenfield. Oxford: Wiley Blackwell, 1987.

MacLennan, D. J. "Conditioned Existence." *Extravolution.com*, April 12, 2015. http://www.extravolution.com/2015/04/conditioned-existence.html.

———. "If I Only Had a Brain." *Extravolution.com*, March 31, 2014. http://www.extravolution.com/2014/03/if-i-only-had-brain.html.

———. "The Massive, Magnanimous Liberty of The Culture." *H+ Magazine*, June 14, 2013. http://hplusmagazine.com/2013/06/14/the-massive-magnanimous-liberty-of-the-culture/.

Martini, Frederic H. *Anatomy and Physiology*. 1st Edition. New Jersey: Prentice Hall, 2005.

Martin, Lee J., Nael A. Al-Abdulla, Ansgar M. Brambrink, Jeffrey R. Kirsch, Frederick E. Sieber, and Carlos Portera-Cailliau. "Neurodegeneration in Excitotoxity, Global Cerebral Ischemia, and Target Deprivation: A Perspective on the Contributions of Apoptosis and Necrosis." *Brain Research Bulletin* 46, no. 4 (July 1998): 281–309. doi:10.1016/S0361-9230(98)00024-0.

Mattson, Mark P. "Hormesis Defined." *Ageing Research Reviews* 7, no. 1 (January 2008): 1–7. doi:10.1016/j.arr.2007.08.007.

Maturana, Humberto R., and Francisco J Varela. *Autopoiesis and Cognition: The Realization of the Living*. Dordrecht, Holland: D. Reidel Publishing Company, 1980.

McCauley, Clark. "Understanding the 9/11 Perpetrators: Crazy, Lost in Hate, or Martyred." In *History Behind The Headlines: The Origins Of Conflict Worldwide*, edited by N.

Matuszak, 5:274–86. New York: Gale Publishing Group, 2002.

McLuhan, Marshall. *Understanding Media: The Extensions of Man*. London: Routledge, 1964.

McSush, and Lunkwill. *Exponential.svg*, 2010.
    http://commons.wikimedia.org/wiki/File:Exponential.svg.

Men, Weiwei, Dean Falk, Tao Sun, Weibo Chen, Jianqi Li, Dazhi Yin, Lili Zang, and Mingxia
    Fan. "The Corpus Callosum of Albert Einstein's Brain: Another Clue to His High
    Intelligence?" *Brain*, September 24, 2013, awt252. doi:10.1093/brain/awt252.

Merkle, Ralph. "The Allocation of Long Term Care Costs at Alcor." *Cryonics*, June 2012.

———. "The Technical Feasibility of Cryonics," 1992.
    http://www.merkle.com/cryo/TheTechnicalFeasibilityOfCryonics.pdf.

Meryman, H. T. "Osmotic Stress as a Mechanism of Freezing Injury." *Cryobiology* 8, no. 5
    (October 1971): 489–500. doi:10.1016/0011-2240(71)90040-X.

Mika, Eric, Arturo Vidich, and Sofy Yuditskaya. "How to Hack Toy EEGs." *Frontiernerds.com*,
    April 7, 2010. http://frontiernerds.com/brain-hack.

Mikulecky, Peter J., Michelle Rose Gilman, and Kate Brutlag. *AP Chemistry For Dummies*.
    John Wiley & Sons, 2009.

Minsky, Marvin Lee. *The Emotion Machine: Commonsense Thinking, Artificial Intelligence, and
    the Future of the Human Mind*. New York: Simon & Schuster, 2006.

———. *The Society of Mind*. New York: Simon and Schuster, 1986.

Mlodinow, Leonard. *The Drunkard's Walk: How Randomness Rules Our Lives*. London:
    Penguin, 2009.

Moore, Alan. "Interview with Alan Moore." In *A Disease of Language*. London: Knockabout
    Comics, 2005.

Moore, Pete. *Enhancing Me: The Hope and the Hype of Human Enhancement*. Chichester,
    England; Hoboken, NJ: Wiley/Dana Centre, 2008.

"Moore's Law (computer Science)." *Encyclopedia Britannica Online*. Encyclopedia Britannica,
    2014. http://www.britannica.com/EBchecked/topic/705881/Moores-law.

Moran, Laurence A. "What Is a Gene?" *Sandwalk*, January 28, 2007.
    http://sandwalk.blogspot.co.uk/2007/01/what-is-gene.html.

More, Max. "Technological Self-Transformation." *Extropy*, 1993.

More, Max, Natasha Vita More, Alexander Chislenko, Anders Sandberg, James Hughes, and
    Nick Bostrom. "Transhumanist FAQ." *Humanityplus.org*, April 28, 2002.
    http://humanityplus.org/philosophy/transhumanist-faq/.

Moscow Times. "Russian Programmer Volunteers for 'Frankenstein' Head Transplant."
    *Moscowtimes.com*, April 8, 2015. http://www.themoscowtimes.com/news/article/russian-
    programmer-volunteers-for-frankenstein-head-transplant/518810.html.

Nagel, Thomas. "What Is It Like to Be a Bat?" *The Philosophical Review* 83, no. 4 (October
    1974): 435. doi:10.2307/2183914.

Neumann, Odmar. "Some Aspects of Phenomenal Consciousness and Their Possible
    Functional Correlates." Paper presented at the conference "The phenomenal mind – how is
    it possible and why is it necessary?," Bielefeld, 1990.

Nishida, Toshisada, and Kenji Kawanaka. "Within-Group Cannibalism by Adult Male

Chimpanzees." *Primates* 26, no. 3 (July 1, 1985): 274–84. doi:10.1007/BF02382402.

Nishimoto, Shinji, An T. Vu, Thomas Naselaris, Yuval Benjamini, Bin Yu, and Jack L. Gallant. "Reconstructing Visual Experiences from Brain Activity Evoked by Natural Movies." *Current Biology* 21, no. 19 (November 10, 2011): 1641–46. doi:10.1016/j.cub.2011.08.031.

Nolan, Christopher. *The Prestige*. Drama, Mystery, Thriller, 2006.

Oberhammer, Rosmarie, Werner Beikircher, Christoph Hörmann, Ingo Lorenz, Roger Pycha, Liselotte Adler-Kastner, and Hermann Brugger. "Full Recovery of an Avalanche Victim with Profound Hypothermia and Prolonged Cardiac Arrest Treated by Extracorporeal Re-Warming." *Resuscitation* 76, no. 3 (March 2008): 474–80. doi:10.1016/j.resuscitation.2007.09.004.

Oe, Kenzaburo. *A Personal Matter*. Grove Press, 1969.

Okonek, S, H Setyadharma, A Borchert, and E G Krienke. "Activated Charcoal Is as Effective as Fuller's Earth or Bentonite in Paraquat Poisoning." *Klinische Wochenschrift* 60, no. 4 (February 15, 1982): 207–10.

Olsen, Tom Skyhøj, Uno Jakob Weber, and Lars Peter Kammersgaard. "Therapeutic Hypothermia for Acute Stroke." *The Lancet Neurology* 2, no. 7 (July 2003): 410–16. doi:10.1016/S1474-4422(03)00436-8.

Orrell, David. *The Future of Everything: The Science of Prediction: From Wealth and Weather to Chaos and Complexity*. New York: Basic Books, 2007.

"Outlier." *Oxford Dictionaries*. Oxford University Press, 2014. http://www.oxforddictionaries.com/definition/english/outlier.

"Paraquat." *Oxford Dictionaries*. Oxford University Press, 2013. http://www.oxforddictionaries.com/definition/english/paraquat.

Parfit, Derek. *On What Matters*. Vol. 1. The Berkeley Tanner Lectures. Oxford; New York: Oxford University Press, 2011.

———. *Reasons and Persons*. Oxford: Oxford University Press, 1987.

Pariser, Eli. *The Filter Bubble: What The Internet Is Hiding From You*. London: Penguin, 2012.

Parker Pearson, Mike, Andrew Chamberlain, Matthew Collins, Christie Cox, Geoffrey Craig, Oliver Craig, Jen Hiller, Peter Marshall, Jacqui Mulville, and Helen Smith. "Further Evidence for Mummification in Bronze Age Britain." *Antiquity* 81, no. 312 (September 2007). http://antiquity.ac.uk/ProjGall/parker/.

Parkinson's UK. "Deep Brain Stimulation." *Parkinsons.org.uk*, February 2013. http://www.parkinsons.org.uk/content/deep-brain-stimulation.

Pearce, David. "Wirehead Hedonism versus Paradise-Engineering," February 1, 2001. http://www.wireheading.com/.

Perlmutter, Dawn. "The Semiotics of Honor Killing & Ritual Murder." *Anthropoetics – The Journal of Generative Anthropology* 17, no. 1 (2011). http://www.anthropoetics.ucla.edu/ap1701/1701Perlmutter.htm.

Petanjek, Zdravko, Miloš Judaš, Goran Šimić, Mladen Roko Rašin, Harry B. M. Uylings, Pasko Rakic, and Ivica Kostović. "Extraordinary Neoteny of Synaptic Spines in the Human Prefrontal Cortex." *Proceedings of the National Academy of Sciences* 108, no. 32

(August 9, 2011): 13281–86. doi:10.1073/pnas.1105108108.

Peters, A., and M. L. Feldman. "The Projection of the Lateral Geniculate Nucleus to Area 17 of the Rat Cerebral Cortex. I. General Description." *Journal of Neurocytology* 5, no. 1 (February 1976): 63–84.

Pettitt, Paul. *The Palaeolithic Origins of Human Burial.* New York: Routledge, 2011.

Pfister, Wally. *Transcendence.* Action, Drama, Sci-Fi, 2014.

Pinker, Steven. *The Language Instinct.* New York: William Morrow, 1994.

Ponomarev, L. I., and I. V. Kurchatov. *The Quantum Dice.* Boca Raton, Florida: CRC Press, 1993.

Porotto, Matteo, Feng Yi, Anne Moscona, and David A. LaVan. "Synthetic Protocells Interact with Viral Nanomachinery and Inactivate Pathogenic Human Virus." Edited by Meni Wanunu. *PLoS ONE* 6, no. 3 (March 1, 2011): e16874. doi:10.1371/journal.pone.0016874.

Porter, Roy. *Blood and Guts: A Short History of Medicine.* London; New York: Penguin, 2003.

Premack, David, and Guy Woodruff. "Does the Chimpanzee Have a Theory of Mind?" *Behavioral and Brain Sciences* 1, no. 04 (1978): 515–26. doi:10.1017/S0140525X00076512.

Price, David. "Fairly Bland: An Alternative View of a Supposed New 'Death Ethic' and the BMA Guidelines." *Legal Studies* 21, no. 4 (2001): 618–43. doi:10.1111/j.1748-121X.2001.tb00183.x.

Quiroga, R. Quian, L. Reddy, G. Kreiman, C. Koch, and I. Fried. "Invariant Visual Representation by Single Neurons in the Human Brain." *Nature* 435, no. 7045 (June 23, 2005): 1102–7. doi:10.1038/nature03687.

Quote Investigator. "Writing About Music Is Like Dancing About Architecture." *Quoteinvestigator.com,* November 2010. http://quoteinvestigator.com/2010/11/08/writing-about-music/.

Rabada, David. "Were There Ritual Burials in the Sima de Los Huesos Outcrop? (Atapuerca Range, Burgos, Spain)." *David Rabada,* July 27, 2012. http://blogs.e-noticies.com/david-rabada/were_there_ritual_burials_in_the_sima_de_los_huesos_outcrop_atapuerca_range_burgos_spain.html.

Raichle, Marcus E, and Abraham Z Snyder. "A Default Mode of Brain Function: A Brief History of an Evolving Idea." *NeuroImage* 37, no. 4 (October 1, 2007): 1083–90; discussion 1097–99. doi:10.1016/j.neuroimage.2007.02.041.

Ramachandran, V. S. "Split Brain with One Half Atheist and One Half Theist." June 3, 2010. http://youtu.be/PFJPtVRlI64.

Rawls, John. *A Theory of Justice.* Reprint of original 1971 edition. Cambridge, Mass: Harvard University Press, 2005.

Reynolds, Alastair. *Chasm City.* London: Gollancz, 2001.

———. *House of Suns.* London: Gollancz, 2008.

———. *Revelation Space.* London: Gollancz, 2000.

Reynolds, Craig. *Boids,* 1986.

Rilling, James K. "Human and Nonhuman Primate Brains: Are They Allometrically Scaled

Versions of the Same Design?" *Evolutionary Anthropology: Issues, News, and Reviews* 15, no. 2 (2006): 65–77. doi:10.1002/evan.20095.

Roach, Mary. *Stiff: The Curious Lives of Human Cadavers.* London: Penguin, 2004.

Robbins, Jim. *A Symphony in the Brain: The Evolution of the New Brain Wave Biofeedback.* New York: Grove Press, 2008.

Rohr, René R. J. *Sundials: History, Theory, and Practice.* Translated by Gabriel Godin. New York: Dover Publications, 1996.

Rosenblatt, F. "The Perceptron: A Probabilistic Model for Information Storage and Organization in the Brain." *Psychological Review* 65, no. 6 (1958): 386–408. doi:10.1037/h0042519.

Saenz, Aaron. "Brain Preservation Technology Prize: A Proposal for Immortality?" *Singularityhub.com*, March 29, 2010. http://singularityhub.com/2010/03/29/brain-preservation-technology-prize-a-modest-proposal-for-immortality-video/.

Sahu, Satyajit, Subrata Ghosh, Kazuto Hirata, Daisuke Fujita, and Anirban Bandyopadhyay. "Multi-Level Memory-Switching Properties of a Single Brain Microtubule." *Applied Physics Letters* 102, no. 12 (March 26, 2013): 123701. doi:10.1063/1.4793995.

Sample, Ian. "Quest for the Connectome: Scientists Investigate Ways of Mapping the Brain." *The Guardian*, May 7, 2012, sec. Science. http://www.theguardian.com/science/2012/may/07/quest-connectome-mapping-brain.

Sample, Ian, and Alex Hern. "Scientists Dispute Whether Computer 'Eugene Goostman' Passed Turing Test." *The Guardian*, June 9, 2014, sec. Technology. http://www.theguardian.com/technology/2014/jun/09/scientists-disagree-over-whether-turing-test-has-been-passed.

Saxe, Rebecca, and Anna Wexler. "Making Sense of Another Mind: The Role of the Right Temporo-Parietal Junction." *Neuropsychologia* 43, no. 10 (2005): 1391–99. doi:10.1016/j.neuropsychologia.2005.02.013.

Saxe, R., and N. Kanwisher. "People Thinking about Thinking People. The Role of the Temporo-Parietal Junction in 'Theory of Mind.'" *NeuroImage* 19, no. 4 (August 2003): 1835–42.

Schrödinger, Erwin. *What Is Life?*. Cambridge: Cambridge University Press, 1944.

Schroeder, Karl. *Ventus.* New York: Tor, 2000.

ScienceDaily. "Scientists Claim Brain Memory Code Cracked." *Sciencedaily.com*, March 9, 2012. http://www.sciencedaily.com/releases/2012/03/120309103701.htm.

Selznick, Brian. *The Invention of Hugo Cabret: A Novel in Words and Pictures.* 1st ed. New York: Scholastic Press, 2007.

Semon, Richard Wolfgang. *The Mneme.* London: George Allen & Unwin, 1921.

Seung, Sebastian. *Connectome: How the Brain's Wiring Makes Us Who We Are.* Kindle Edition. Penguin, 2012.

Shatz, Carla J., and Corey S. Goodman. "Developmental Mechanisms That Generate Precise Patterns of Neuronal Connectivity." *Cell* 72, Supplement (January 1993): 77–98. doi:10.1016/S0092-8674(05)80030-3.

Shay, J. W. "Aging and Cancer: Are Telomeres and Telomerase the Connection?" *Molecular*

*Medicine Today* 1, no. 8 (November 1995): 378–84.

Shermer, Michael. "Nano Nonsense & Cryonics." Blog. *Michaelshermer.com*, September 2001. http://www.michaelshermer.com/2001/09/nano-nonsense-and-cryonics/.

Shubin, Neil. *Your Inner Fish: The Amazing Discovery of Our 375-Million-Year-Old Ancestor.* London: Penguin, 2009.

Silk, Joseph. *The Infinite Cosmos: Questions from the Frontiers of Cosmology.* Oxford; New York: Oxford University Press, 2006.

Simmons, Dan. *The Endymion Omnibus.* London: Gollancz, 2005.

———. *The Hyperion Omnibus.* London: Gollancz, 2004.

Sironi, Vittorio A. "Origin and Evolution of Deep Brain Stimulation." *Frontiers in Integrative Neuroscience* 5 (August 18, 2011). doi:10.3389/fnint.2011.00042.

Skloot, Rebecca. *The Immortal Life of Henrietta Lacks.* London: Pan Books, 2010.

Society for Venturism. "Kim Suozzi Charity." *Venturist.info*, March 15, 2011. http://venturist.info/kim-suozzi-charity.html.

Sommer, Jeffrey D. "The Shanidar IV 'Flower Burial': A Re-Evaluation of Neanderthal Burial Ritual." *Cambridge Archaeological Journal* 9, no. 01 (1999): 127–29. doi:10.1017/S0959774300015249.

Soon, Chun Siong, Anna Hanxi He, Stefan Bode, and John-Dylan Haynes. "Predicting Free Choices for Abstract Intentions." *Proceedings of the National Academy of Sciences*, March 18, 2013, 201212218. doi:10.1073/pnas.1212218110.

Spalding, Kirsty L, Olaf Bergmann, Kanar Alkass, Samuel Bernard, Mehran Salehpour, Hagen B Huttner, Emil Boström, et al. "Dynamics of Hippocampal Neurogenesis in Adult Humans." *Cell* 153, no. 6 (June 6, 2013): 1219–27. doi:10.1016/j.cell.2013.05.002.

Sporns, Olaf, Giulio Tononi, and Rolf Kotter. "The Human Connectome: A Structural Description of the Human Brain." *PLoS Computational Biology* 1, no. 4 (September 2005). doi:10.1371/journal.pcbi.0010042.

Steiner, Susie. "Top Five Regrets of the Dying." *The Guardian*, February 1, 2012, sec. Life and style. http://www.theguardian.com/lifeandstyle/2012/feb/01/top-five-regrets-of-the-dying.

Stein, Rob. "'Vegetative' Woman's Brain Shows Surprising Activity." *The Washington Post*, September 8, 2006, sec. Nation. http://www.washingtonpost.com/wp-dyn/content/article/2006/09/07/AR2006090700978.html.

Stix, Gary. "Cool Aid: Drug That Lets Body Temperature Drop Could Save Stroke Victims." *Scientificamerican.com*, February 15, 2012. http://www.scientificamerican.com/article/cool-aid-drug-that-causes/.

"Strong's Greek: 1453. Ἐγείρω (egeiró) – to Waken, to Raise up." Accessed October 15, 2011. http://biblehub.com/greek/1453.htm.

Taleb, Nassim Nicholas. *Antifragile: How to Live in a World We Don't Understand.* London: Allen Lane, 2012.

———. *The Black Swan: The Impact of the Highly Improbable.* London: Penguin Books, 2008.

Taubes, Gary. *The Diet Delusion.* London: Vermilion, 2007.

Taylor, Jill Bolte. *My Stroke of Insight.* London: Hodder & Stoughton, 2008.

Taylor, M. J., J. E. Bailes, A. M. Elrifai, S.-R. Shih, E. Teeple, M. L. Leavitt, J. G. Baust, and J. C. Maroon. "A New Solution for Life Without Blood: Asanguineous Low-Flow Perfusion of a Whole-Body Perfusate During 3 Hours of Cardiac Arrest and Profound Hypothermia." *Circulation* 91, no. 2 (January 15, 1995): 431–44. doi:10.1161/01.CIR.91.2.431.

Taylor, Timothy. *The Buried Soul: How Humans Invented Death.* London: Fourth Estate, 2003.

———. "What Questions Have Disappeared?" *Edge.org,* 2001. http://www.edge.org/response-detail/12054.

Tegmark, Max. "Consciousness as a State of Matter." *arXiv:1401.1219 [cond-Mat, Physics:hep-Th, Physics:quant-Ph],* January 6, 2014. http://arxiv.org/abs/1401.1219.

———. "Importance of Quantum Decoherence in Brain Processes." *Physical Review E* 61, no. 4 (April 2000): 4194–4206. doi:10.1103/PhysRevE.61.4194.

Teilhard de Chardin, Pierre. *The Phenomenon of Man.* Translated by Bernard Wall. London; New York: Collins, Harper & Row, 1959.

Thera, Narada. "Buddhism in a Nutshell – Anatta." *Buddhanet.net,* n.d. http://www.buddhanet.net/nutshell09.htm.

———. "Buddhism in a Nutshell – Is Buddhism a Religion." *Buddhanet.net,* n.d. http://www.buddhanet.net/nutshell03.htm.

"Thomas K. Donaldson." *Wikipedia, the Free Encyclopedia,* January 19, 2014. http://en.wikipedia.org/w/index.php?title=Thomas_K._Donaldson.

Thomson, Helen. "First Human Head Transplant Could Happen in Two Years." *Newscientist.com,* February 25, 2015. http://www.newscientist.com/article/mg22530103.700-first-human-head-transplant-could-happen-in-two-years.html#.VPA-zI7m7ke.

———. "Suspended between Life and Death." *New Scientist* 221, no. 2962 (March 29, 2014): 8–9. doi:10.1016/S0262-4079(14)60610-2.

Thuras, Dylan. "Sokushinbutsu of Dainichi Temple." *Atlas Obscura,* June 27, 2007. http://www.atlasobscura.com/places/sokushinbutsu-dainichi-temple.

Tononi, Giulio. "Consciousness as Integrated Information: A Provisional Manifesto." *The Biological Bulletin* 215, no. 3 (December 1, 2008): 216–42.

Turing, A. M. "Computing Machinery and Intelligence." *Mind* LIX, no. 236 (1950): 433–60. doi:10.1093/mind/LIX.236.433.

———. "On Computable Numbers, with an Application to the Entscheidungsproblem." *Proceedings of the London Mathematical Society* s2–42, no. 1 (January 1, 1937): 230–65. doi:10.1112/plms/s2-42.1.230.

"Turing Machine (computing Device)." *Encyclopedia Britannica Online.* Encyclopedia Britannica, 2014. http://www.britannica.com/EBchecked/topic/609750/Turing-machine.

Turkheimer, Eric. "Three Laws of Behavior Genetics and What They Mean." *Current Directions in Psychological Science* 9, no. 5 (October 1, 2000): 160–64. doi:10.1111/1467-8721.00084.

Tyler, Mitchell, Yuri Danilov, and Paul Bach-Y-Rita. "Closing an Open-Loop Control System: Vestibular Substitution through the Tongue." *Journal of Integrative Neuroscience* 2, no. 2 (December 2003): 159–64.

UCL. "Human Brain Project Wins Major EU Funding." *UCL Website*, January 28, 2013. http://www.ucl.ac.uk/news/news-articles/0113/130128-human-brain-project-wins-major-EU-funding.

Uexküll, Jakob von. *Umwelt und Innenwelt der Tiere*. Berlin: Springer, 1909.

University of Cambridge. "Surprising Solution to Fly Eye Mystery." *Cam.ac.uk*, October 11, 2012. http://www.cam.ac.uk/research/news/surprising-solution-to-fly-eye-mystery.

Van Gennep, Arnold. *The Rites of Passage*. Translated by Monica B. Vizedom and Gabrielle L. Caffee. First published in 1909. Chicago: University of Chicago Press, 1960.

Varki, Ajit, and Tasha K. Altheide. "Comparing the Human and Chimpanzee Genomes: Searching for Needles in a Haystack." *Genome Research* 15, no. 12 (December 1, 2005): 1746–58. doi:10.1101/gr.3737405.

Vesalius, Andreas. *De Humani Corporis Fabrica*, 1543.

Villazon, Luis. "What Percentage of My Body Is the Same as Five Years Ago?" *BBC Focus*, January 2013.

Vinge, Vernor. "The Coming Technological Singularity," 1993. http://www-rohan.sdsu.edu/faculty/vinge/misc/singularity.html.

Von Hagens, Gunther. "Plastination." *Bodyworlds.com*, 2006. http://www.bodyworlds.com/en/plastination/idea_plastination.html.

"Von Neumann Machine (computer science)." *Encyclopedia Britannica Online*. Encyclopedia Britannica, 2014. http://www.britannica.com/EBchecked/topic/1252440/von-Neumann-machine.

Voytek, Bradley, and Paul King. "How Do We Know That There Are 100 Billion Neurons in the Brain?" *Eyewire*, January 4, 2013. http://blog.eyewire.org/how-do-we-know-that-there-are-100-billion-neurons-in-the-brain/.

Wachowski, Andy, and Lana Wachowski. *The Matrix*. Science Fiction, 1999.

Wade, Andrew, Andrew Nelson, and Gregory Garvin. "Another Hole in the Head? Brain Treatment in Ancient Egyptian Mummies." *Anthropology Presentations*, April 1, 2010. http://ir.lib.uwo.ca/anthropres/5.

Waldrop, M. Mitchell. "Computer Modelling: Brain in a Box." *Nature* 482, no. 7386 (February 22, 2012): 456–58. doi:10.1038/482456a.

Walpoth, Beat H., Beyhan N. Walpoth-Aslan, Heinrich P. Mattle, Bogdan P. Radanov, Gerhard Schroth, Leonard Schaeffler, Adam P. Fischer, Ludwig von Segesser, and Ulrich Althaus. "Outcome of Survivors of Accidental Deep Hypothermia and Circulatory Arrest Treated with Extracorporeal Blood Warming." *New England Journal of Medicine* 337, no. 21 (1997): 1500–1505. doi:10.1056/NEJM199711203372103.

Ware, Bronnie. *The Top Five Regrets of the Dying: A Life Transformed by the Dearly Departing*. London: Hay House, 2012.

Wedekind, Johann Kaspar, and Christoph Helbigk. *Dissertatio inauguralis medica de alkahest ...* Typis Groschianis, 1685.

Wheeler, Mark. "Hyperactivity in Brain May Explain Multiple Symptoms of Depression." *UCLA Newsroom*, February 27, 2012. http://newsroom.ucla.edu/portal/ucla/hyperactivity-in-brain-may-explain-228954.aspx.

Wiener, Norbert. *Cybernetics; Or, Control and Communication in the Animal and the Machine.* New York: J. Wiley, 1948.

Willis, Roy. *World Mythology: The Illustrated Guide.* London: BCA, 1993.

Wilson, Andrew. "Iain Banks Interview." *Textualities*, 1994. http://textualities.net/andrew-wilson/iain-banks-interview.

Wise, Jeff. "About That Overpopulation Problem." *Slate*, January 9, 2013. http://www.slate.com/articles/technology/future_tense/2013/01/world_population_may_actually_start_declining_not_exploding.html.

Witelson, Sandra F, Debra L Kigar, and Thomas Harvey. "The Exceptional Brain of Albert Einstein." *The Lancet* 353, no. 9170 (June 1999): 2149–53. doi:10.1016/S0140-6736(98)10327-6.

Wolpert, Daniel. "The Real Reason for Brains." Speech presented at the TED conference, July 2011. https://www.ted.com/talks/daniel_wolpert_the_real_reason_for_brains.

Woodruff, Robert, and Ferris N. Pitts. "Monozygotic Twins with Obsessional Illness." *American Journal of Psychiatry* 120, no. 11 (May 1, 1964): 1075–80.

Yeats, William Butler. "Among School Children." In *The Tower*. London: Macmillan, 1928.

———. "Sailing to Byzantium." In *The Tower*. London: Macmillan, 1928.

Zeithamova, Dagmar, Margaret L. Schlichting, and Alison R. Preston. "The Hippocampus and Inferential Reasoning: Building Memories to Navigate Future Decisions." *Frontiers in Human Neuroscience* 6 (March 26, 2012). doi:10.3389/fnhum.2012.00070.

Zhang, Jian-Min, Yan Sheng, Yong-Zhi Cao, Hong-Yan Wang, and Zi-Jiang Chen. "Cryopreservation of Whole Ovaries with Vascular Pedicles: Vitrification or Conventional Freezing?" *Journal of Assisted Reproduction and Genetics* 28, no. 5 (May 2011): 445–52. doi:10.1007/s10815-011-9539-3.

# INDEX